Proceedings of the
Engineering Foundation Conference on

IMPROVED HYDROLOGIC FORECASTING WHY AND HOW

William P. Henry,
Conference Chairman

Asilomar Conference Grounds
Pacific Grove, California
March 25-30, 1979

CO-SPONSORED BY
American Society of Civil Engineers
American Meteorological Society

AMERICAN SOCIETY OF CIVIL ENGINEERS FOUNDED 1852

PUBLISHED BY
American Society of Civil Engineers
345 East 47th Street
New York, New York 10017

ISBN 0-87262-203-7

CONTENTS

*Not available for publication.

OPENING REMARKS

William P. Henry, P.E.
Manager, Industrial Projects
CH2M Hill, San Francisco, CA

On behalf of the Engineering Foundation and the conference organizing committee, we welcome you to Asilomor and this conference on "Improved Hydrologic Forecasting - Why and How."

This conference was structured to achieve several purposes. High on the list was to share lessons that were learned during the drought of 1976 and 1977 with practicing hydrologists. During that drought, methods that were developed during wet periods became unsatisfactory. Project forecasting for hydropower, for irrigation supply and for municipal and industrial supply came under more scrutiny than ever before. Hydrologic forecasters were forced to improve their techniques and many of these improvements will be shared with you during the conference.

A second purpose of the conference is to remind the practicing hydrologists of the wide variety of purposes for which the numbers he develops can be used. Too often the hydrologist sees only part of a project and does not recognize the full impact of his input. However, in a water resource project, the hydrologist has the overriding role in project sizing. His information is the basis for decisions that commit large amounts of capital. Because of this, the hydrologist must understand the need for accuracy in various forecasts. But accuracy alone is often not enough. Time is also an important factor, particularly in urban flooding problems. Sophisticated forecasts which may take weeks to prepare, will be useless when the response time of a community is in the neighborhood of 1 hour or less.

As in other engineering disciplines, the analytical techniques used in hydrologic forecasting are constantly changing. We are fortunate to have with us during the week not only practicing hydrologists, but meteorologists and mathematical modellers of the rainfall-runoff process. This collection of talent allows us to have a full discussion of some of the emerging techniques in hydrologic forecasting that may and can be used to augment traditional analyses.

There is a final purpose for the conference. We will engage in a little crystal ball gazing as we attempt to raise a few thoughts about what the future may hold for the practicing hydrologist.

I feel very strongly that this is your conference. As such, responsibility is with you to make the most of it. The sessions have been scheduled to allow more time for questions and answers and discussion than for formal presentations. We encourage you to make full use of these discussion periods.

As in all Engineering Foundation conferences, where the emphasis is on dialogue, any one of you may be free to enter into discussions which you do not want reported. Merely signify that your comments are "off the record" and they will not be made a part of the proceedings published from this conference.

Thank you for attending. The organizing committee and I hope that you find the week here a valuable one in your professional careers.

CONFERENCE ORGANIZING COMMITTEE

John J. Cassidy
Washington Water Research Center
Washington State University

Catalino B. Cecilio
Pacific Gas and Electric Company

William P. Henry, Chairman
CH2M HILL

D. Earl Jones
U.S. Department of Housing and
Urban Development

Jerry Knapp
CH2M HILL

Eugene L. Peck
National Weather Service

Flood Hydrograph and Peak Flow Frequency Analysis

Arlen D. Feldman*

The accurate prediction of streamflows is essential to the planning of our water resource systems. This paper addresses the practical state-of-the-art of techniques to predict flood peaks and their associated frequency of occurrence; and techniques for predicting critical flood hydrographs (or series of hydrographs) and their frequencies of occurrence. Statistical relationships, empirical equations, and watershed models will be investigated as means for predicting the peak discharges and flood hydrographs.

Peak discharge information is required to determine the appropriate size of water conveyance systems such as natural channels, diversion canals, storm drains, bridge openings, etc. The frequency of the peak discharges is necessary to determine how often the conveyance system capacity is exceeded. Criteria for sizing conveyance systems are derived from socio-economic responses to the inconveniences associated with the exceedence of system conveyance capacities and the cost of providing those systems.

Flood control studies usually base flood damages on peak discharges as representative of damage due to several associated flood problems. It is especially convenient to be able to express flood damages in terms of discharge (stage), however, other factors such as flow velocities and duration of flooding may need to be considered separately. Flood control measures may take the form of increasing the capacity of conveyance systems or regulating the flood waters through storage, diversions, or local control measures. Flood damage reduction measures include these items plus nonstructural measures such as flood proofing structures, etc.

The tradeoff between conveyance capacity and storage in a flood control system is a classic consideration. This analysis is appropriate for small urban drainage systems (storm and combined sanitary) up to the large river/reservoir networks. Many techniques have been developed, tested, and implemented for sizing these systems. In general, the larger or more complex the drainage system becomes, the more the analysis shifts from predicting peaks to predicting the whole hydrograph. The

*"Flood Hydrograph and Peak Flow Frequency Analysis", Arlen D. Feldman, Chief, Research Branch, U.S. Army Corps of Engineers, The Hydrologic Engineering Center, Davis, California

techniques addressed in this paper are separated into the following catagories:

Frequency analysis of historical streamflows
Statistical equations
Empirical formulae
Single event watershed models
Continuous watershed models

The single event watershed models are further broken down into techniques using hypothetical storms and historical storms. Continuous watershed models are discussed in terms of relatively simple models and also the more complex complete soil moisture accounting models. In one of the proposed methods of analysis, a simple continuous model is used to screen the historical rainfall-runoff record to determine important individual events which are then simulated in more detail with a single event model. A distinction will also be made between techniques which predict peak flows from urban and nonurban areas. Much emphasis has been placed on urban runoff in recent years and several techniques have been developed to meet those needs (1).

The statistical equations and empirical formulae are best used to predict peak flow rates for small areas, less than 50 mi^2. When analyzing larger areas, the storage and routing effects in the basin usually require the use of a watershed model for adequate definition of the hydrograph.

Peak Flow Estimation Techniques.--Peak flows may be estimated directly as functions of historical streamflow records or statistical/empirical relationships. The peak flow techniques referred to in this paper are those techniques which predict only the peak flow - not including the whole hydrograph. The techniques which predict the whole hydrograph or series of hydrographs also compute a peak flow but they will be discussed in the Watershed Modeling sections. The peak flow techniques are functions of rainfall intensity or runoff frequency and various geographic characteristics of the basin. Usually the annual peak flow frequency curve is derived either directly from an equation or by estimating a series of flood peaks which are then analyzed with standard frequency techniques.

Frequency Analysis of Historical Streamflows.--Historical streamflow records may be used directly to estimate discharges at various frequencies. If adequate streamflow records exist and the watershed has remained relatively unchanged during the course of that record, then those observed streamflows are probably the best indicator of the potential flood responses of the watershed in its present condition. The Water Resource Council's guidelines (2) describe the currently recommended techniques. Those guidelines describe the use of the Pearson Type III distribution and associated topics of high and low outliers, generalized skew, two-station comparisons, mixed populations, confidence limits, flood estimates from precipitation data, and equivalent accuracy for independent estimates for analysis of historical flood peaks.

If it is desired to predict the magnitude-frequency of streamflows under some future watershed land use development or regulated condition, then the historical streamflow records cannot be used directly. In this case one must usually resort to a watershed model. The same requirement arises where long term historical streamflow records exist but the watershed has undergone significant changes during that time. Thus, a nonstationary streamflow series exists and cannot be used directly in the frequency analysis. The nonstationary series problem can also be unraveled through the use of watershed models (3).

If a stationary series of data is available, but not at the specific locations of interest, then a regional frequency analysis may be undertaken (4). The regional analysis allows one to transfer the parameters of the flood frequency distribution at gaged locations to other locations of interest. This is accomplished through relating frequency parameters to geographic and meteorologic characteristics which are known at the gaged and ungaged locations.

Water Resource Council's Ungaged Areas Flood Frequency Study.--The Water Resources Council - Hydrology Committee, work group for peak flow frequency for ungaged areas, has recently begun a study of the following eight flood frequency estimation techniques.

Statistical estimation of Qp
Statistical estimation by moments
Index flood method
Transfer method
Empirical equations
Single storm
Multiple discrete events
Continuous simulation

The first phase of the WRC study has just been completed for selected watersheds in the northwestern and central U.S. and they expect to publish a report in November 1979 (personal communication with John Miller, National Weather Service). The methods reported in the pilot tests are: USGS Equations, FHWA, Reich, Snowmelt, Index Flood, Rational Formula, TR55, (RP149), TR55 (TC), TR20, and HEC-1. They are encouraging the widest possible review of this work before going on with similar applications in the southwestern and southeastern U.S. A later phase of the WRC studies will include urban areas.

Preliminary results of the WRC study show there to be a fair amount of variation within the application of the same method on the same watershed by different participants. This was observed even with the apparently straight forward methods such as the USGS State Equations and FHWA where all that is needed is drainage area and other simple geographic location parameters. The results varied even more, as one would expect, as more judgement/experience factors were required to use the methods such as SCS' TR-20 and The Corps of Engineers' HEC-1. The above observations were made from a very preliminary review of the pilot study raw data. The WRC Ungaged Watershed Group is now in the process of editing and analyzing that data and preparing their report.

Statistical Flood Peak Estimation Techniques.--Statistical flood peak estimation techniques predict instantaneous peak flows of prescribed frequencies through a regression analysis of geographic variables affecting the flood runoff. An excellent discussion of drainage basin and meteorologic characteristics which can be used to explain behavior of streamflows is given in Thomas and Benson (21). They analyzed over twenty characteristics and discuss the relative affects of each. Drainage basin area and normal annual precipitation were among the most significant. Certainly some of the most widely available examples of these techniques are the U.S. Geological Survey (USGS) "State Regression Equations," (5).

Patterson and Gamble (5), developed relationships between drainage area and mean annual flood in different hydrologic areas. The mean annual flood is reduced in proportion to the lake storage in the basin if applicable, and the peak flows for recurrence intervals from 1.1 to 50 years may be determined from a graph of recurrence interval vs. ratio of discharge to mean annual flood as in figure 1.

A generalized procedure is also used by the Federal Highway Administration (FHWA), (6) for small rural watersheds, generally less than 100 square miles. In this procedure the 10-year event is determined as a function of the drainage area, an iso-erodent factor, and a difference in elevation in the watershed. Separate equations are given for each of twenty four hydrophysiographic zones in the U.S. and Puerto Rico. A fixed relationship is given between the 2-year, 100-year peaks and the 10-year peak.

Other techniques use watershed runoff characteristics and precipitation intensities to predict the peak runoff rates. A time on concentration and infiltration index may be determined directly from watershed characteristics such as length and elevation change of main channel, soil characteristics, and land cover. The x-minute rainfall intensity for the desired frequency is obtained from TP40 or HYDRO-35 (7). A peak discharge per square mile may be derived from the time-of-concentration, infiltration index, and peak 30-minute intensity. Adjustments can be made for antecedent precipitation. This method is subject to the standard criticisms of assuming the frequency of the runoff is the same as the rainfall.

Empirical Equations.--The most popular and long lasting of the empirical equations is the Rational formula (8). Despite the many more sophisticated methods available today, the rational formula is still popular because of its easy and economic use. The regression equations previously discussed are similar to this except that the coefficients in the equations are determined by a minimum error statistical techniques. Application of the Rational method has even been inconsistent (37).

Runoff vs. Rainfall Based Methods.--Many of the statistical estimation techniques for peak flow are directly streamflow based and do not go through the rainfall-to-runoff analysis. The flow estimates are determined by analyzing streamflows of known frequencies in a hydrologic region and relating them to basin characteristics, primarily drainage area and sometimes general meteorologic measures such as average annual

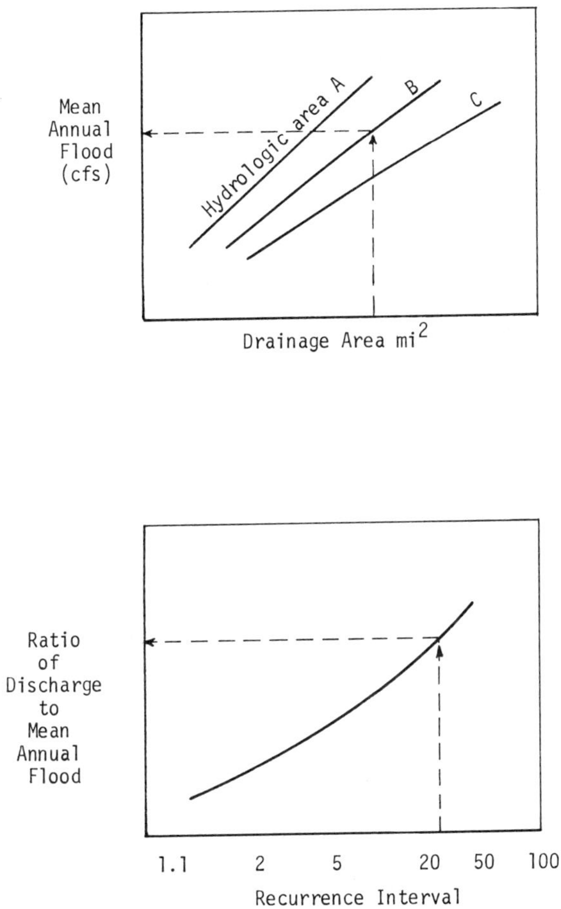

Figure 1. Area Regression Equation Method

precipitation. It is generally more difficult to develop these relationships in urbanizing basins because of the nonhomogeneous nature of the runoff series.

As urbanization occurs the rainfall runoff response function changes and additional parameters must be brought into the relationship to explain that variation. Usually the percent of impervious area and watershed conveyance factors are found to be suitable measures of urbanization (9). Rainfall is also brought into these relationships so that impact of changed precipitation loss rates can be analyzed directly instead of trying to reflect change only in the routing parameters.

There are two general classes of rainfall based flood prediction methods: 1) runoff frequency is assumed to be the same as the rainfall frequency, and 2) the runoff frequency is computed independently of the rainfall frequency. The assumption that runoff frequency equals rainfall frequency is generally agreed to be undesireable (10) but is oftentimes used because it simplifies the required analysis. Precipitation frequency analysis is discussed in a later section. Rainfall of some frequency can be applied to several different antecedent moisture conditions in the same watershed and largely different runoff may result. As the frequency of the event becomes more rare, the runoff is less affected by the antecedent moisture condition and the ensuing loss rates. The single event models can be used on many historic events and then the runoff peaks ranked and the frequencies determined by standard methods. Continuous simulation models take the final step of analyzing the entire precipitation runoff record maintaining consistency with respect to soil moisture storages. The continuous process modelers claim to have the most realistic basis for computing flood frequencies.

Watershed Modeling.--When is it necessary and/or desirable to use a watershed model instead of the simplified statistical and empirical techniques? The watershed models are generally required when: an entire hydrograph is desired; analyzing complex areas; or when the past or proposed future watershed response functions are changing. Watershed models are particularly desirable when analyzing the effect of various water management schemes.

Watershed models range widely in complexity. Brandstetter (1), compares many different models for urban storm runoff and many of these models are equally good for nonurban cases. Some are nothing more than simple empirical equations within a subbasin network routing/combining framework. Others perform a complex accounting of soil moisture and water in various stages of runoff. The following discussion of watershed modeling looks at the practical state-of-the-art in single event and continuous models and combinations thereof. The attendent precipitation analyses required with both techniques is also discussed.

Hydrographs are necessary when storage projects (reservoirs of different forms) are being investigated as flood control measures. A frequency analysis of runoff can be made to determine the expected frequency of various flow durations (4). If a certain flow-duration relationship is desired for project design, it can be determined separately, as just mentioned, and then the simulated hydrograph can be balanced (11) to conform to that flood duration. Continuous event simulation models are often preferred for storage analysis if it is a sequence of storms which cause the flood problem as opposed to one large, single event.

Single Event Models.--A single event model is one that is used primarily for individual storm events, although it may be of long duration and multi-peaked. Two factors usually constrain their use to single events: the continuity of soil moisture (loss rates) is not simulated, and/or the model simulates in such detail and requires time consuming computations so that it is not economical to run over long periods. Many of the single event models can be used equally well in urban and nonurban areas but have usually been developed for a specific purpose and then generalized to meet more needs. Some of the most widely used single event models are:

HEC-1: Flood Hydrograph Package (11)

TR-20: Computer Program for Project Formulation Hydrology (12)

MITCAT: MIT Catchment Model (13)

USGS Rainfall-Runoff Simulator (14)

SWMM: Storm Water Management Model (15)

Many other models exist and some contain more advanced representation of various aspects of the precipitation-runoff process. Many comparisons of such models have been made (1, 16). Few models are more comprehensive and/or widely used than those above. These models are generally well supported by government agencies or private engineering consultants and are continually being improved to meet new needs. The HEC-1 model, for example, goes beyond the basic rainfall/snowmelt runoff simulation process and has special options for computation of expected annual flood damages, automatically sizing components of a flood control system for maximum net benefits, and simulation of dam overtopping and failure per the requirements of the National Dam Safety Inspection Program.

The current tendency in watershed modeling, both single event and continuous, is to incorporate parameters with realistic relationships to the physical process and that can be determined directly from readily available geographic data. Because of this strong interest in relating watershed model parameters to geographic characteristics, the Soil Conservation Service's (SCS) curve number technique has received much increased interest and usage. The SCS curve number technique is the only one in which both the precipitation loss rate and the water excess-to-runoff transformation (unit hydrograph) can be determined from readily available geographic data. The data used are: land cover, hydrologic soil type, average slope of the watershed, and length of the main water course. Curve numbers have been recommended for various land cover - hydrologic soil group combinations in both urban and nonurban areas (17) as shown in Table 1. Calibration with observed rainfall runoff data is still required. The curve number technique, although a rather simplistic representation of the runoff process, appears to work well in many cases. Further research is being currently undertaken by the HEC and Espey, Huston & Assoc. to test the validity of the technique in urbanizing watersheds.

Table 1. Runoff Curve Numbers for

Selected Land Uses (from Table 2.2 (17))

Land Use	Hydrologic Soil Group			
	A	B	C	D
Cultivated land	72	81	88	91
Pasture or range land	68	79	86	89
Wood or Forest land	25	55	70	77
Residential, 1/4 acre lots	61	75	83	87
Commercial (85% imperv.)	89	92	94	95
Industrial (72% imperv.)	81	88	91	93
Paved parking lots, roofs, driveways, etc.	98	98	98	98

A particularly interesting and powerful benefit of using geographically related watershed parameters is that of interconnecting watershed models with geographic information systems. This concept was used in a project for Fairfax County, VA in which a geographic grid cell information system was used as the basis for computation of watershed model parameters (18). This study made use of the MITCAT watershed model and parameters were estimated from a geographic data bank of land use, etc.

The Hydrologic Engineering Center, (19), made a practical application of a geographic information system for automatically computing hydrologic and economic parameters in the Oconee River Expanded Flood Plain Information study, figure 2. This concept has been expanded into a comprehensive flood plain planning tool (20), and is now being implemented as a regular tool in Corps of Engineers' project investigation.

As single event models became more geographically based and capable of easily predicting starting conditions, (initial values of model parameters), the less necessary continuous watershed models would appear to be. With this capability, the single event model could be started before every significant event. Statistical analysis of the output peak flows and volumes could be performed to make predictions for design purposes. Using a single event model for many storm events and using the resulting frequency curve, overcomes the common criticism of single event models - that runoff frequency equals rainfall frequency.

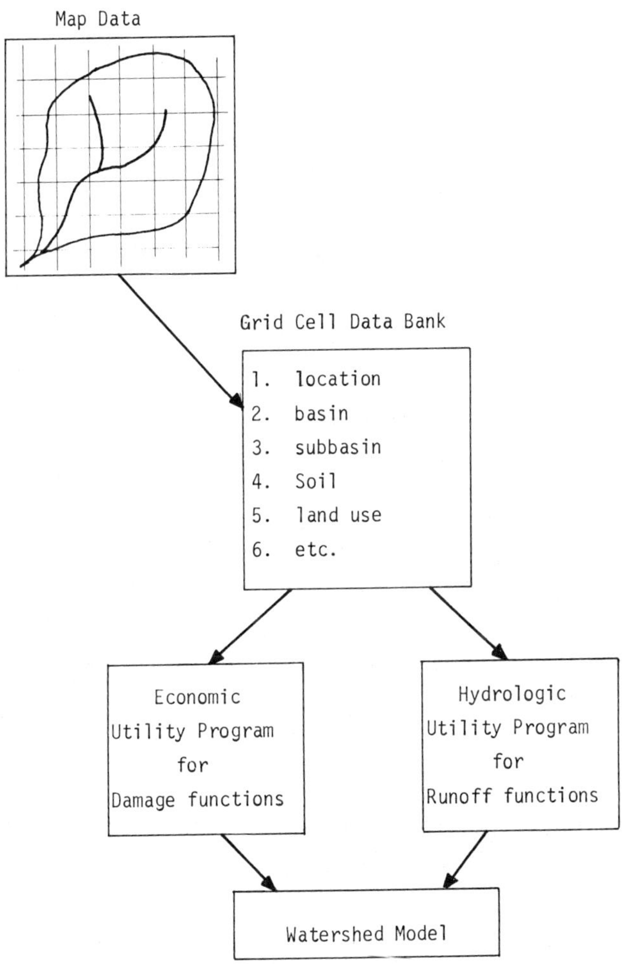

Figure 2. Spatial Data Management and Hydrologic Analysis System

Analysis of Flood Control Measures and Land Use Changes.--The hydrologic engineer is often asked to determine the impact of land use changes or flood control management measures on specific design floods and the entire flow frequency curve. This can be accomplished with either the single event or continuous watershed model. Both methods require the ability to change watershed parameters to reflect new watershed response functions.

Watershed modelers use many different characteristics affecting the runoff process with which to predict the parameters of the model. The common procedure is to establish a relationship between the model parameters, say loss rates, and runoff transformations (unit graph and kinematic wave), and basin characteristics. Basin characteristics are discussed in relation to runoff production by Thomas and Benson (21). Urbanization factors are included if the basin has been or is being developed (9).

For evaluating flood control management alternatives, the watershed modelers simply run the model in the with and without project control modes to determine the impact of the project. In the continuous models, one usually simulates the entire record with and without the modified land use and/or flood control projects. The annual peak flows, for each case, are subjected to traditional frequency analysis and the modified frequency curve is obtained.

Single event models can be used to develop a modified frequency curve by simulating several storms, of varying magnitude, under each development condition (11). A base case frequency curve is required and can be developed by any preferred method. Usually a frequency curve is "adopted" which may be some specific curve or combination of curves. The frequency of the base case computed peak flow, for each storm magnitude, is determined from the adopted frequency curve, figure 3. The same storm is simulated again under the modified conditions and the frequency of the runoff is assumed to be the same. The modified frequency curve is determined as shown in figure 3. A potential fallacy with this approach is that the storm runoff may change in frequency for the modified watershed development. That is, the ranking of peaks flows might change under the modified condition.

Precipitation Frequency Analysis.--Many studies have been made of precipitation frequency and critical design events such as the probable maximum precipitation and various frequency intensity-duration relationships of the National Weather Service (NWS) (7). A recent analysis by Marsalek (10) reviewed the Chicago and Illinois design storm methods and compared them with results obtained from continuous simulation. Marsalek reinforced those common feelings about the pitfalls of design storms, at least for the limited geographic area analyzed.

Nevertheless, design storms are a commonly used tool and must be given serious consideration, especially because of their economic attractiveness. Several single event watershed modelers have developed their own rainfall frequency analysis techniques (22) and link these techniques directly with their watershed models. These techniques as well as the NWS (7) publications can be used effectively as long as the

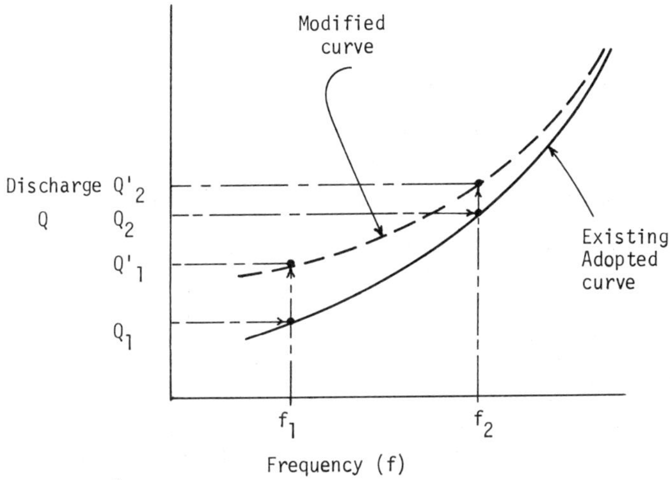

Q_i = peak discharge for design storm i

f_i = frequency of Q_i from given curve

Q'_i = peak discharge from design storm i under modified watershed condition

Figure 3. Modifying Frequency Curves

impact of the antecedent precipitation assumptions are fully realized. A study by Yen (23) demonstrates the use of synthetic storms in designing projects for the Federal Highway Administration. One of the major problems occurs when the particular sequence of precipitation events causes the critical flood situation as opposed to the magnitude of any one part of the multiple events. This type of problem leads one to prefer the analysis found in the continuous models.

The use of continuous watershed models does not solve all of one's rainfall analysis problems. Granted, it is the most comprehensive analysis of the hydrology of the basin and much is to be gained from that insight. These models have a major dependence upon the precipitation measured or synthetically generated; and the precipitation data are usually the least well known part of the runoff process. The difficulty comes in the spatial variation of precipitation; point measurements are made and spatial averages are inferred.

The construction of a long-record precipitation series is a difficult task. As one goes back in time, the observation stations become fewer, and one must make more and more assumptions about the spatial and temporal variation of the precipitation. The National Weather Service maintains tape files of daily precipitation records since 1948 and shorter interval precipitation measurements for selected stations and time periods. Before 1948, most precipitation data were not in computer compatible format and thus, extensive preparation by the analyst is required. The Hydrologic Engineering Center estimated that approximately 4 to 6 person-months of effort would be required to construct the precipitation record from 1900 to 1948 for a 130 mi^2 basin near Chicago, IL.

Continuous Watershed Models.--Most of today's highly sophisticated continuous watershed models are derived from the Stanford Watershed Model (24). Another model, developed at about the same time, is the SSARR model of the Corps of Engineers (25). The SSARR model does not have all of the complexity of the Stanford derived models, but has been shown to be comparable in results with the more comprehensive models (26).

The Stanford Watershed Model has been elaborated upon at several universities: Kentucky (27); Texas (28); Ohio (20); and others. Notable among these is the Kentucky version, entitled OPSET, where the parameters of the model are derived automatically by an optimization routine. The National Weather Service also used the Stanford Watershed Model as the basis for its NWSRFS model (30). The National Weather Service Sacramento Model (31) has more comprehensive soil moisture accounting algorithms, but may be considered less sophisticated in its runoff transformation via linear unit graphs and the fact that it does not route stream flows in a comprehensive river system.

One of the most highly developed versions of the Stanford Watershed Model existing today is the Hydrocomp HSP Model (32). The HSP system of programs incorporates the precipitation-runoff model as one piece of an array of study tools ranging from water quality simulation to unsteady flow dam break flood routings. The technical analysis tools all link together with a comprehensive data management system which arranges input and saves output for further analysis.

Efficient data management in continuous watershed simulation models is an exceedingly important requirement. The HSP system is probably the most advanced and comprehensive in this regard. The continuous models are frequently criticized for their enormous appetite for data. In fact, the cost of assembling the necessary data often negates the use of these models in all but the most comprehensive studies which require a well orchestrated analysis of competing interdisciplinary uses of water.

One of the simplest and most economical to run continuous watershed models is the STORM program (33). The model was originally developed by Water Resource Engineers, Inc. in connection with stormwater runoff in the city of San Francisco. The original model was essentially a long term hyetograph analysis with a simple rational formula type transformation to runoff. Long term, say 50 years or more, of hourly precipitation data can easily be analyzed at an affordable computer cost. The original model was limited to a single subbasin analysis but later versions incorporate multibasin routing and combining.

The STORM program was studied by Brandstetter (1) and its characteristics can easily be compared with other models in that report. Another comparison of several continuous and single event watershed models was recently published by the ASCE Urban Water Resources Research Program (16). While this comparison was not an exhaustive testing of the models, it does give good insight into the relative performance, attributes, and difficulties one may find in these models. Another comparison of continuous models was made by Lumb (34).

The Hydrologic Engineering Center has undertaken a detailed analysis of the HSP watershed model (35). The purpose of this analysis was to see how well and practically a comprehensive continuous simulation model could be used in a standard Corps of Engineers flood frequency study. The HSP model was chosen as a state-of-the-art model and applied to the DuPage River Basin near Chicago, IL. This study drew several conclusions:

1) The model can produce reasonable results when properly calibrated. Annual flood events, when analyzed together, exhibited characteristics similar to recorded flows, although individual years were significantly different from the observed.

2) The model can account for urbanization but more in theory than was able to be accomplished in practice. The application was begun with five land uses for runoff production, but this was soon reduced to two, nonurban and urban, together with an impervious area. Without runoff data to distinguish the urban and nonurban contributions to runoff it was furthermore not possible to make that land use distinction. The final model was constituted of an urban/rural mixture land use and an impervious area. The theory of the model would lead one to believe that several different land use runoff segments could be used but in the end only one could be realistically used.

3) The model is relatively easy to operate in terms of input instructions, file organization and manipulation.

4) The model is difficult to calibrate because of the large number of parameters and the mass of data processing. Many of the parameters were found to be more empirical than one would expect and their best values could only be determined by calibration. It was difficult to know the starting values for several parameters, but this would be easier with experience. It would certainly be desirable to have parameter values more directly derivable from geographic data.

5) The data requirements are extensive both in quantity and in the labor necessary for preprocessing.

6) The use of such a model produces a major benefit of becoming very knowledgeable about the full complexity of the watershed's hydrologic cycle; that knowledge will surely permit a greater number/variety of water resource management alternatives to be considered.

7) The general conclusion of this application was that continuous simulation models can be used for the flood frequency study, but, considering the amount of work and data required, it is doubtful that they are the best tool for such limited studies. These models are best suited for comprehensive river basin studies requiring analysis of both high and low flows.

Hybrid Single Event/Continuous Watershed Modeling.--The previous discussion has noted that single event models used with design storms have certain difficulties in estimating the frequency of the runoff. Single event models can be used on historical storms and could be taken to the limit of analyzing all storms of record. That task would be quite expensive and receive the same criticism as the data and manpower intensive comprehensive continuous simulation models. Is it possible, then, that a blend of the single event and continuous models could fulfull this need?

The Hydrologic Engineering Center has recommended the use of a simple continuous simulation model to screen the history of precipitation runoff events in a watershed (36). The STORM model or other similar simple model is used to simulate the entire historic record, figure 4. Different amounts of storage in a simple linear reservoir model could also be used to see if basin storage, natural and/or artifical, would bring about different series of critical events (22).

The simulated runoff events are then ranked according to magnitude and/or volume. A conventional flood frequency analysis is performed for those simulated flood events. This yields a gross estimate of the flood frequency curve. This rough estimate of the frequency curve is then improved with detailed runoff analysis by a single event model.

Several representative storm events are selected across the range of computed floods. These storm events are then simulated in detail using models such as MITCAT or HEC-1. The antecedent moisture conditions prior

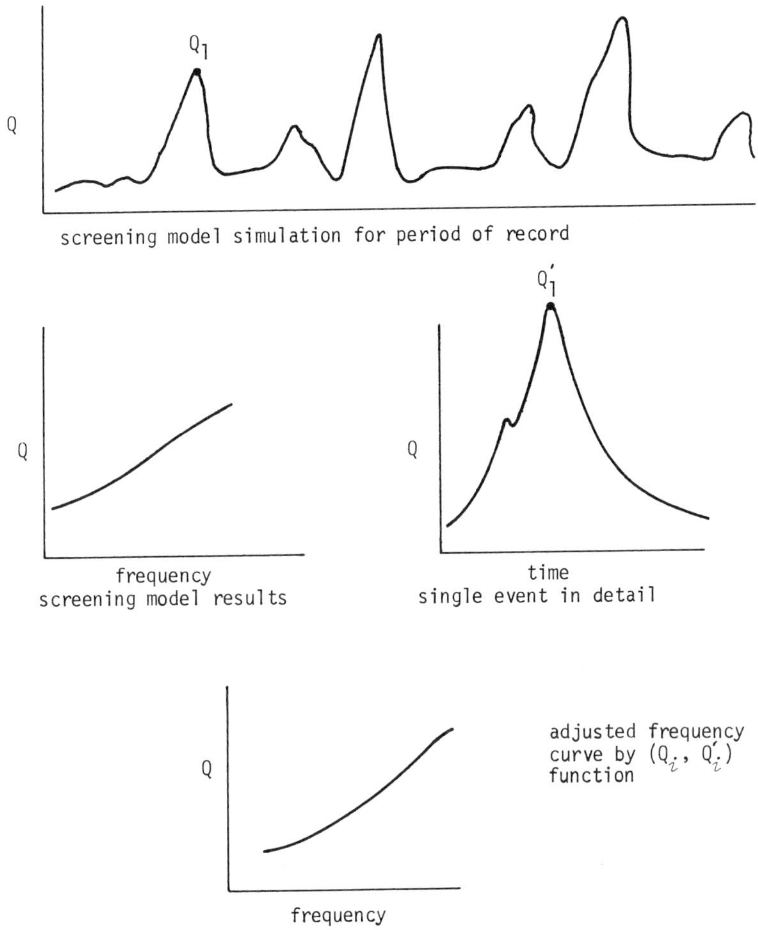

Figure 4. Hybrid Single Event/Continuous Model Analysis

to each storm can be estimated from the continuous simulation results. A statistical analysis of the simple continuous model peak flows and the detailed single event model's peak flows is made and a regression equation is developed. The regression equation is used to determine a better value for the other simple continuous model peak flows not simulated in detail. This results in an adjusted frequency curve based on the detailed simulation of a few significant flood events, figure 4.

Conclusions.--There are numerous techniques for predicting peak flood discharges/volumes of prescribed frequencies. The budget of one's study and one's familiarity with different analytical techniques usually determine which approach is used.

Statistical and empirical flood peak estimation techniques may be good for small areas where river routing/storage effects are not significant. For larger watersheds and studies requiring analysis of alternative flood control/watershed management procedures, the watershed simulation model is the best tool. The advantages and disadvantages of single event models with design storms and complex continuous simulation models have been discussed. A hybrid approach using a simple continuous model to identify significant events and a single event model to analyze those events in detail is a promising method of analysis.

Watershed models are tending to become more directly based on readily measurable geographic parameters. The need for continuous accounting of soil moisture conditions would appear to be less necessary as one becomes better able to predict watershed model parameters from direct geographic/meteorologic measurements. With this capability, single event models theoretically could be easily started for any event in question.

Geographic information systems and utility programs to compute automatically the watershed model parameters are a promising technology for comprehensive river basin studies. This technique is particularly powerful in analyzing many land use and watershed and flood damage reduction management alternatives.

If one uses a subjective measure of flood severity, such as general public inconvenience, then projects may be sized by methods which are not dependent on a true estimate of a flood frequency. That is, the projects are designed by some consistent method, say design storms, and the severity of the storms is changed dependent upon the public's reaction to flooding inconvenience. This is generally the case in urban storm sewer design.

If an objective estimate of expected economic damage is to be the design criterium, then realistic flood frequencies must be computed. Projects based on specific flood frequencies, say the 100-year flood of the flood insurance studies, would require consistent, but not necessarily precise, estimates of flood frequencies. In that case, insurance premiums could be adjusted, as actuarial experience indicates, to made the project viable. The landowners, however, would argue for the use of a precise frequency.

References

1. Brandstetter, Albin, "Assessment of Mathematical Models for Storm and Combined Sewer Management," for Environmental Protection Agency, Technology Series, EPA-600/2-76-175a, August 1976.

2. Water Resources Council, "Guidelines for Determining Flood Flow Frequency," Bulletin 17A, Hydrology Committee, June 1977.

3. Gundlach, David L., "Adjustments of Peak Discharge Rates for Urbanization," Irrigation and Drainage Journal, American Society of Civil Engineers, Vol. 104, No. IR3, September 1978.

4. Beard, Leo R., "Statistical Methods in Hydrology," U.S. Army Corps of Engineers, Sacramento District, January 1962.

5. Patterson, James L. and C.R. Gamble, "Magnitude and Frequency of Floods in the United States, Part 5. Hudson Bay and Upper Mississippi River Basins," U.S. Geological Survey, Water Supply Paper 1678, Washington, D.C. 1968.

6. Trent, R.E., "FHWA Method for Estimating Peak Flow Rates of Runoff from Small Rural Watersheds," Federal Highway Administration, Offices of Research and Development, Environmental Design and Control Division, Washington, D.C., 1978.

7. U.S. Weather Bureau, "Rainfall Frequency Atlas of the United States," Dept. of Commerce, Technical Paper No. 40, May 1961, and National Weather Service, "Five to 60-Minute Precipitation Frequency for the Eastern and Central United States," NOAA Technical Memo NWS HYDRO-35, Silver Spring, MD, June 1977.

8. Gregory, R.L., and C.E. Arnold, "Rational Runoff Formulas," Transactions, American Society of Civil Engineers, Vol. 96, 1932.

9. Espey, William H. Jr., Duke G. Altman, and Charles B. Graves, Jr., "Nomographs for Ten-Minute Unit Hydrographs for Small Urban Watersheds," American Society of Civil Engineers Urban Water Resources Research Program Technical Memo No. 32, New York, 1977.

10. Marsalek, Jiri, "Research on the Design Storm Concept," American Society of Civil Engineers Urban Water Resources Research Program, Technical Memo No. 33, New York, 1978.

11. Hydrologic Engineering Center, "Flood Hydrograph Package, HEC-1," U.S. Army Corps of Engineers, California, 1973.

12. Soil Conservation Service, "Computer Program for Project Formulation Hydrology," Technical Release No. 20, U.S. Dept. of Agriculture, Washington, D.C., 1965.

13. Resource Analysis, Inc., "MITCAT Catchment Simulation Model," Description and Users Manual, Version 6, Massachusetts, 1975.

14. Dawdy, David R., Robert W. Lichty, and James M. Bergmann, "A Rainfall-Runoff Simulation Model for Estimation of Flood Peaks for Small Drainage Basins," U.S. Geological Survey Professional Paper 506-B, Washington, D.C., 1972.

15. Heaney, James P. and Wayne C. Huber, "Storm Water Management Model: Refinements, Testing, and Decision-Making," University of Florida, 1973.

16. Abbott, Jesse W., "Testing of Several Runoff Models on an Urban Watershed," American Society of Civil Engineers Urban Water Resources Research Program Technical Memo No. 34, New York, October 1978.

17. Soil Conservation Service, "Urban Hydrology for Small Watersheds," Technical Release No. 55, U.S. Dept. of Agriculture, Washington, D.C., January 1975.

18. Michel, Henry L. and William P. Henry, "Flood Control and Drainage Planning in the Urbanizing Zone: Fairfax Co. Virginia," in _Urban Runoff Quantity and Quality_, Engineering Foundation and American Society of Civil Engineers, New York, 1974.

19. Hydrologic Engineering Center, "Phase 1 Oconee Basin Pilot Study, Trail Creek Test," U.S. Army Corps of Engineers, California, 1975.

20. Davis, Darryl W., "Comprehensive Flood Plain Studies Using Spatial Data Management Techniques," Water Resources Bulletin, Vol. 14, No. 3, June 1978.

21. Thomas, H. and M. Benson, "Generalization of Streamflow Characteristics," U.S. Geological Survey Water Supply Paper No. 1975, Washington, D.C., 1970.

22. Resource Analysis, Inc., "Rainfall Analysis and Generation Model Description and Theoretical Background," Boston, MA, March 1975.

23. Yen, Ben Chie and Ven Te Chow, "Feasibility Study on Research of Local Design Storms," Prepared for Federal Highway Administration, Report No. FHWA-RD-78-65, Washington, D.C., November 1977.

24. Crawford, Norman H. and Ray K. Linsley, "Digital Simulation in Hydrology: Stanford Watershed Model IV," Stanford University, Civil Engineering Technical Report No. 39, July 1966.

25. Rockwood, David M., "Streamflow Synthesis and Reservoir Regulation," U.S. Army Engineer Division, North Pacific, Technical Bulletin No. 22, January 1964.

26. World Meteorological Organization, "Intercomparison of Conceptual Models Used in Operational Hydrological Forecasting," Operational Hydrology Report No. 7, WMO-No. 429, Geneva, 1975.

27. James, L. Douglas, "An Evaluation of Relationships Between Streamflow Patterns and Watershed Characteristics Through the Use of OPSET," Research Report No. 36, Water Resources Institute, University of Kentucky, Lexington, 1970.

28. Claborn, B.J., and W.L. Morre, "Numerical Simulation of Watershed Hydrology," Hydrologic Engineering Laboratory, Civil Engineering Dept., University of Texas, Technical Report HYD-14-7001, 1970.

29. Ricca, V.T., "The Ohio State University Version of the Stanford Streamflow Simulation Model," Water Resources Research Center, Ohio State University, 1972.

30. National Weather Service, "National Weather Service River Forecast System Forecast Procedures," NOAA Technical Memo, NWS HYDRO-14, Silver Spring, MD, 1972.

31. Burnash, Robert J.C., R. Larry Ferral and Richard McGuire, "A Generalized Streamflow Simulation System," Joint Federal-State River Forecast Center, Sacramento, CA 1973.

32. Hydrocomp, Inc., "Hydrocomp Simulation Programming Operations Manual, Fourth Edition," Palo Alto, California, January 1976.

33. Hydrologic Engineering Center, "Storage, Treatment, Overflow, Runoff Model, STORM," U.S. Army Corps of Engineers, California, 1976.

34. Lumb, Alan M., "Comparison of the Georgia Tech, Kansas, Kentucky, Stanford, and TVA Watershed Models in Georgia," Georgia Institute of Technology, January 1976.

35. Cermak, Robert J., "Continuous Hydrologic Simulation of the West Branch DuPage River Above West Chicago: An Application of Hydrocomp's HSP," Hydrologic Engineering Center, U.S. Army Corps of Engineers, California, Draft, November 1978.

36. Peters, John C., "Techniques for Developing Frequency Curves in Urban Areas," Lecture Notes, Hydrologic Engineering Center, U.S. Army Corps of Engineers, California 1978.

37. Ardis, Colby V., K.J. Dueker, and A.T. Lenz, "Storm Drainage Practices of Thirty-Two Cities," American Society of Civil Engineers, Journal of the Hydraulics Division, Vol. 95, No. 1, January 1969.

TECHNIQUES FOR THE GENERATION OF LONG STREAMFLOW SEQUENCES

by
K.M. Leytham and D.D. Franz[1]

WHY GENERATE?

Over the past 20 years we have seen considerable advances in computer hardware development, in simulation and data generation techniques, and in the concepts and methods used for assessing the benefits derived from water resource systems. These advances, together with the spectacular and continuing dive in computational costs, make it not only feasible, but advantageous to study the design and operation of water resource systems in considerable detail.

Let us loosely define the "response" of a water resources system as being the benefits derived from the operation of the system over its economic time horizon. This response is controlled by interactions between the physical characteristics of the system (reservoir storage, irrigated area), the system operating policy, and the hydrometeorology (rainfall, streamflow, evaporation) experienced over the future life of the facility. The great unknown in determining the response is the future rainfall or streamflow sequence. The historic record may be of the same length or longer than the economic life of the project and this may be used in simulations to determine one possible response from the system. However, the historic record is quite unlikely to be repeated in the future and its use alone may result in a response which is quite unrepresentative of possible future responses. To make useful decisions about the design or operation of a facility in the face of uncertain streamflow or rainfall sequences we wish to know at least the expected or mean response and its variability. Ideally, we should determine the complete distribution of the response.

Fortunately, it is often reasonable to assume that the future streamflow or rainfall record will have the same statistical properties as the observed historic record. Streamflow records with similar statistical properties may appear quite different and may result in very different responses from the system under investigation. Thus using some suitable mechanism we can generate many equally likely future streamflow sequences which maintain the important statistical features of the historic record. Simulation of the systems response using these synthetic sequences then allows the designer to determine the approximate distribution of the response and hopefully derive a more appropriate operating policy or system configuration.

[1] K. Malcolm Leytham, Hydrologist, Hydrocomp, Inc., Palo Alto, CA.
Delbert D. Franz, Vice President, Hydrocomp, Inc., Palo Alto, CA.

The analysis process outlined above makes use of data generation techniques to obtain synthetic sequences, simulation techniques to determine the behavior of the system, and methods for quantifying the benefits derived from operation of the system. This paper considers one aspect of the process - data generation.

PRINCIPLES FOR DATA GENERATION

The goal of data generation in the present context is to generate streamflow sequences which replicate certain important features or statistics of the historic record. The goal is purposely vague at this stage. The features to be preserved in the synthetic sequence may depend both on the type of study being undertaken and on details of the system's physical characteristics and operation. In all cases, we would wish to preserve the mean and variance of the historic streamflow. For storage problems (reservoir analysis) preservation of the historic skew and correlation structure would also be required, and in some cases, especially in long-term storage problems, it may be necessary to reproduce the "Hurst" effect.

Historic data provides us with sample statistics complete with all the problems of small sample bias and sampling error. However, these are the statistics which will be preserved, within statistical bounds, by the generator; from the point of view of the generator they are population statistics. It is thus extremely important that we have good estimates of the statistics of the historic data and that we understand the implications of uncertainties in the statistics.

Two questions then have to be answered. Firstly, what features of the historic record should be preserved, implying a choice of model or generator. Secondly, how can one avoid or alleviate the problems of uncertainty in the statistics of the historic record.

CHOICE OF MODEL OR GENERATOR

Model choice should ideally be based on considerations of the following items:

- the features or statistics to be preserved (a function of the nature of the study)
- the extent and quality of historic data
- the type of distribution followed by historic data (e.g. gamma, log normal)
- the effects of uncertainty in parameter levels
- the implications of failing to preserve certain statistics or features

In applications in a commercial (as opposed to academic) environment however there are obviously also a set of pragmatic considerations which could influence model selection:

- availability of models
- ease of application of models
- availability of trained personnel
- cost of model application
- future applications in other studies

A variety of models or generators have been presented in the literature and a number of comparative studies have been performed to evaluate several of these models. In general, the models fall into two categories:

1. those that preserve short-term persistence (e.g. Markov)
2. those that preserve long-term persistence (i.e. Hurst effect)

We will restrict our considerations in the first category to lag-one Markov models which largely differ in their abilities to preserve skewness and correlation structures. The various approaches to preserving skewness are based on assumptions about what distribution the data comes from (various possibilities are discussed by Fiering and Jackson (1971) and by Srikanthan and McMahon (1978a)). However, historic records are generally so short that arguments about the distribution are inconclusive, and in practise other considerations may intervene. Generation of data at more than one station may require preservation of the cross-correlation between events at the different stations. The most suitable, and possibly only approach to the multisite problem is to sample from a multivariate normal distribution. Use of a log-normal transformation permits preservation of the skewness. This approach, as presented by Matalas (1967), is capable of preserving mean, variance, skewness, lag-one serial correlation, and lag-one cross correlation for both single-site and multi-site problems. In our opinion, in the commercial environment the Matalas approach has no serious rivals if one only wishes to preserve short-term persistence features. As additional reassurance, Srikanthan and McMahon (1978a) present a review of various lag-one Markov models used for single-site generation of annual streamflows. They suggest that log-normal transformations be used for coefficients of variation (CV) greater than 0.2, and that sampling from a gamma distribution only becomes attractive for CV less than 0.2.

If one wishes to preserve the long-term persistence features of a streamflow record (i.e. Hurst effect) then one has to turn to the Fractional Gaussian Noise (FGN) models (e.g. Mandelbrot 1971) or the Broken Line model developed by Mejia, et al. (1972). Burges and Lettenmaier (1977) used the Mandelbrot FGN model, and the Matalas Markov model to study the effects of parameter uncertainty in annual streamflow sequences on the sizing of storage reservoirs under a variety of demand conditions. They conclude that the Hurst effect is most important for high demand levels. For low demand levels, correct preservation of skewness and coefficient of variation seem to be much more important than correct modeling of the Hurst number. In a wide range of practical storage problems, correct modeling of the Hurst effect is unimportant. Burges and Lettenmaier also point out that constraints on feasible combinations of correlation and Hurst number for the FGN model, impose severe limitations on its use for the generation of annual streamflows. Srikanthan and McMahon (1978b) also compared Factional Gaussion Noise models with Markov models. They found that although the FGN models preserved the Hurst coefficient in some cases, other parameters were not preserved very well. They concluded with the rather alarmist statement that "for operational purposes the model (FGN) is very expensive to run relative to other generating techniques and its use is not justified."

Our immediate interest in data generation was for studying the operation of a system of reservoirs for power production. A review of the literature as outlined above suggested that a multi-site lag-one Markov model as presented by Matalas (1967) would be the most appropriate method for our needs. Such a model appears to have a wide range of applicability but its inability to preserve the Hurst-coefficient makes it unsuitable for long-term, high-demand storage problems. The remainder of the paper is devoted to a rather detailed discussion of the application of a lag-one Markov model. Particular emphasis is placed on the difficulties encountered and methods available to alleviate or circumvent such difficulties.

MECHANICS FOR GENERATION - LAG-ONE MARKOV MODEL

The model outlined in this section is a multi-season multi-site lag-one Markov model designed for the generation of monthly data. The model is derived from that developed by Matalas (1967). Complete details of the model can be found in the Matalas paper.

The generation scheme is:

$$(X_{i+1} - MEAN_{i+1}) = A*(X_i - MEAN_i) + B*RANDOM_{i+1} \qquad (1)$$

where

X_i = (m x 1) matrix whose Pth element represents the time series value in month i at site P

$MEAN_i$ = (m x 1) matrix whose Pth element represent the mean of the time series in month i at site P

A = (m x m) conditional lag-one correlation matrix

B = (m x m) conditional lag-zero correlation matrix

$RANDOM_{i+1}$ = (m x 1) matrix whose elements are independent random normal deviates with mean of zero and variance of 1

m = number of sites

The (m x m) matrices A and B are given by:

$$A = M_{i+1,i}*M_{i,i}^{-1} \qquad (2)$$

$$B*B^T = M_{i+1,i+1} - A*M_{i+1,i}^T \qquad (3)$$

where

$M_{i+1,i}$ = (m x m) correlation matrix expressing the dependence of values in month (i+1) on those in month i

The elements of the M and MEAN matrices are obtained from the statistics of the historic monthly data (mean, variance, lag-zero, and lag-one serial, and cross correlations). For the generation scheme of Equations 1-3, the historic data is assumed to be from a multi-variate normal distribution. The generated synthetic events will then be sampled from a multi-variate normal distribution and will resemble the historic events in terms of mean, variance, and lag-zero and lag-one serial, and cross correlations.

If the population skewness cannot be assumed to be zero then skewness may also be preserved by using a three-parameter log-normal transformation of the historic data:

$$y = \log(x - SHIFT) \tag{4}$$

where

y = transformed data
x = original time series
SHIFT = constant shift

The use of such a transformation implies an assumption that the historic data follow a log-normal distribution. The transformed variables Y thus follow a normal distribution so the generation scheme in Equation 1 may be used. Matalas (1967) and Charbeneau (1978) discuss the mathematically correct methods of parameter estimation for preservation of the historic statistics.

In solving Equations 2 and 3 to determine matrices A and B, it should be noted that matrices $M_{i,i}$ and $B*B^T$ are symmetric. $B*B^T$ may be solved for B using a Choleski decomposition (Pinder and Gray 1977).

The model as presented is quite general; it can be applied to any time series for which the basic assumptions are reasonably met. For example, we can use the model to generate monthly rainfall amounts, streamflow volumes, water quality indices, traffic flow, etc. At this point, it is worth summarizing the assumptions and requirements inherent in the model. These are:

1. Stationarity of the time series.
2. For a lag-one Markov model, the lag-k serial correlation coefficient is ρ_i^k where ρ_i is the lag-one serial correlation. This relationship is shown in Figure 1 for $\rho_i = 0.5$.
3. Data is sampled from a multi-variate normal or log-normal distribution.
4. Correlation matrices $M_{i,i}$ must be positive definite.
5. Matrix $B*B^T$ must be positive-definite.

Problems of stationarity and the inherent Markovian correlation structure will be addressed in the next section of the paper. These problems do not at least prevent data generation even though the generated data may be invalid.

On the other hand if matrix $B*B^T$ (Equation 3) is not positive definite then it cannot be decomposed and the data generation scheme of Equation 1 cannot be used. The covariance matrix $M_{i,i}$ may be "inconsistent" or "non-positive definite" if the statistics for multi-site generation are obtained from historic sequences of different lengths (Fiering 1968). Matalas and Wallis (1974) show that even if $M_{i,i}$ is positive definite it does not follow that $B*B^T$ will also be positive definite. The correlations obtained from the data must satisfy a set of constraints imposed by the use of the multi-variate normal generation. Violating the constraints may show that the assumption of a multi-variate normal distribution is simply invalid for the population or sample in question. Whatever the reason for the inconsistency, matrices $M_{i,i}$ or $B*B^T$ must be adjusted to make

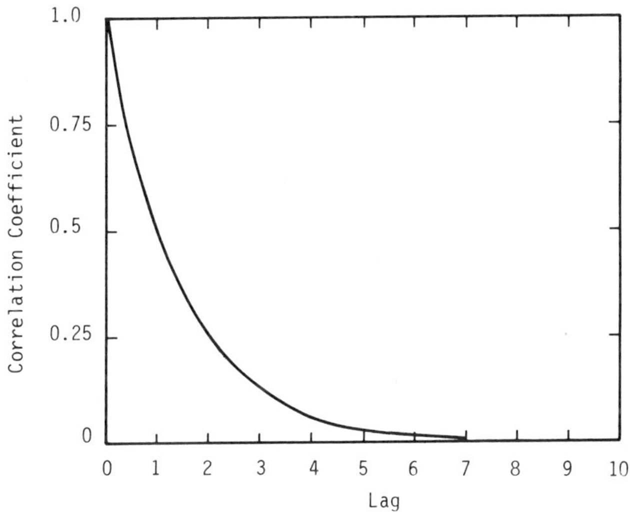

Figure 1. Correlogram for lag-1 Markov Model (ρ=0.5)

them positive definite before data generation can proceed. Fiering (1968) presents two methods for adjusting inconsistent matrices. We successfully used Fiering's eigen space approach for adjusting matrices, but did not rigorously attempt to minimize the difference between the original inconsistent matrix and the adjusted positive-definite matrix.

DIRECT GENERATION OF STREAMFLOW DATA

We will now consider direct generation of streamflow data. By this we mean the analysis of existing streamflow records to estimate population statistics and use of these statistics to generate additional monthly streamflow data using the Markov model discussed above.

Streamflow generation has its own peculiar set of problems. Let us first look at a basic problem related to multi-site generation on a stream network such as that of Figure 2. We wish to generate flows at points A,B,C, and D and we come across a physical requirement not addressed by the model. For monthly streamflow volumes we generally require:

$$\text{Flows at C} > \text{Flows at A} \tag{5}$$

$$\text{Flows at D} > \text{Flow at C} + \text{Flow at B} \tag{6}$$

i.e. downstream points have greater monthly volumes than upstream points. Although there may be a strong cross correlation between the various points, preservation of the correlation does not guarantee that the generator will always produce flows such that (5) and (6) are met. The random component of the model may for example produce flows at A greater than C for some months in the synthetic trace. We have not studied the implications of this feature of streamflow modeling, but a convenient way of getting around the problem is to create time series containing the incremental flows between the points of interest. In the network of Figure 2, we would create the following time series:

A, C-A, B, D-C-B

These flows are then divided by the tributary areas between the various points to produce four time series of land surface runoff per unit area. Data can then be generated to preserve the characteristics of the land surface runoff. Manipulation and summation of the generated series gives us the required streamflow sequences at points A,B,C, and D.

As pointed out earlier, use of the lag-one Markov model implies a fixed relationship between the lag-one serial correlation and the lag-k serial correlation (Figure 1). In areas with a pronounced dry season, such as California, summer streamflows are controlled by groundwater levels at the end of the winter and by the recession rate. The multiple-lag correlation structure of the historic data may be quite different from that implied by use of a lag-one model. To preserve the historic correlation structure, we would have to use a

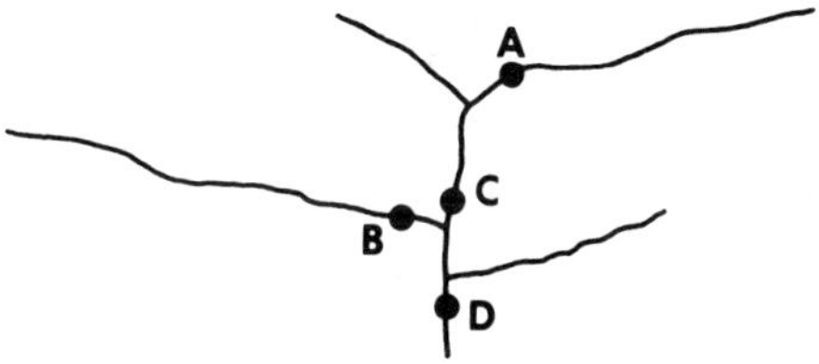

Figure 2. Stream network

multi-lag model (Jackson and Fiering 1971) or alternatively generate annual streamflow data and then find some mechanism for disaggregation to monthly amounts.

A further problem with streamflow generation is that of stationarity of the data. Streamflow volumes are particularly sensitive to changes of land use within the watershed. Urbanization of a watershed or conversion from forest to farmland will produce trends in the streamflow data. The data should be corrected to remove the trends and the adjusted data should reflect conditions representative of the future. If rainfall data is available this problem may be solved by use of a rainfall/runoff model to produce a streamflow time series reflecting current or future land use. However this introduces a very attractive alternative to direct streamflow generation; that is generation of rainfall data followed by a rainfall/runoff transformation to produce the synthetic streamflow sequences.

RAINFALL GENERATION FOLLOWED BY RAINFALL/STREAMFLOW TRANSFORMATION

Generation of streamflow by first generating rainfall and then transforming rainfall to streamflow offers a number of significant advantages over direct streamflow generation; it also has a few occasionally fatal flaws.

Let us first assume that our raingage network gives an accurate description of monthly areal and temporal variations in rainfall and that the calibrated rainfall/streamflow model accurately simulates streamflow volumes produced by the input rainfall. Use of a deterministic rainfall/streamflow transformation with detailed description of the stream network firstly permits generation of streamflow at any number of points within a stream network, including ungaged sites. The generated flows are moreover always consistent in that downstream points will experience greater flows than upstream points.

In the cases we have studied, lag-one correlations of monthly rainfall data have always been quite small (< 0.2) and the multiple-lag correlations insignifcant. It appears that the correlation structure of monthly rainfall data in general fits that implied by a Markov model much better than does streamflow data. A correct rainfall/streamflow model should maintain such features as a dry season recession in streamflows and the corresponding correlation structure.

A third point, previously mentioned in reference to direct streamflow generation, was stationarity. Assumptions of stationarity in rainfall data are usually much better founded than the same assumption for streamflow data, though the reasons for trends are less well understood. As pointed out earlier, some account may also be made in the rainfall/streamflow transformation for future changes in land use or channel configuration in the watershed.

The new and potentially insurmountable problem introduced by this approach is clearly in the rainfall/streamflow transformation. For the approach to be acceptable, the rainfall/streamflow transformation has to be "reasonably accurate." Unfortunately it is very difficult to make general quantitative statements about the accuracy achieved by the various models for various climate types, raingage densities and soil characteristics etc. However we do know from experience that the transformation is difficult and inaccurate in areas prone to either intense localized rainfall (convective storms) or in areas where streamflow is produced by snowmelt. This streamflow generation technique should not be used in such areas. Beyond those guidelines questions of accuracy can only be answered after attempts have been made to calibrate the rainfall/streamflow model for the area in question. However other questions of accuracy are also pertinent. How do errors in the statistics of the historic rainfall affect the generated streamflow. For example, if the variance of the historic monthly rainfall amounts is changed by 10 percent, how would this affect the generated streamflow? Unfortunately these questions have not received much attention, but they lead to another extremely important topic, both for rainfall generation and direct streamflow generation; that of parameter uncertainty.

PARAMETER ESTIMATION

Two questions should be asked when estimating the parameters or historic statistics used in the data generation scheme:

1. What reliability can I place on the parameters obtained from a record of a given length?
2. What is the effect of parameter uncertainty on the results of my study?

In general, rainfall records are longer than streamflow records and so one would expect the rainfall statistics to be more reliable than those from the streamflow data; but let us consider each of the parameters in turn.

Mean

Estimates of the mean by the method of moments are unbiased. It would seem that for most rainfall and streamflow sequences a twenty year record would give reasonably accurate estimates of the mean. Wallis, et al. (1974), using Monte Carlo simulation, show the effect of record length on the variance of the mean for samples from several distributions. This data could be used to give a general indication of uncertainty in the mean.

Variance and Skewness

Small sample bias of variance and skewness estimates has been studied in detail by Wallis, et al. (1974). Wallis used Monte Carlo simulation techniques to investigate parameter bias for small samples taken from a variety of distributions. For the log-normal, which we used exclusively in our work, Wallis shows bias in both skewness and standard deviation as a function of skewness and sample size. The bias factors Wallis computed are shown in Figures 3 and 4. Notice the dramatic increase in bias for sample sizes less than 20, and the increase in bias as skewness increases. Wallis actually derived complete distributions for small sample estimates of standard

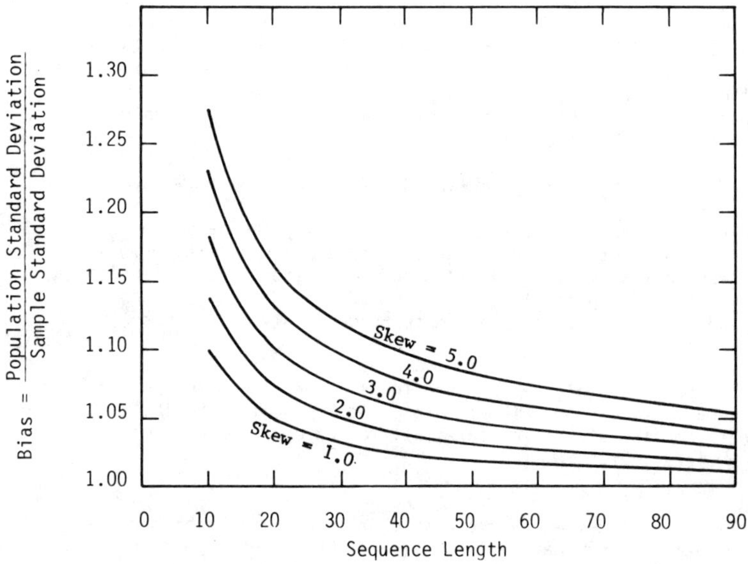

Figure 3. Short sample bias factors in standard deviation for samples from a log normal distribution (Wallis, et al. 1974)

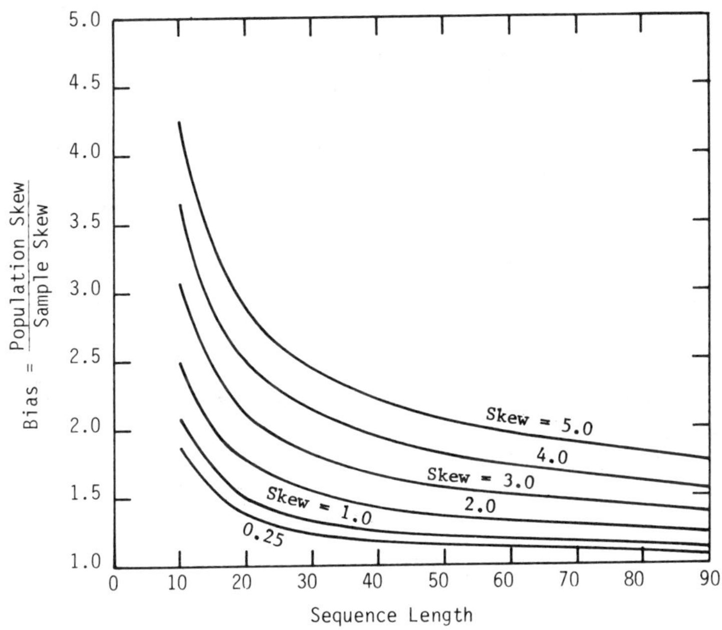

Figure 4. Short sample bias factors in the skew for samples from log normal distribution (Wallis, et al. 1974)

deviation and skewness, so the Wallis paper also gives indication of the variability in the parameter estimates. Of course we cannot say exactly what distribution our sample comes from but Wallis' work gives a good indication of the magnitude of the small sample problem. His bias factors should be used to adjust estimates of statistics from historic sequences.

It is also interesting to note that there is an algebraic bound on the sample skewness and coefficient of variation (Kirby 1974). In other words for a given length sample, the sample skewness cannot exceed a certain value irrespective of the skewness of the population. These bounds are shown in Figure 5.

The effect of parameter uncertainty on the result of any study is difficult to generalize and some form of sensitivity analysis should undoubtedly be done. However, Burges and Lettenmaier (1977) do present some results showing the effect of changes in the skewness and coefficient of variation of the inflow sequence on the sizing of storage facilities. It appears that an increased skewness will decrease storage requirements whereas an increased coefficient of variation will of course increase storage requirements for a given demand level.

Serial Correlation

There are several formulae for estimating the correlation coefficient. Since each approach has different short sample bias characteristics, care has to be taken in choosing an appropriate estimator. Different methods for computing correlation coefficient are discussed by Wallis and O'Connell (1972). They performed Monte Carlo simulations to investigate the short sample bias of the most efficient estimators. Specifically they sampled from normal data generated by a lag-one Markov process. Typical results are shown in Figure 6. Note that for high population correlations, the sample statistics may be grossly underestimated for short sequence lengths. For streamflow sequences underestimation of serial correlation coefficients will underestimate storage requirements.

It is of interest to compare the range of the statistics exhibited by monthly rainfall and streamflow sequences. In almost all situations the lag-one serial correlation of rainfall will be considerably less than that of the streamflow. For monthly data in Minas Gerais, Brazil for example the lag-one correlation coefficient in the dry season is about 0.9 for streamflow and less than 0.1 for rainfall. The lag-one correlation coefficient is thus an important parameter for direct streamflow generation and is subject to a large short sample bias. The coefficient appears to be relatively unimportant for rainfall generation.

The situation with other statistics is unfortunately not so obvious. Rainfall generally exhibits higher skewness and higher coefficient of variation than streamflow. On the other hand, the longer records generally available for rainfall data could result in much more accurate estimates of statistics than those obtained from a shorter

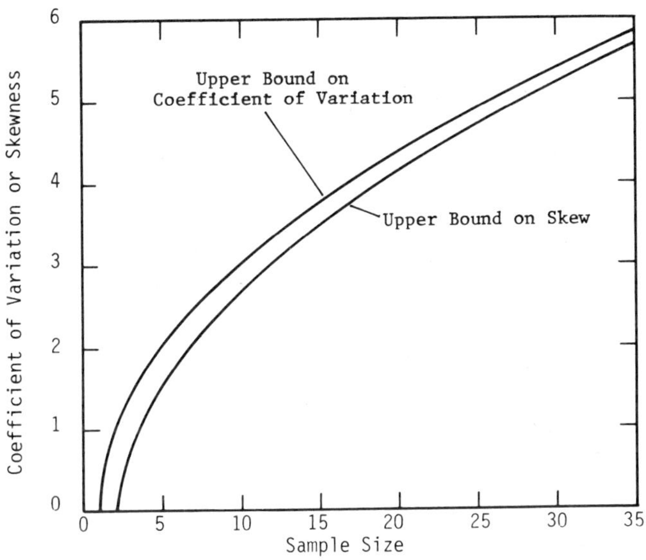

Figure 5. Upper bound on sample statistics

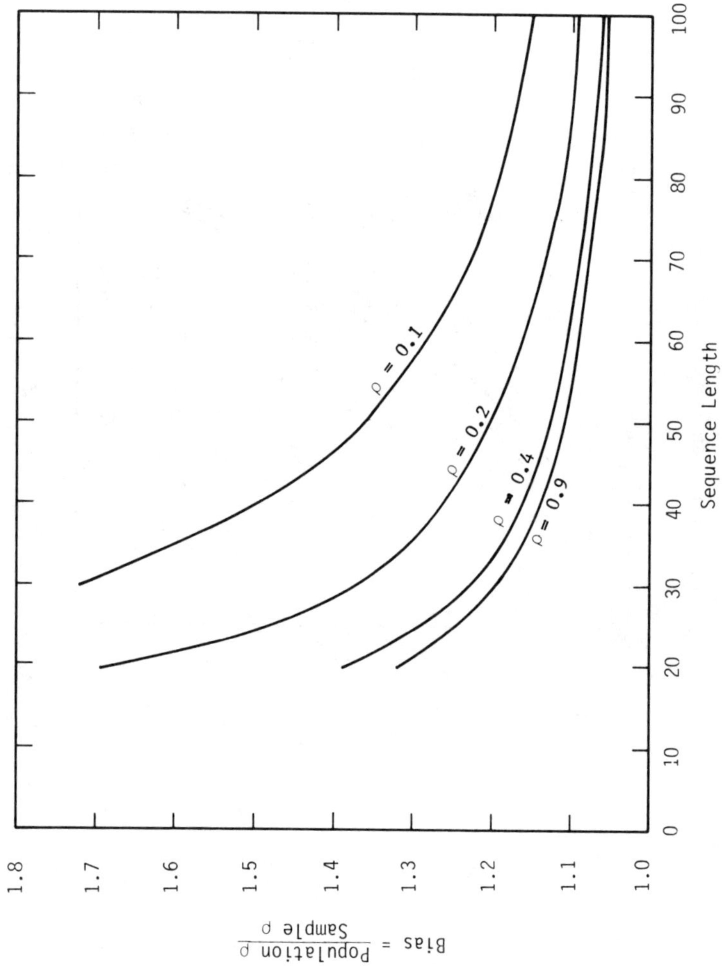

Figure 6. Short sample bias factors in lag one correlation for samples from a lag 1 Markov process (Wallis and O'Connell, 1972)

streamflow record (Figures 4 and 5 give some indication of this).

SUMMARY - DIRECT OR INDIRECT STREAMFLOW GENERATION?

Some of the advantages and disadvantages of direct and indirect streamflow generation are summarized in Table 1. The decision as to which method to use in a particular case hinges on data availability, the potential accuracy of the rainfall/streamflow transformation, and some judgement as to the improvment of results possible with the indirect method vis-a-vis the increased effort which may be involved. Note that although the indirect method involves calibration of a rainfall/streamflow model, it does greatly facilitate streamflow generation at multiple sites.

One important point about the indirect approach requiring further investigation is that we do not yet know how errors in rainfall statistics affect the streamflow quantities produced.

EXAMPLE OF INDIRECT STREAMFLOW GENERATION

Monthly rainfall generation followed by a rainfall/streamflow transformation was used on a watershed of 29300 km^2 in Brazil (Hydrocomp 1978). The intention was to generate additional streamflow data at a number of existing streamgage sites for use in operation studies on a proposed series of reservoirs. The watershed, the Upper Paranaiba, and the location of rainfall and streamflow gages is shown in Figure 7. Streamflow data was to be generated at the four streamgage sites Porto Jacinto, Porto da Barra, Cachoeira da Sertao, and Ponte Festelita Campos. Rainfall data was available for a period of 18 years at seven sites within the basin. The gages and length of record are shown in Table 2. Note that the streamflow record is particularly short at Porto Jacinto.

Despite the poor siting of the raingages for simulation, it was felt that in view of the short streamflow records, the indirect generation approach would produce more reliable results. The first step was the generation of rainfall at the seven gaged sites. Ninety-nine years of data was generated; the statistics of recorded and generated data for a typical site are shown in Figures 8, 9, and 10. The statistics of the generated data agree well with those of the recorded data (within the approximate 95% confidence limits). The principle difficulty encountered was generation of data for the dry season months June through August. There were a substantial number of months in which no rainfall was recorded and the data undoubtedly are from a mixed distribution and not from a log-normal distribution as we assumed. The effect of the zero data was to introduce inconsistent (non-positive definite) $B{*}B^T$ matrices. These were adjusted to allow generation using Fiering's eigen space approach.

Transformation of rainfall to streamflow was achieved using a version of HSP (Hydrocomp 1976) calibrated for the Paranaiba watershed. Figure 11 compares mean monthly simulated and recorded runoff at Ponte Estelita Campos for the 13 year period of record, using the calibrated model. Annual volumes agree well, but there is a bias in the monthly

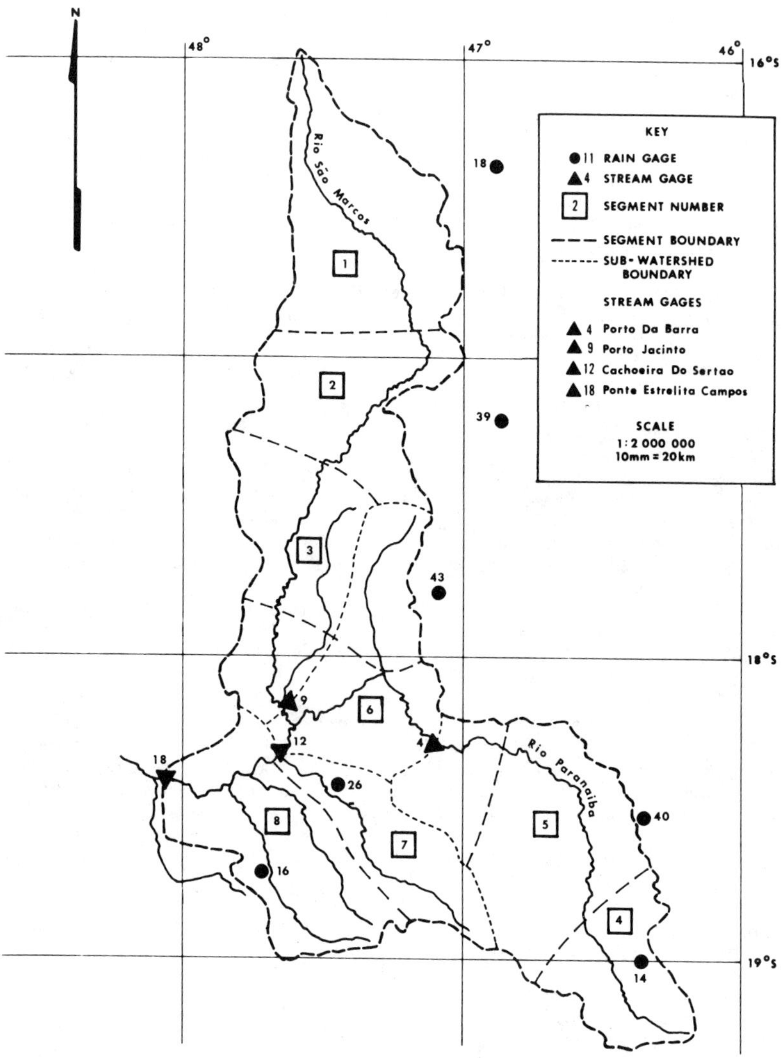

Figure 7. Rio Paranaiba watershed

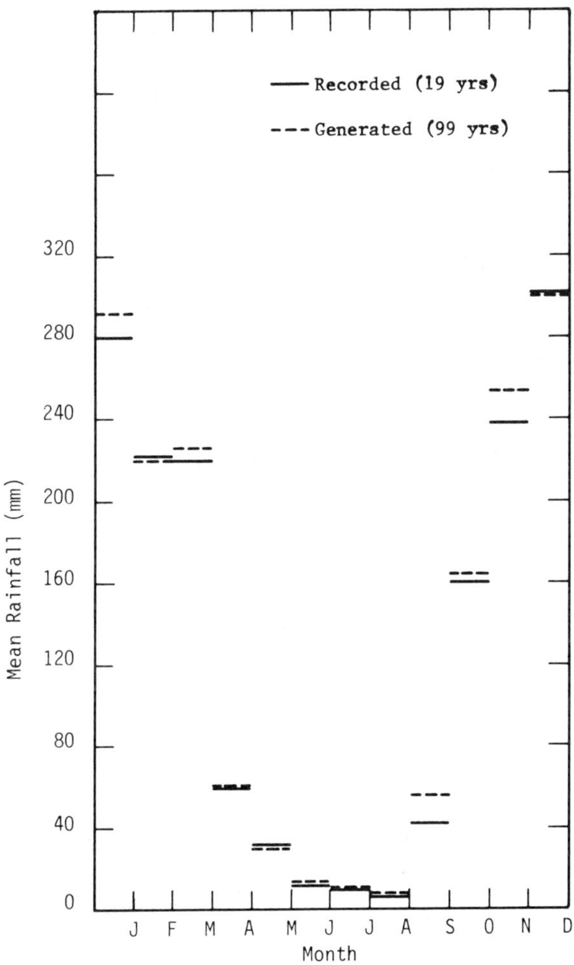

Figure 8. Mean monthly rainfall at station 16

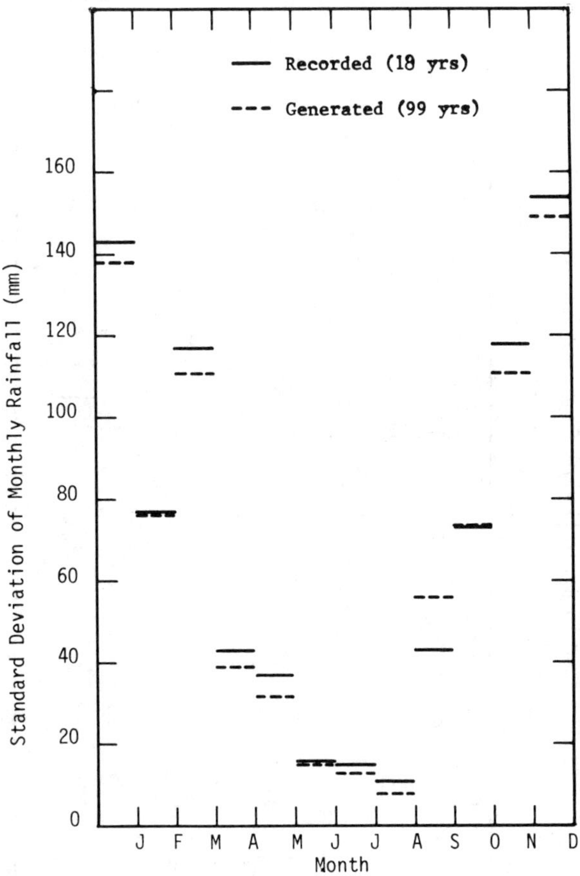

Figure 9. Standard deviation of monthly rainfall at station 16

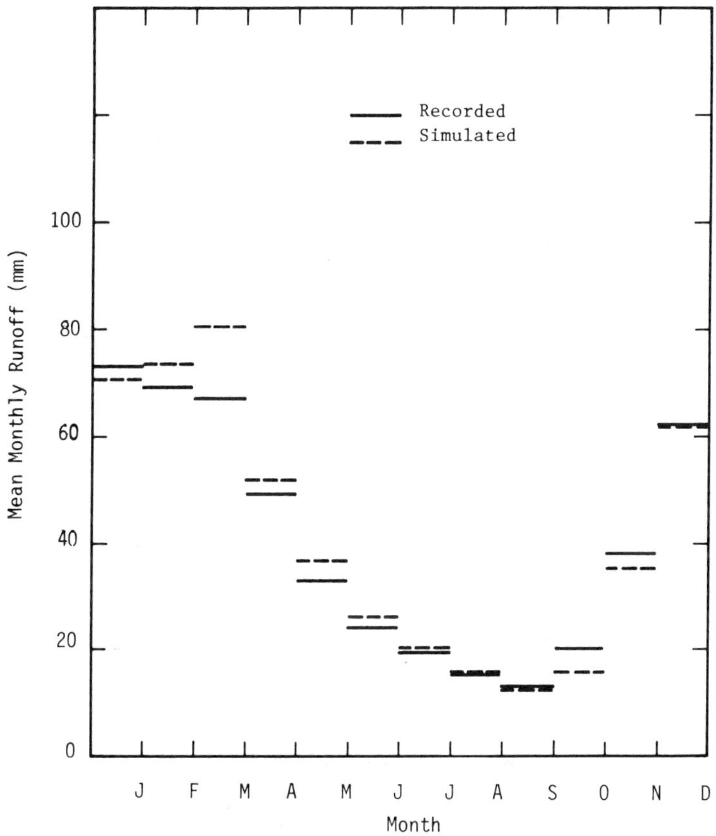

Figure 11. Recorded and simulated runoff at Ponte Estelita Campos

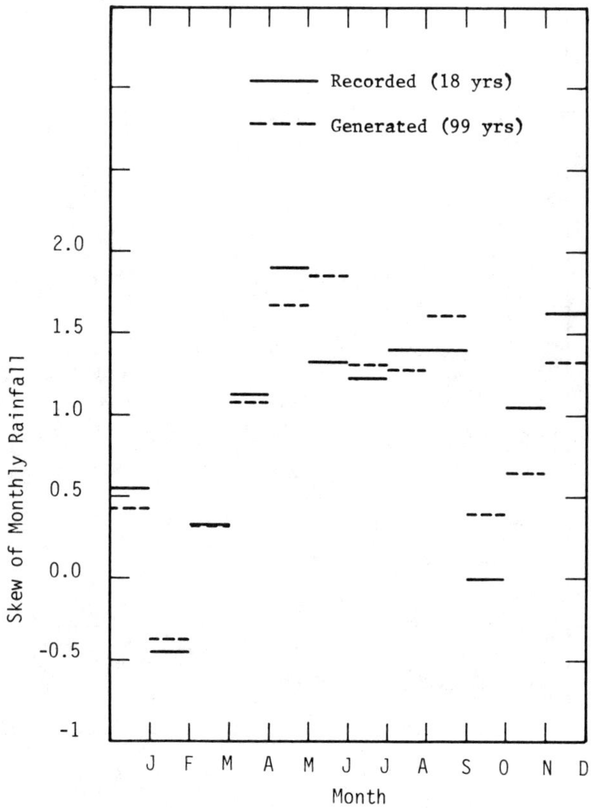

Figure 10. Skew of monthly rainfall at station 16

Table 1. COMPARISON OF DIRECT AND INDIRECT STREAMFLOW GENERATION

Topic	Direct Streamflow Generation	Rainfall Generation Followed by Rainfall/ Streamflow Transformation
Consistency of multi-site flow data (downstream upstream).	Inconsistencies possible.	Streamflow data always consistent.
Ease of flow generation at multiple sites.	Computational effort increases significantly. Difficulties at ungaged points.	Rainfall generated at a small number of sites can be easily used to generate streamflow at many sites, including ungaged sites.
Impact of imposed multi-lag correlation structure.	Occasionally distorts generated data.	Unimportant.
Impact of stationarity requirement.	May require data adjustment if changes have occured in the watershed during the period of record.	Probably unimportant.
Parameter uncertainty.	Largely a function of record length. Rainfall records are often larger than streamflow records.	
Errors in rainfall/ streamflow transformation.	Not Applicable.	Largely a function of climate and raingage density. Errors probably untenable in snowmelt regimes and in areas with considerable areal variations in rainfall.

Table 2. RECORD LENGTHS

Station		Period of Record
Precipitation	Station 14	January 1958 - January 1977
Precipitation	Station 16	January 1958 - January 1977
Precipitation	Station 18	January 1959 - January 1977
Precipitation	Station 26	January 1958 - January 1977
Precipitation	Station 39	July 1958 - February 1977
Precipitation	Station 40	July 1958 - January 1977
Precipitation	Station 43	July 1958 - January 1977
Flow	Porto da Barra	January 1958 - May 1977
Flow	Porto Jacinto	April 1972 - May 1977
Flow	Cachoeira do Sertao	May 1967 - May 1977
Flow	Ponte Estelita Campos	October 1963 - May 1977

simulation with slight oversimulation in the dry season recession and undersimulation at the beginning of the wet season.

The 99 years of synthetic rainfall data were used in the calibrated rainfall/streamflow model to produce a corresponding period of synthetic streamflow. The recorded and generated flow statistics at Ponte Estelita Campos are shown in Figures 12 and 13. The record at the gage was for approximately 13 years; 1963 to 1977. The rainfall data however covered the period 1958 to 1977 with years 1961 and 1962 being amongst the wettest on record. The streamflow record at Ponte Estelita Campos was extended to include years 1958 to 1977 using simulation.

Figures 12 and 13 show statistics for the original record, the extended record, and the generated data. Notice that extension of the record often has a substantial effect on both the mean and standard deviation of the monthly flows.

CONCLUSIONS

Despite the difficulties encountered in calibration on the Paranaiba, the results from indirect streamflow generation are quite encouraging. Comparison of the statistics of the record and extended record in Figures 12 and 13 suggest that indirect generation produced a more reliable synthetic sequence than could be obtained from the existing record at Ponte Estelita Campos. However, it is appparent that bias in the calibration will also bias the synthetic streamflow. Until we gain more experience using this approach it is difficult to say at what point calibration difficulties should force us to use direct streamflow generation.

REFERENCES

Burges, S.J. and D.P. Lettermaier. 1977. Comparison of annual streamflow models. Journal of the Hydraulics Division, ASCE, 103(HY9):91-1006.

Charbeneau, R.J. 1978. Comparison of the two- and three-parameter log normal distributions used in streamflow synthesis. Water Resources Research 14(1):149-150.

Fiering, M.B. 1968. Schemes for handling inconsistent matrices. Water Resources Research 4(2):291-297.

Fiering, M.B. and B.B. Jackson. 1971. Synthetic streamflows. Water Resources Monograph 1. American Geophysical Union, Washington, D.C. pp. 98.

Hydrocomp, Inc. 1976. Hydrocomp simulation programming operations manual. Palo Alto, California.

Hydrocomp, Inc. 1978. Hydrologic simulation of the Rio Paranaiba, Brazil. Palo Alto, California. pp. 178.

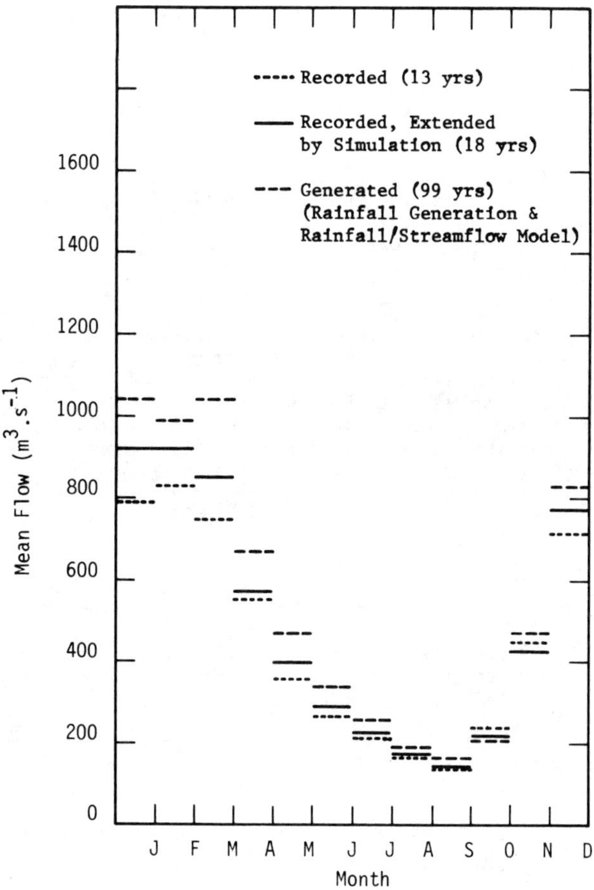

Figure 12. Mean monthly flows at Ponte Estelita Campos

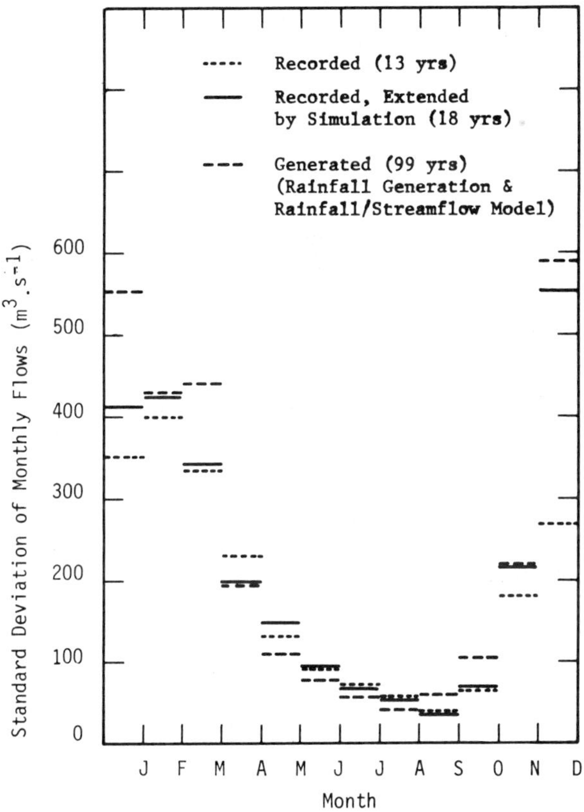

Figure 13. Standard deviation of monthly flows at Ponte Estelita Campos

Kirby, W. 1974. Algebraic boundedness of sample statistics. Water Resources Research. 10(2):220-222.

Mandelbrot, B.B. 1971. A fast fractional gaussian noise generator. Water Resources Research, 7(3):543-553.

Matalas, N.C. 1967. Mathematical assessment of synthetic hydrology. Water Resources Research, 3(4):937-945.

Matalas, N.C. and J.R. Wallis. 1974. Correlation constraints for generating processes, in Mathematical Models in Hydrology. Proceedings of the Warsaw Symposium, July 1971. IAHS/UNESCO/WMO, vol. 1, pp. 372-376.

Mejia, J.M., I. Rodriguez-Iturbe, and D.R. Dawdy. 1972. Streamflow simulation, 2. The broken line process as a potential model for hydrologic simulation. Water Resources Research, 8(4):931-941.

Pinder, G.F. and W.G. Gray. 1977. Finite element simulation in surface and subsurface hydrology. Academic Press, New York.

Srikanthan, R. and T.A. McMahon. 1978a. A review of lag-one Markov models for generation of annual flows. Journal of Hydrology, 37:1-12.

Srikanthan R. and T.A. McMahon. 1978b. Generation of annual streamflow using a filtered fractional gaussian noise model. Journal of Hydrology, 37:13-21.

Wallis, J.R. and P.E. O'Connell. 1972. Small sample estimation of Water Resources Research. 8(3):707-712.

Wallis, J.R., N.C. Matalas, and J.R. Slack. 1974. Just a moment! Water Resources Research. 10(2):211-219.

PROBLEMS IN OPERATIONAL HYDROLOGIC FORECASTING

Jack F. Hannaford and Roderick L. Hall[1/]

STATEMENT OF PROBLEM

As with other critical natural resources, the water use in California is continually increasing, and there is a continuing and growing need for water management skills and refined operational tools to make the best use of available water supply as well as to control flow during periods of excess supply or rate of flow. This paper addresses the problem of forecasts and forecast error, primarily from the viewpoint of water supply forecasting, but the same type of problems and errors persist for short range forecasts, such as flood forecasts, which may in many cases go only a few hours or days into the future.

Historically, water supply forecasting, the assessment of potential future volumetric runoff, has been one of the most valuable operational tools for water management, particularly in those watersheds where the lag between winter snowfall and spring runoff provides an opportunity for determination and prediction of volumetric water supply. Winter accumulation of snowpack and snowmelt during spring and early summer in California's Sierra Nevada provides a situation in which very effective application of water supply forecasting has been made.

Many techniques have been utilized to relate the measured quantity of water stored in the watershed, particularly in the form of snowpack, to resulting runoff volume. The intended use of the forecast in operation usually dictates the type of forecast procedure and relative level of refinement required. All forecast procedures have some degree of error associated with the relationship between runoff and its causitive parameters. Sources of this "forecast error" have been examined by many investigators, and the relative magnitude of error in the forecast "procedures" is well understood by those working in the field.

Unfortunately, water management decisions can not always wait after the major portion of the seasonal precipitation has occurred

1/ Engineering Consultants, Sierra Hydrotech, Placerville, Ca.

and snowpack has accumulated. Often decisions must be made in February, January or even earlier, committing seasonal water delivery which is anticipated from precipitation occurring subsequent to the date of forecast and decision. Error related to the inability to accurately predict future weather conditions may have a major impact on estimates of future runoff. This "weather related" error is dominant early in the season (prior to the time when the major portion of the seasonal snowpack accumulation can be measured). Recognition of the probability of "weather related" error is absolutely essential to application of early season forecasts to water management decisions. "Weather related" error constitutes the major limitation on the use of water supply forecasts in general, and a thorough and accurate understanding of "error probability" is absolutely essential for effective application of any water supply forecasting technique to operational decisions.

RELATING RUNOFF TO CAUSUAL PARAMETERS

Precipitation is the primary source of water for streamflow runoff. The characteristics of the watershed govern losses within the watershed and the portion of that precipitation resulting in surface runoff as well as the time-distribution of runoff. Various techniques may be used to relate precipitation to corresponding runoff from a given watershed. These techniques vary in complexity, but are primarily dependent upon the intended use of the results and the historic data available to develop or calibrate the procedure. As a general rule, the shorter the time period of runoff to be simulated (i.e., daily runoff as opposed to seasonal runoff), the more complex and sophisticated the procedure must be.

Hydrologic Models

The nomenclature "model" or "hydrologic model" has been overworked in scientific publications for a number of years. Many "models" have been developed to represent natural and man-made processes. Mathematical models range from the most simplified representation of a process to the very complicated representation of many processes to simulate a desired result.

A "model" represents a technique for mathematical simulation of physical processes or relationships. Perhaps the least sophisticated forecasting "model" to relate precipitation and runoff would be a simple plot of measured precipitation (as an index to basinwide precipitation) against runoff for the historic period of record. Figure 1 relates observed precipitation at Caribou Power House in the Feather River basin to observed unimpaired runoff of Feather River near Oroville (Figure 2). Two things are immediately apparent from this plot. First, there appears to be a substantial amount of scatter,

FIGURE 1

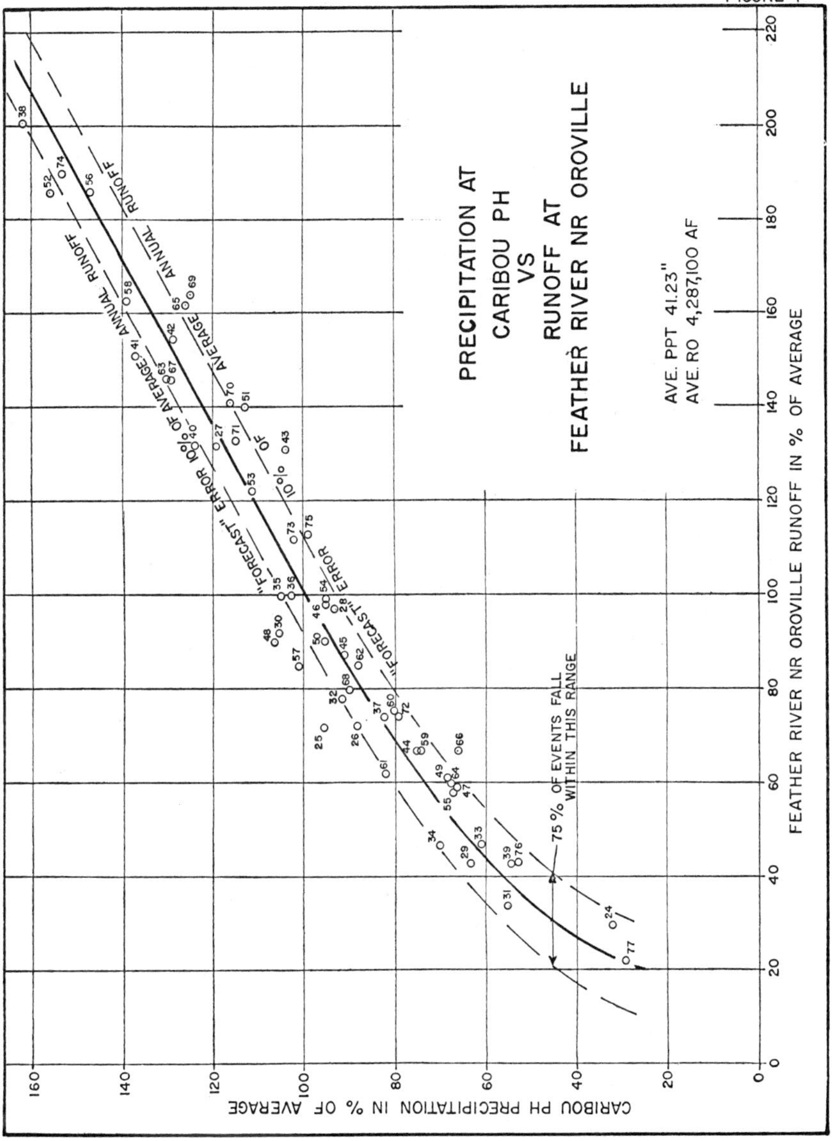
PRECIPITATION AT
CARIBOU PH
VS
RUNOFF AT
FEATHER RIVER NR OROVILLE
AVE. PPT 41.23"
AVE. RO 4,287,100 AF
"FORECAST" ERROR 10% OF AVERAGE ANNUAL RUNOFF
75 % OF EVENTS FALL WITHIN THIS RANGE
FEATHER RIVER NR OROVILLE RUNOFF IN % OF AVERAGE
CARIBOU PH PRECIPITATION IN % OF AVERAGE

suggesting that in using this very simple approach, errors as great as 20 percent could be expected in estimating runoff, even when all precipitation data were known. Second, there appears to be curvature at the lower end of the relationship suggesting that a "straight line" relationship between precipitation and runoff should not be expected for all ranges of flows, and that caution must be exercised in extrapolating procedures for estimating flows under very extreme conditions. In addition, if this procedure were to be used operationally, it would become apparent that only minimal information is available to the forecast user on the time-distribution of flow and since the procedure relates only to water year runoff, any forecast error would be concentrated in the remaining runoff subsequent to date of forecast. These are serious deficiencies in flexibility for operational forecasting.

Complex hydrologic models to describe every detail of the hydrologic cycle are at the opposite extreme of techniques for relating precipitation to runoff. There are many very complex hydrologic models currently in operation intended to perform many different tasks. Various academic organizations, operating agencies, and hydrologic specialists have developed models applicable to certain tasks, including operational forecasting. At the present time, the most sophisticated hydrologic models may be of academic interest only, since many of the data required for their operation are unavailable on an operational basis. It should also be pointed out that even a sophisticated conceptual hydrologic model is not entirely free from restriction of the basic data and must be calibrated to relate to observed conditions. Conditions beyond the limits of historical data may require special consideration, regardless of the technique used to relate runoff to causual parameters.

Most techniques in current use to relate future runoff to its causitive parameters fall intermediate between these extremes. Operational techniques represent compromises involving available data, complexity of development, complexity of operation, economic aspects, and intended results. Most techniques used are very dependent upon the range of historic data and the skill of the hydrologist for extrapolation under extreme conditions. With the advent of the electronic computer, there may often be a tendency to rely too greatly or completely on the computer to determine the shape and characteristics of the hydrologic relationships. The computer itself is no substitute for the trained hydrologist in developing techniques which will provide hydrologically sound results under all conditions.

More generally, a typical hydrologic model intended for operational forecasting would consist of numerous mathematical relationships representing the various hydrologic processes occurring within the watershed. A typical hydrologic model might be intended to simulate daily flow from a watershed over some extended time period utilizing various hydrologic, climatologic, and meteorologic parameters including precipitation, snowpack, and temperature. Although developed primarily as a mathematical simulation of daily runoff, such a model might have application as a hydrologic forecast tool to enable project operators to make timely and pertinent analysis regarding operational decisions. A good hydrologic model represents a tool to enable the hydrologist to evaluate the effect upon runoff of various sequences of assumed or observed climatologic or meteorologic events which might represent either historical or hypothetical conditions.

One of the most serious deficiencies in many forecast procedures for operational purposes is lack of flexibility to meet the needs for operational investigation of critical situations on a real time basis as the operational season unfolds. Perhaps the greatest advantage obtainable in water supply forecasting from utilization of a hydrologic model with at least some intermediate degree of sophistication is the flexibility which can be made available to permit the hydrologist and water manager to answer specific questions regarding probability and time-distribution of runoff. It should be emphasized that familiarity with the capabilities and limitations of the technique for hydrologic simulation and capability of interpretation of results are perhaps even more important than the particular model or technique used to relate runoff to significant parameters.

FORECAST ERROR

Hydrologic forecasts, especially those made early in the water year, seldom verify precisely. In fact, one would not expect them to verify precisely when a significant portion of the season during which precipitation is anticipated remains in the future. An understanding of the failure of forecasts to verify precisely and the potential sources and magnitude of this forecast "error" is essential to application of water supply forecasts to operational decision making. This understanding should include not only a philosophical ("of course I know a forecast can't be 100 percent accurate") understanding, but also knowledge of the impact on project operations that would result should forecast "error" occur at any level of probability.

Sources of Error

Although there are any number of sources to which forecast "error" may be attributed, there are three basic sources of forecast error (ignoring human error). These are 1) basic data error, 2) forecast procedure error, and 3) weather-related error.

Basic Data Error

A portion of error in relating runoff to pertinent hydrologic parameters may be the result of error in the historic basic data from which the forecast procedure has been developed. There may be error in recorded historic parameters which has not or can not be eliminated even by careful editing and review. The magnitude of such error is usually indeterminant, but in many cases should not be considered insignificant. Care must be taken in the acquisition, editing and preanalysis of historic data to prevent inclusion of errors in the historic data file.

During the forecast season, certain basic data may be erroneous, unavailable, or at least not available on a timely basis so that estimates must be made of the effect of erroneous or missing data on the hydrologic forecast at the time the forecast is prepared. Careful editing or incoming data may be required to detect erroneous information before it is used in forecast preparation. In general, qualified personnel with good background in the hydrology of watersheds being forecasted can minimize the consequences of erroneous or missing data at the time of forecast. Although each season has a substantial amount of missing data due to equipment failures, storm activities, and other factors, this is probably not a serious problem in most water supply forecasts. Nevertheless, care must be taken to assure availability of adequate data during forecast preparation.

A problem partially related to basic data and partially related to forecast procedure development is that of representativeness of data. For example, how representative is measured snowpack water content at a group of snow courses within the watershed of the volume of water stored in the snowpack and available to produce snowmelt runoff? How representative is precipitation measured at the lower elevation, manned stations of the unobserved precipitation falling in the higher elevation, water producing areas of the basin? Compromises must always be to utilize those data which are actually available or practical to obtain. Representativeness of new data or increased or revised data networks would fall into this category. Problems involving representativeness of current data with respect to past data from the same source may also at times exert influence on forecast accuracy.

Forecast Procedure Error

Forecast procedure error is that error related to inability to completely describe the dependent variable, runoff, using the various hydrologic parameters in a forecast procedure. This error may be partially related to lack of true representativeness in the historic data from which the procedure has been developed or failure of the technique used to relate runoff and other factors to truly simulate the hydrologic relationships in the watershed.

All forecast procedures or hydrologic models intended to develop runoff forecasts are subject to procedure-type errors related to representativeness of both data and technique. Degree of error may vary between watersheds, and, to some extent, the level of error may be controlled during procedure development. However, a very high level of refinement necessary to supress error could result in unjustifiably large expenditures of time and effort in procedure development, data collection, and operational forecasting.

Weather Related Error

Under most conditions, the largest contributor to operational forecast error is that error related to inability to accurately predict weather conditions subsequent to the date of forecast. There is a high degree of variability in precipitation from month to month, season to season, and year to year throughout the Sierra Nevada, which in turn contributes to an even more exaggerated variability in runoff. Figure 2 represents the degree of variability for water year precipitation at Caribou Power House in the Feather River basin and water year runoff of Feather River near Oroville. Values are expressed in terms of percentage of the 50 year averages. Note that there may appear to be a tendency for years to group in terms of quantities, but a predictable pattern is not readily discernible.

If the data in Figure 2 are arrayed by magnitude, they may be plotted against their probability of occurrence during the period of historic record. Figure 3 represents observed water year precipitation and runoff plotted against probability of being exceeded. For example, there is a 90 percent chance that precipitation at Caribou PH will be 60 percent of average or greater (and a corresponding 10 percent chance that it will be less than this value) in any given season, based on the assumption that seasonal amounts of precipitation are random by season. There is a 50 percent chance that precipitation will be 98 percent of average or greater. The 50 percent value is the "median". In most hydrologic data, there is a tendency for the median to fall somewhat below the average, and generally in the case of precipitation and runoff from the same area, the median runoff will fall further below average than the median precipitation. In the drier areas of California's Sierra Nevada, there is a tendency in both precipitation and runoff records for median values to fall further below average values than in the wetter areas.

Probability analysis may be made for precipitation subsequent to any given date of forecast as well as for an entire water year. Such analysis is usually predicated on the supposition that precipitation

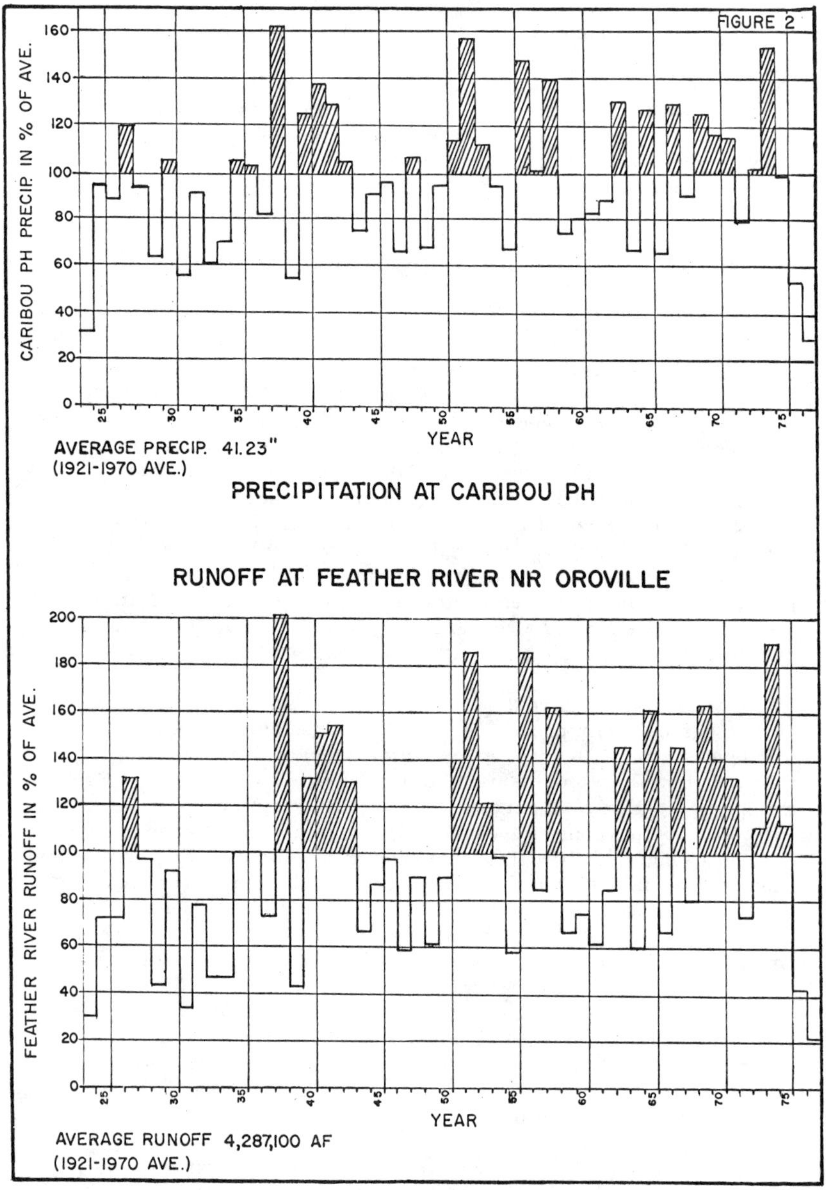
FIGURE 2
CARIBOU PH PRECIP. IN % OF AVE.
YEAR
AVERAGE PRECIP. 41.23"
(1921-1970 AVE.)
PRECIPITATION AT CARIBOU PH
RUNOFF AT FEATHER RIVER NR OROVILLE
FEATHER RIVER RUNOFF IN % OF AVE.
YEAR
AVERAGE RUNOFF 4,287,100 AF
(1921-1970 AVE.)

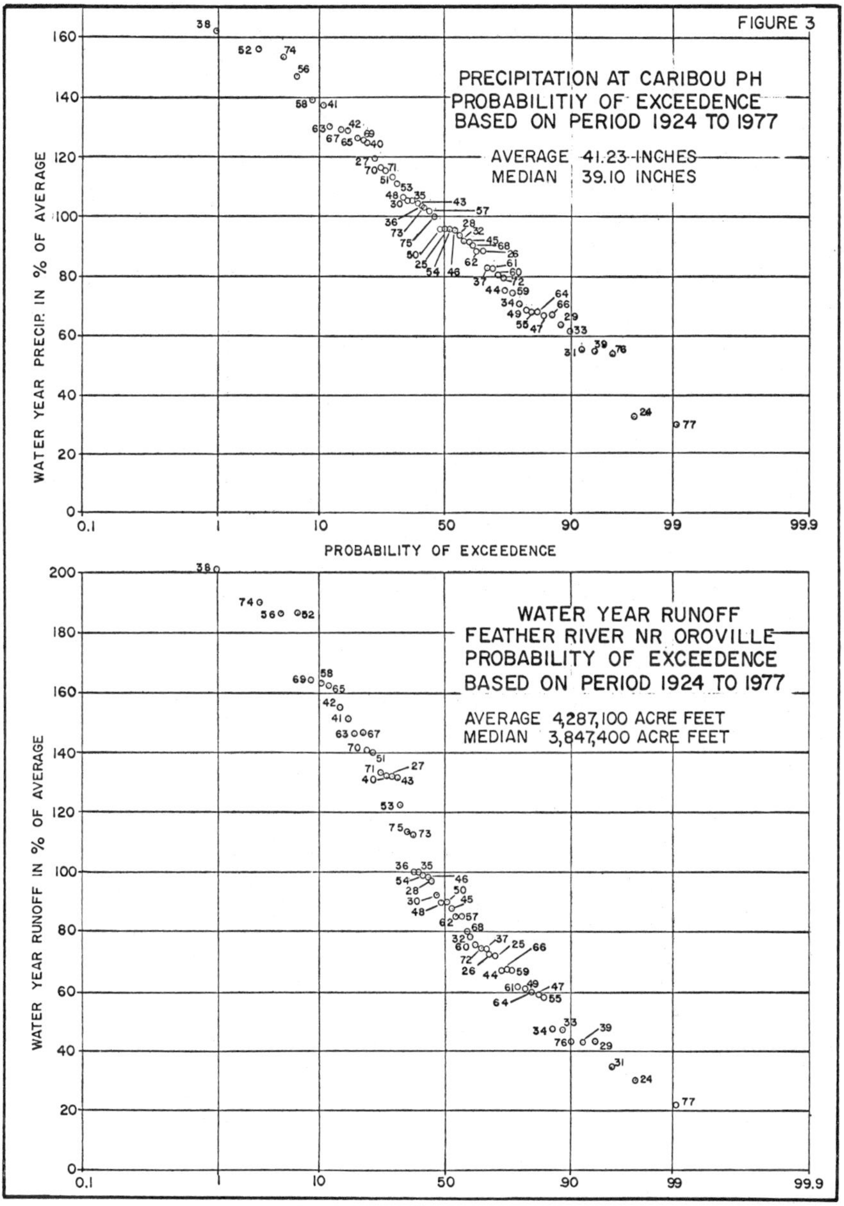
FIGURE 3
PRECIPITATION AT CARIBOU PH
PROBABILITIY OF EXCEEDENCE
BASED ON PERIOD 1924 TO 1977
AVERAGE 41.23 INCHES
MEDIAN 39.10 INCHES
WATER YEAR PRECIP. IN % OF AVERAGE
PROBABILITY OF EXCEEDENCE
WATER YEAR RUNOFF
FEATHER RIVER NR OROVILLE
PROBABILITY OF EXCEEDENCE
BASED ON PERIOD 1924 TO 1977
AVERAGE 4,287,100 ACRE FEET
MEDIAN 3,847,400 ACRE FEET
WATER YEAR RUNOFF IN % OF AVERAGE

prior to and precipitation subsequent to the date of forecast are independent of each other. That is, quantity of precipitation to a given date would not necessarily be related to the quantity of precipitation after that date. Although it is recognized that there may be some element of carryover or serial correlation between precipitation amounts prior to and subsequent to a given date during the season, data do tend to exhibit a relatively high degree of independence for practical analysis. In Figure 4, observed precipitation during given water years prior to date of forecast has been plotted against observed precipitation subsequent to date of forecast. Plots have been made for "forecast" dates of January 1 and March 1. The points exhibit a high degree of scatter and no relationship is readily discernible between precipitation prior to and subsequent to date of forecast. It must not be assumed that no relationship exists, and in practice, adjustment may be made for any degree of interdependence or carryover effect exhibited in the historic data.

This all points up the fact that perhaps the greatest increase in operational hydrologic forecasting, particularly early in the water year (say up through April 1) would be achieved through development of skills in long-range meteorologic projections, and not necessarily in sophistication of forecast procedures. However, degree of skill and numerical description of error have yet to be demonstrated in long-range meteorological forecasting.

Combined Error

Conventional hydrologic procedures for water supply forecasting are based upon hydrologic and meteorologic conditions as they appear on the date of forecast plus the assumption of certain specified climatic conditions subsequent to the date of forecast. Generally, forecasts are based on the assumption of either median or average climatic conditions, but any desired or required condition subsequent to the date of forecast could be assumed. Knowing the historic occurrence of climatic conditions subsequent to date of forecast (for a period of perhaps 50 years), the magnitude and probability of forecast error may be defined by conventional methods and applied to operational problems. Forecast error as of a given date is related principally to the three independent error types described above.

Error-Probability Diagram

A typical relationship between date of forecast and probability of forecast error for the Kings River basin in the southern Sierra Nevada is delineated on Figure 5. This diagram is intended to illustrate error in remaining volume of runoff from date of forecast through the end of the water year. The forecasting procedure for which this diagram was developed employs ongoing evaluation of data through June 1.

FIGURE 4

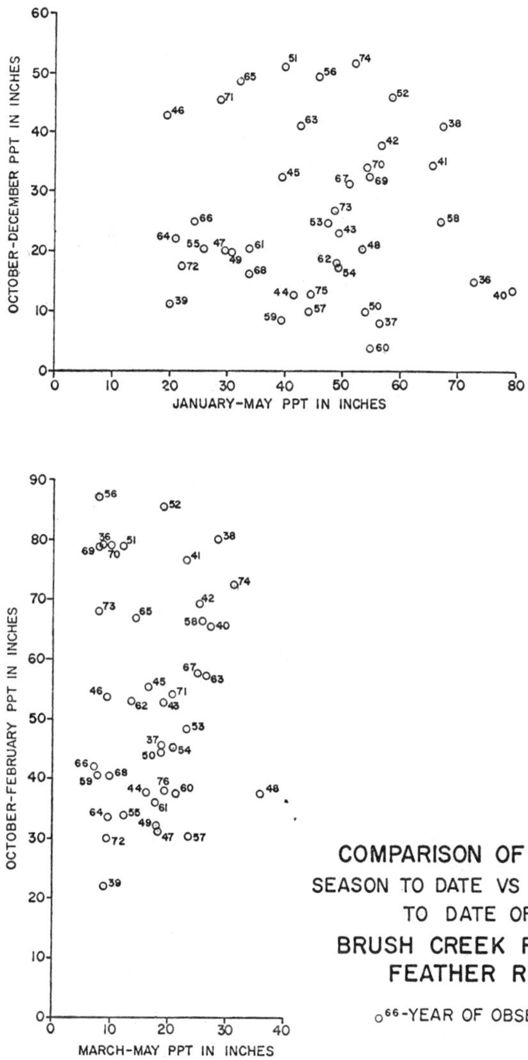

COMPARISON OF PRECIPITATION

SEASON TO DATE VS SEASON SUBSEQUENT TO DATE OF FORECAST

BRUSH CREEK RANGER STATION FEATHER RIVER BASIN

o[66]-YEAR OF OBSERVATION

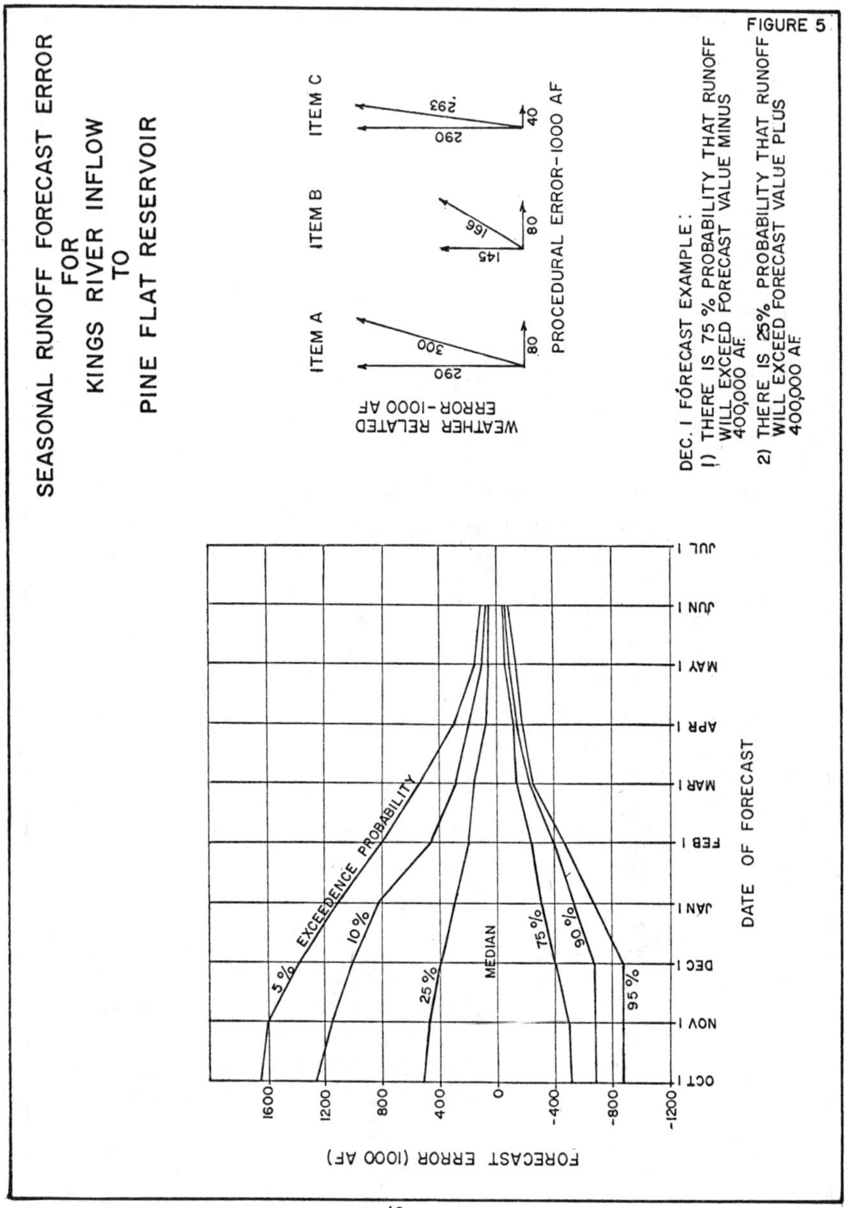
SEASONAL RUNOFF FORECAST ERROR
FOR
KINGS RIVER INFLOW
TO
PINE FLAT RESERVOIR
ITEM A
ITEM B
ITEM C
WEATHER RELATED ERROR-1000 AF
290
300
80
145
166
80
290
293
40
PROCEDURAL ERROR-1000 AF
DEC. 1 FORECAST EXAMPLE:
1) THERE IS 75 % PROBABILITY THAT RUNOFF WILL EXCEED FORECAST VALUE MINUS 400,000 AF.
2) THERE IS 25% PROBABILITY THAT RUNOFF WILL EXCEED FORECAST VALUE PLUS 400,000 AF.
FIGURE 5
FORECAST ERROR (1000 AF)
1600
1200
800
400
0
-400
-800
-1200
5 %
10 %
EXCEEDENCE PROBABILITY
25 %
MEDIAN
75 %
90 %
95 %
OCT 1
NOV 1
DEC 1
JAN 1
FEB 1
MAR 1
APR 1
MAY 1
JUN 1
JUL 1
DATE OF FORECAST

12

Forecast error, positive (observed greater than forecast) and negative (observed less than forecast), is represented on the vertical axis while date of forecast is represented on the horizontal axis. Exceedence probability is represented by lines forming a funnel shape across the diagram. Exceedence probability indicates the percentage probability that observed runoff will exceed a forecast runoff based on median conditions subsequent to date of forecast by the intercept read from the vertical scale, noting the algebraic signs. As the season progresses, the likelihood of a forecast error of given magnitude decreases until June 1, and, as far as this particular procedure is concerned, remains at the June 1 level for the remainder of the water year. An example of computation of forecast ranges and probability is indicated on Figure 5 for a December 1 forecast.

It should be noted that in operational practice, Figure 5 is actually three dimensional, with the third dimension (not delineated on Figure 5) representing the character of the season being forecast in terms of runoff volume. Drier seasons are bounded on the negative side by zero, since it is not possible to have less than zero precipitation (or runoff) subsequent to the date of forecast. Wetter seasons, on the other hand, might tend to produce a greater amount of incremental runoff from each additional inch of precipitation than would drier seasons. For complete flexibility, error-probability analysis must consider the overall size of the season being forecast.

Examination of the Kings River diagram will reveal that the lower or negative side of the diagram has less "spread" than the positive side, particularly at extreme values. This "skew" results from the fact that the negative side of the diagram is limited by zero precipitation subsequent to the date of forecast, while the extremes on the positive side might result from precipitation several times the median value or more. Note that this situation is different than, but not completely independent of, the problem described in the paragraph above. The skew appears to be even more pronounced on the lower elevation, higher precipitation basins of the northern Sierra, such as the Feather River, than on the relatively high elevation Kings River basin (Figure 6). The degree of skew is considerably less as median is approached from either the negative or positive side.

The error probability diagrams have been presented here primarily for demonstration. Using an electronic computer with a "library" of historic climatologic and hydrologic probability parameters and an adequate forecast procedure, an error probability analysis can be generated responsive to observed conditions. This approach removes the requirement for the "three dimensional" error probability diagram which would, in practice, be a difficult procedure to generate and utilize. The technology is available to perform error probability analysis on the basis of 1) climatologic probability from an

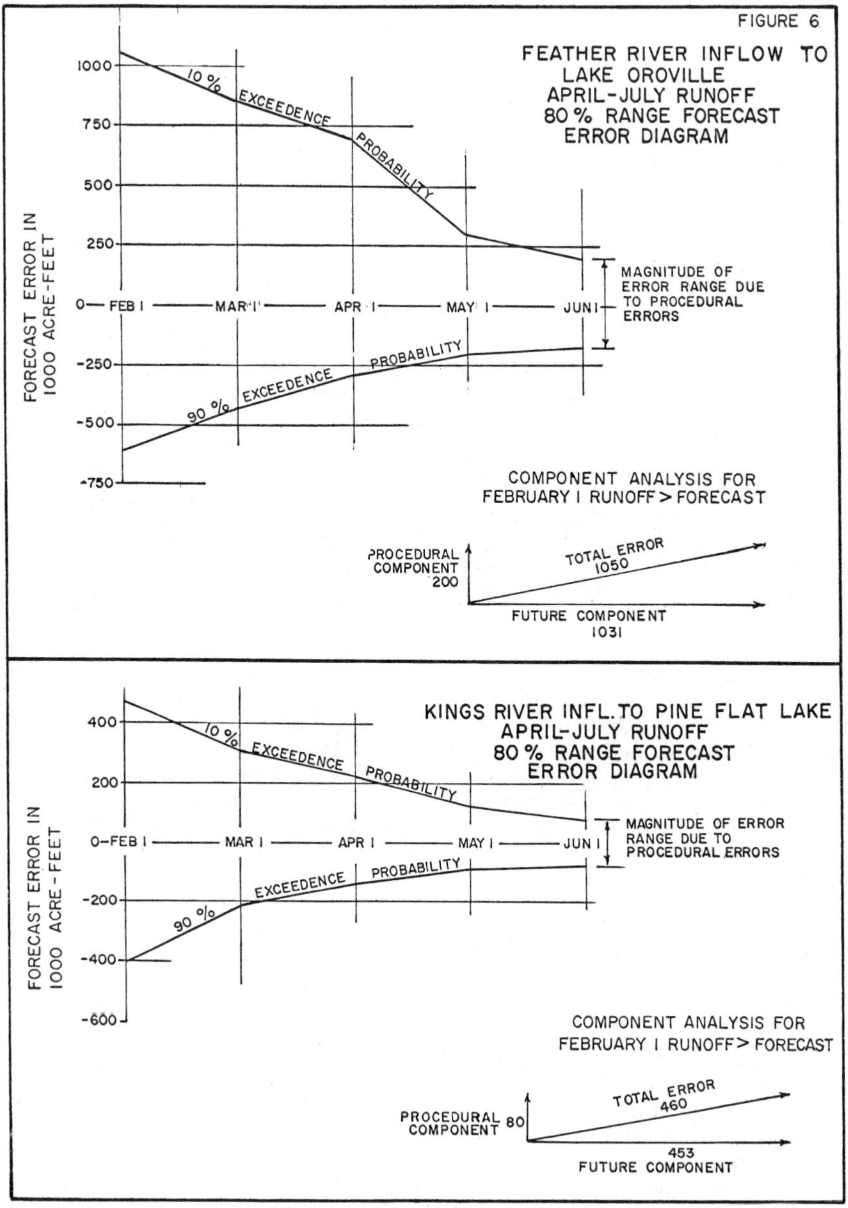
FEATHER RIVER INFLOW TO
LAKE OROVILLE
APRIL-JULY RUNOFF
80 % RANGE FORECAST
ERROR DIAGRAM
FORECAST ERROR IN
1000 ACRE-FEET
1000
750
500
250
0
-250
-500
-750
FEB 1
MAR 1
APR 1
MAY 1
JUN 1
10 % EXCEEDENCE PROBABILITY
90 % EXCEEDENCE PROBABILITY
MAGNITUDE OF
ERROR RANGE DUE
TO PROCEDURAL
ERRORS
COMPONENT ANALYSIS FOR
FEBRUARY 1 RUNOFF > FORECAST
PROCEDURAL
COMPONENT
200
TOTAL ERROR
1050
FUTURE COMPONENT
1031
KINGS RIVER INFL. TO PINE FLAT LAKE
APRIL-JULY RUNOFF
80 % RANGE FORECAST
ERROR DIAGRAM
FORECAST ERROR IN
1000 ACRE-FEET
400
200
0
-200
-400
-600
FEB 1
MAR 1
APR 1
MAY 1
JUN 1
10 % EXCEEDENCE PROBABILITY
90 % EXCEEDENCE PROBABILITY
MAGNITUDE OF ERROR
RANGE DUE TO
PROCEDURAL ERRORS
COMPONENT ANALYSIS FOR
FEBRUARY 1 RUNOFF > FORECAST
PROCEDURAL
COMPONENT
80
TOTAL ERROR
460
453
FUTURE COMPONENT
FIGURE 6

historic record, 2) estimated probability of error in long-range meteorologic projections, or 3) any other practicable system for estimating the future weather-related error.

Error-Type Comparison and Combination

Forecast error as delineated on Figure 5 is composed of two independent elements. The first independent type of error is "procedure error" which results from inability of the basic data and forecast procedure to portray observed runoff when all parameters (except the dependent -- remaining runoff) are available. Procedure error, for purposes of this discussion, includes both "forecast procedure error" and "basic data error" described above. This type error would represent that error which could potentially be reduced by use of more refined data, new parameters or better techniques for analysis. The second type of error is caused by inability to predict future weather conditions subsequent to date of forecast at the time a forecast is made. In the illustration above (Figure 5) on June 1 the entire error can be considered as procedure error (for this particular illustration). Prior to June 1, the combined error includes a factor related to weather subsequent to date of forecast which varies substantially with time of year. Since the two error types may be considered as independent, they are not directly additive, but can be combined vectorially at any given date and level of probability to obtain an estimate of the combined error. Item A in the inset on Figure 5 shows the relative levels of prcedural and weather-related error combined in the total error (diagonal line) for March 1 at the 10 percent exceedence level. If the error resulting from inability to forecast March, April and May precipitation could be cut in half, the reduced total error would be represented by Item B, Figure 5. For example, the March 1 weather related error of 290,000 acre-feet and procedural error of 80,000 acre-feet would combine to give a total March 1 error of 300,000 acre-feet at the 10 percent exceedence level (Item A). Reducing the procedure error by 50 percent would result in an overall March 1 error of 293,000 acre-feet (Item C), while reducing error attributable to weather by 50 percent would result in a combined error of 166,000 acre-feet (Item B), illustrating the relative importance of these two error types at this date. Even total elimination of procedure error would reduce total error only to 290,000 acre-feet. By June 1, however, cutting the procedure error in half would also cut the "combined error" in half. Figure 7 illustrates forecast improvement obtainable at the 10 percent and 90 percent exceedence levels by completely eliminating procedure error for all forecast dates, demonstrating the diminishing effect of improving forecast procedure error for forecasts made early in the forecast season.

It should be noted that that error probability diagram illustrated was developed on the basis of combined error for about 45 years of

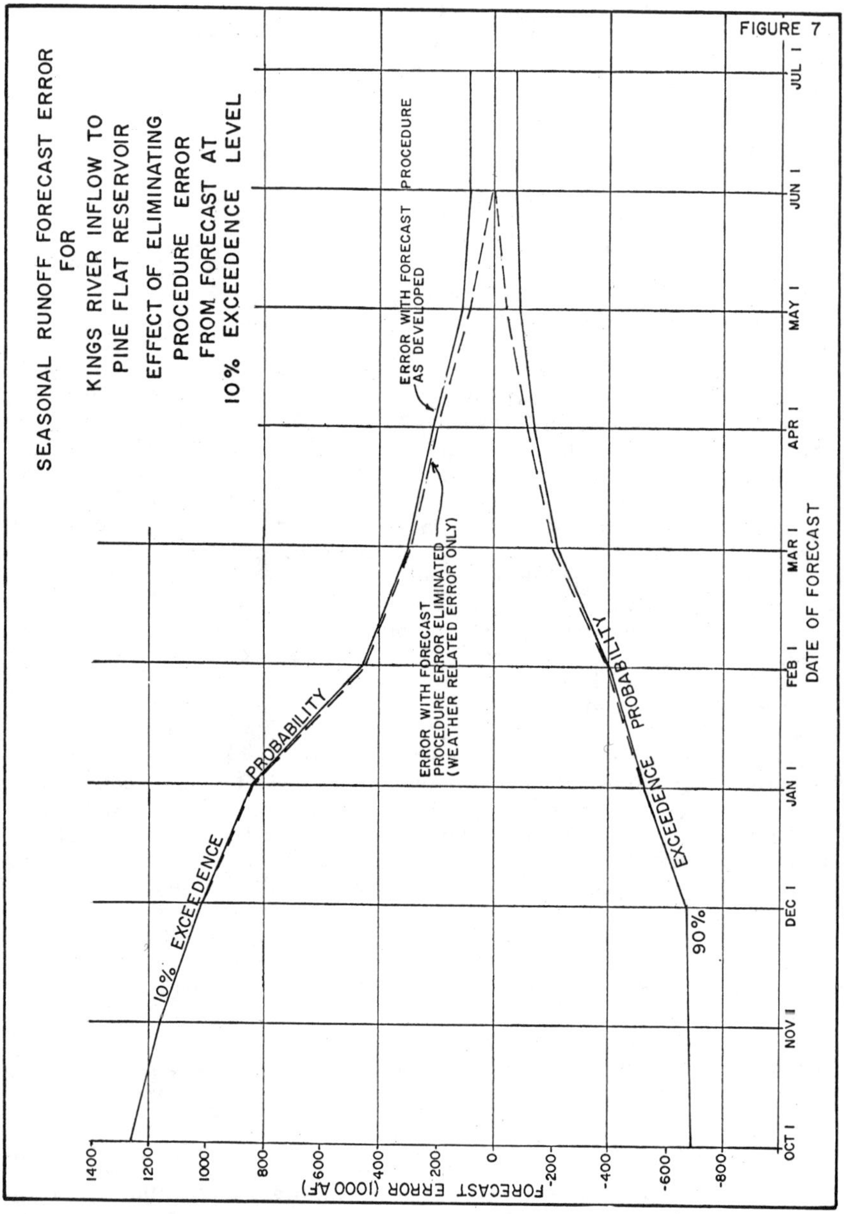
SEASONAL RUNOFF FORECAST ERROR
FOR
KINGS RIVER INFLOW TO
PINE FLAT RESERVOIR
EFFECT OF ELIMINATING
PROCEDURE ERROR
FROM FORECAST AT
10% EXCEEDENCE LEVEL
FIGURE 7
ERROR WITH FORECAST PROCEDURE AS DEVELOPED
ERROR WITH FORECAST PROCEDURE ERROR ELIMINATED (WEATHER RELATED ERROR ONLY)
10% EXCEEDENCE PROBABILITY
90% EXCEEDENCE PROBABILITY
FORECAST ERROR (1000 AF)
1400
1200
1000
800
600
400
200
0
-200
-400
-600
-800
OCT 1
NOV 1
DEC 1
JAN 1
FEB 1
MAR 1
APR 1
MAY 1
JUN 1
JUL 1
DATE OF FORECAST

historical record. The reference to different error types is for illustrative purposes only, but represents an accurate picture of the relative importance of these error types as the season progresses.

The described relationship between procedural and weather-related errors is typical of most western snowmelt streams and although the relative importance of error types may vary depending upon basin hydrology and climatic conditions, the basic problem still persists. Lower elevation watersheds, such as the Feather River where snowmelt season runoff may account for less than 40 percent of water year runoff, may have a proportionally greater weather-related vector than the higher elevation Kings River where snowmelt season runoff averages 75 percent of water year runoff and a large portion of the snowmelt season water supply is stored as snowpack even as early as February 1. Figure 6 illustrates the relative difference in characteristics of error probability diagrams for the Feather and Kings River basins for the February 1 through June 1 forecasts of April-July runoff.

Application of refined techniques to estimate runoff volume may be effective in reducing procedure error of conventional volumetric water supply forecasts. The potential for achieving a major decrease in forecast error on April 1 or earlier is limited primarily by physical, climatological and statistical considerations related to variability in future weather conditions, and only to a very minor extent by the sophistication of the procedure to relate observed forecast parameters to resulting runoff. In California, even the most optimistic reduction in procedure error through major changes or refinment in techniques or data would not provide a material increase in forecast accuracy in the early season. Forecast procedures used currently in most of the Sierra are relatively good, and probably any increase in accuracy attributable to refinement in techniques or data would be overshadowed by weather related error prior to the beginning of spring snowmelt.

SUMMARY AND CONCLUSIONS

There are many techniques, procedures, and hydrologic models which have been developed pertinent to water supply forecasting representing methodologies currently in use. The objectives of a forecast procedure and the methods for interpretation of forecast results must of necessity be closely related to the intended use of the forecast and the specific forecast problems. Forecasts of water supply which are intended for public distribution through the media may be very different in concept and presentation from those intended for water management decision making and operational planning. A single forecast expressed in acre-feet or percentage of average may be satisfactory for general public distribution, but a thorough understanding and application of error probability is a prerequisite for many areas of water management operation.

Forecast methodology for operational forecasting is extremely dependent on the intended use of the forecast. In general, the shorter the time period of runoff being forecast (i.e. days or weeks as opposed to season or water year), the more sophisticated the technique required. No single forecasting technique has a clear cut advantage over all others. The adequate application of water supply forecasting techniques to operational management is not so much related to type of procedure used as to the ability of the techniques to respond to the operational needs of the water manager. Hydrologic models do, for the most part, represent a more sophisticated and possibly more realistic approach to simulation of streamflow than to index correlation techniques. Such models, if properly designed, can provide the flexibility for operational decision making.

Forecast error, particularly early in the forecast season (say prior to April 1 in California's Sierra Nevada) is highly dependent upon variability of weather conditions subsequent to the date of forecast. This weather related error decreases as the season progresses. The magnitude of weather related error in the early season is so great when compared with the magnitude of error introduced by currently used water supply forecast procedures that even complete elimination of procedure error would have only minor effect on overall water supply forecast accuracy until well into the period of snowmelt runoff on most California snowmelt streams.

An accurate and realistic understanding of "forecast error", its magnitude, composition, and characteristics, is extremely important in application of hydrologic forecasts to decision making in water management. This applies to all types of hydrologic forecasting, whether water supply or other types of hydrologic forecasts. A large component of the error in early season is related to weather conditions subsequent of the date of forecast and exerts a major influence on the forecast result whether the technique used to relate runoff to its causitive parameters is a simplified relationship or a relatively sophisticated hydrologic model. The water manager must understand the necessity for developing operational plans contingent upon the probability of certain climatologic and hydrologic conditions materializing. He must recognize that extreme conditions could be more critical than those observed in limited past record and although the probability of such extremes being encountered might be low, certain operational restraints may require consideration of such extremes. Misunderstanding or misinterpretation of the nature, magnitude and time-relationship of forecast error may be one of the most critical problems faced in water management at the present time.

Tropical Storm Surge Forecasting in the National Weather Service

by

Chester P. Jelesnianski[1]

INTRODUCTION

An abnormal rise of water generated by a tropical storm is called a storm surge. The surge is a long gravity wave and not to be confused with short period wind waves, surf, swell, etc. By storm surge we mean the still water heights with short period waves filtered out, that is, measurements with stilling wells such as tide gages. Along a coastline, the significant surge has a reach of about 100 miles with a duration of about 6 hours. There are notable exceptions to this reach and duration depending on storm size, residence time of the storm on the continental shelf, storm path, and speed of the storm along its path. For low-lying, onshore terrain, there can be massive inland flooding for many miles. Inland bodies of water such as bays, estuaries and lakes can be tilted to dangerously high levels; also, there is added storage of sea water from inland flooding and flow of sea water through mouths or channels connected to the sea.

In any one particular geographical area, observed surge data is scarce due to rarity of storm events. Consequently, appeals are made to dynamic surge modeling to ascertain flooding from tropical storms. The models can be used for either planning or forecasting purposes. By planning we mean the simulation of storm surges with hypothetical storms for design criteria; examples are maximum probable surge for specialized power plants, frequency of surge heights for insurance purposes, and disaster preparedness. By forecasting we mean computing a surge, for a specified lead time, with a real, in-situ storm provided by nature.

All dynamic surge models can deal with both planning and forecasting. There are basic differences in application however, such as the output format of surge values and the way basic input data is used. In planning, an outside arbiter distinguishes which storms are used as input to the model; also, a local geographic area only is usually con-

---

[1]Tropical Storm Surge Forecasting in the National Weather Service.
Chester P. Jelesnianski
Techniques Development Laboratory
Systems Development Office
National Weather Service, NOAA
Silver Spring, Md.

sidered. In forecasting, a storm exists and the user must predict any changes in the storm's character during the lead time of the forecast; also, the model cannot be limited to a particular region and all possible areas affected by any natural storm must be handled.

In this report, we substitute hindcasting for forecasting. By hindcasting we mean inputting actual observed historical storms to surge models; that is, we assume a perfect or adequate forecast has been made for any required lead time and run the model in a forecast mode.

A dynamic surge model will numerically solve a set of partial differential equations of motion, according to input data. By input data we mean, (1) storm driving forces, (2) pre-set values for undetermined coefficients in the equations of motion, (3) representation of real geography or basin data on a grid net, and (4) the initial and quiescent surface height of the basin waters before the storm affects the basin area. Some modelers concentrate on the grid net and numerical approaches and treat the input data cavalierly. This will not do for users of the model for no matter the sophistication of a dynamic surge model the final product is strongly dependent on the quality of the input data.

To arrive at storm driving forces, storm models must be developed; this is not an easy task. It is just as important to develop an adequate storm model as a dynamic surge model. A storm model is characterized by meteorological parameters and then there are undetermined coefficients to be set. It is desirable to have a storm model, to generate surges in a surge model, not sensitive to small changes or errors in storm parameters or coefficient values; if the surge is sensitive, then the final surge computations are suspect no matter the surge model's sophistication.

The undetermined coefficients in equations of motion can be set through calibration procedures so that observed and computed surges for historical storms have reasonable agreement. It may be dangerous to set these coefficients from a one storm event. This is so because meteorological observations of storm parameters (central pressure, storm size, storm track and speed along the track) are replete with errors and the coefficient values are then suspect. It is desirable to set the coefficients in a universal manner by some form of best fit calibration with many storms in many geographical areas.

No surge model is complete without basin data. Preparing basin data is an art and not readily automated. Vertical, rather than horizontal, accuracy for topography is a must; especially so with barrier discontinuities such as roads, levees, spoil banks, etc. The placement of barriers along grid points can be subjective, and exhaustive sensitivity tests need to be performed. Surge generation is especially sensitive to topographic basin data for inland flooding, such as overtopping of barriers and flow through channels connecting bodies of water.

When using a surge model in a forecast mode it must be complete in the sense of storing prepared basin data, pre-set values for coefficients in the equations of motion, and the availability of an adequate storm model. The only input data required from a user are, (1) simple storm parameters for the required lead time of an in-situ storm, and (2) the initial and quiescent height or datum of water surfaces in the absence of a storm.

It is well to know that surge models do not deal directly with short period wind waves. An exception is the long-period, wave set-up at coasts from breaking wave activity. Some modelers merely add a hypothetical height to account for this effect; the NWS models account for it through the fixed coefficients. Usually, the wind waves break offshore with only residual effects inland; however, it waters are deep at the coast and the surge is substantial, then there can be tumultuous destruction from breaking wave activity at the coast.

SURGE MODELS USED BY THE NATIONAL WEATHER SERVICE (NWS)

The NWS uses two surge models, (1) SPLASH (Jelesnianski, 1972, 1974, 1976), and (2) SLOSH (Jelesnianski and Chen, 1979).

The SPLASH model is restricted to surge computations on a basin truncated from a large sea, including a 600 mile length of coastline. The equations of motion are simplified since inland inundation and extreme shallow depths are not considered. The output format of the model are (1) a line envelope of highest waters on the coast during the storm's history, and (2) a time-history of coastal surges, that is, a time-history of surges for any point on the coast. The model has been field tested by the National Hurricane Center at Miami, Florida for eight hurricane seasons. All required basins along the eastern seaboard and Gulf of Mexico, from the Mexican to the Canadian borders, for over 3500 miles of U.S. coasts, have been incorporated in the model. The real model time is 18 hours, requiring about 1 minute of CPU machine time for computations.

The SLOSH model is more comprehensive. It allows overtopping of barriers and coastal dunes, inland flooding from the sea, channel and other sub-grid size flow. The equations of motion are more complicated with non-linear flow to account for inland inundation, extreme shallow depths, and overtopping of barriers. The output format are (1) a surface envelope of highest waters on the entire basin during a storm's existence, (2) snapshots of water heights on the entire basin at selected time intervals, and (3) continuous time histories of surges at selected points on the basin. The model is now undergoing field tests at the National Hurricane Center and other Weather Offices, and various test runs have been made for planning purposes. The real time is 36 hours, requiring about 3 minutes of CPU time for computations.

The SLOSH model is relatively new and only a very few basins with bays, estuaries and lakes have been prepared for operational runs. The size of the basins are variable, according to the geography of inland water bodies relative to a coast; the basin must be large enough to hold inland water bodies, low lying terrain and the continental shelf. The preparation of basin data is an on-going project in the NWS and it will be several years before basin data is available for all bays, estuaries, lakes and continental shelves along the U.S. coastlines.

The only input to the models by a user are storm parameters and quiescent surface heights before the storm affects the basin area. In particular, boundary values with time are not an input requirement; the models use simple boundary conditions located at large distances from regions of interest.

The undetermined coefficients of the equations of motion are set identical in both surge models as universal and fixed constants. They do not vary in time, nor for particular basins, nor for particular storms. They have been set as best fit constants by comparing computed and observed surges for over 50 historical storms striking the U.S. coasts. The storm coefficients were similarly set from comparisons of observed and computed winds for only a very few storms. Some specialized coefficients in the SLOSH model for highly non-linear flow, such as overtopping of barrier and channel flow, have been tentatively set as fixed constants using only a limited set of observed surge data. In the NWS surge models, the computed surge is not sensitive to small changes in coefficient values. All computer runs use the coefficients as fixed constants in the same way; that is, they are not allowed to vary arbitrarily in particular basins or local regions, nor for a particular storm.

The same storm model, by Jelesnianski and Taylor (1973), is used in both models. The storm model balances forces on the water surface, including surface friction. Although the computed wind can be grossly in error compared to observed winds, there are compensating effects so that the generated surge is not sensitive to the computed winds, providing the input storm parameters are not in error. The maximum wind of the storm is not an input parameter; it is computed by the storm model. The storm model gives two sets of wind, (1) for a large sea, and (2) for bays, estuaries, lakes, and inland inundation; the second set of winds is computed the same way for all inland bodies of water. It is assumed a storm traversing a large sea has a weak memory of its past history because of the unvarying nature of a sea, whereas over small bodies of water the storm has a strong memory of its immediate past history, after having traversed over complicated terrain features. For this reason the storm model, in the SLOSH surge model, discriminates by computing a weaker (but more convergent) wind field over inland bodies of water and terrain than over a large sea. The storm model is used in the same way, with the same fixed coefficients, for all basins and all storms. The following inout parameters, (1) central pressure and storm size a function of time (2) storm track, and (3) speed of motion of the storm along its track as a function of time, must all be accurate.

OUTPUT OF THE SPLASH MODEL FOR OPERATIONAL FORECASTING

The present SPLASH model uses a sheared coordinate system to accommodate curved coastlines; for details see Jelesnianski (1976). The model computes surges on the coastline itself, assuming inland inundation extends for only a small distance inland. To run the model with a given storm, an input data deck of storm parameters is inserted at the end of a small program deck (Jelesnianski, 1972, 1974). The small program deck calls in the SPLASH program for execution. The storm track, specified in the data deck, determines which basin--from a library of 35 basins-- is to be used in the model computations.

An example of a programmed computer output is shown in Fig. 1. The graph displayed is the envelope of coastal high water heights for a storm reaching land.

The printed message beginning the output is a resume of storm parameters selected by the programmer. For information, we also print out the computed maximum wind of the storm. This is a sustained, average wind for a stationary storm; as such, it should not be used for meteorological forecasting. The initial height of the water surface before the storm affects the basin is assumed to be zero, unless the user wishes to enter an alternate value.

The next message gives factors to update surges for a variation of the pressure drop, ΔP. This factor corrects heights on the surge envelope if alternate pressure drops are used. The range of updated pressure is ± 10 mb; it is incremented in 2-mb steps and added to the original pressure drop. We remark that it is unnecessary to rerun the program for small errors in pressure drop; the user merely multiplies the envelope surge heights with the correction factor for the updated pressure estimates, providing there are no explosive pressure changes with time.

The next message is a printout of astronomical tide levels for selected stations for 12 hours before and 12 hours after the predicted time of landfall, at 1-h intervals. The stations correspond to cities along the basin's coast. Tides are predicted in feet above MSL.

```
YOU HAVE SPECIFIED THE FOLLOWING STORM PARAMETERS
LANDFALL POINT IS ----------------------------------------    16 MILES  LEFT OF  GULFPORT
THE STORM IS MOVING TO THE -------------------------------   NNW AT  12.0 MPH
THE PRESSURE DROP IS -------------------------------------   100 MBS.
THE RADIUS OF MAXIMUM WIND IS ----------------------------  15.0 STATUTE MILES
THE MAXIMUM (SUSTAINED) WIND IS -------------------------- 136.7 MPH
THE MAX WIND IS DERIVED FOR A (STATIONARY) STORM
     ON USING THE INPUT STORM PARAMETERS

IF YOU WANT TO UPDATE YOUR  100 MB PRESSURE DROP, USE THE FOLLOWING CORRECTION FACTORS ON THE COMPUTED SURGE VALUES
```

UPDATED PRESSURE DROP	90	92	94	96	98	100	102	104	106	108	110
CORRECTION FACTOR	.89	.92	.93	.96	.97	1.00	1.01	1.04	1.05	1.08	1.10

```
TIDE HEIGHTS ARE FT ABOVE MEAN SEA LEVEL .
HOURLY VALUES ARE PRINTED 12 HRS BEFORE TO
12 HRS AFTER ESTIMATED TIME OF LANDFALL.
ESTIMATED LANDFALL TIME 1300 LCL STD TIME, 17 AUG 1969      ESTIMATED TIME OF LANDFALL
```

LCL STD TIME	22	23	24	1	2	3	4	5	6	7	8	9	10	11	12	13	14	15	16	17	18	19	20	21	22
PRT ST JOE	.2	.3	.3	.4	.4	.3	.3	.2	.2	.1	.1	.1	.1	.2	.2	.3	.3	.3	.3	.3	.3	.3	.3	.4	.4
ST ANDR CH	.3	.3	.3	.3	.3	.3	.2	.2	.1	.1	.1	.1	.2	.2	.3	.3	.3	.3	.3	.3	.3	.3	.4	.4	.5
PNSCL B EN	.2	.2	.3	.3	.3	.3	.2	.2	.1	.1	.1	.1	.1	.2	.2	.2	.2	.2	.2	.2	.2	.2	.3	.3	.4
MOBILE PT	.1	.2	.2	.2	.2	.2	.1	.1	.0	-.0	-.0	-.0	.0	.1	.1	.2	.2	.2	.2	.2	.2	.2	.2	.2	.3
PASCAGOULA	.3	.3	.4	.4	.4	.4	.3	.2	.2	.1	.1	.1	.2	.2	.3	.3	.3	.3	.3	.3	.3	.3	.3	.4	.5
BILOXI	.3	.4	.4	.5	.5	.5	.4	.3	.2	.2	.1	.1	.2	.2	.3	.3	.4	.4	.4	.4	.4	.4	.4	.5	.5
BRETON IS	.2	.3	.3	.3	.3	.3	.3	.2	.2	.1	.1	.1	.1	.2	.2	.2	.3	.3	.3	.3	.3	.3	.3	.3	.4
SW PS,M,PV	.3	.3	.3	.3	.3	.2	.2	.1	.1	.1	.1	.2	.2	.3	.3	.3	.3	.3	.3	.3	.3	.3	.4	.5	.5
TIMBALR IS	.2	.3	.3	.3	.3	.3	.2	.2	.1	.1	.1	.1	.1	.2	.2	.2	.3	.3	.3	.2	.2	.3	.3	.3	.4
EUGENE IS	-.3	-.4	-.3	-.1	.1	.3	.5	.6	.6	.4	.2	-.1	-.4	-.5	-.5	-.3	-.1	.1	.3	.5	.5	.5	.4	.2	.1

* PRIMARY TIDE PREDICTION SITE - IF AVAILABLE

Figure 1. An example of the computer output by SPLASH. The graph is the computed envelope of open coast high waters generated by hurricane Camille, 1969. The inset compares computed versus observed surges on the coast.
(Figure 1 continued on next page)

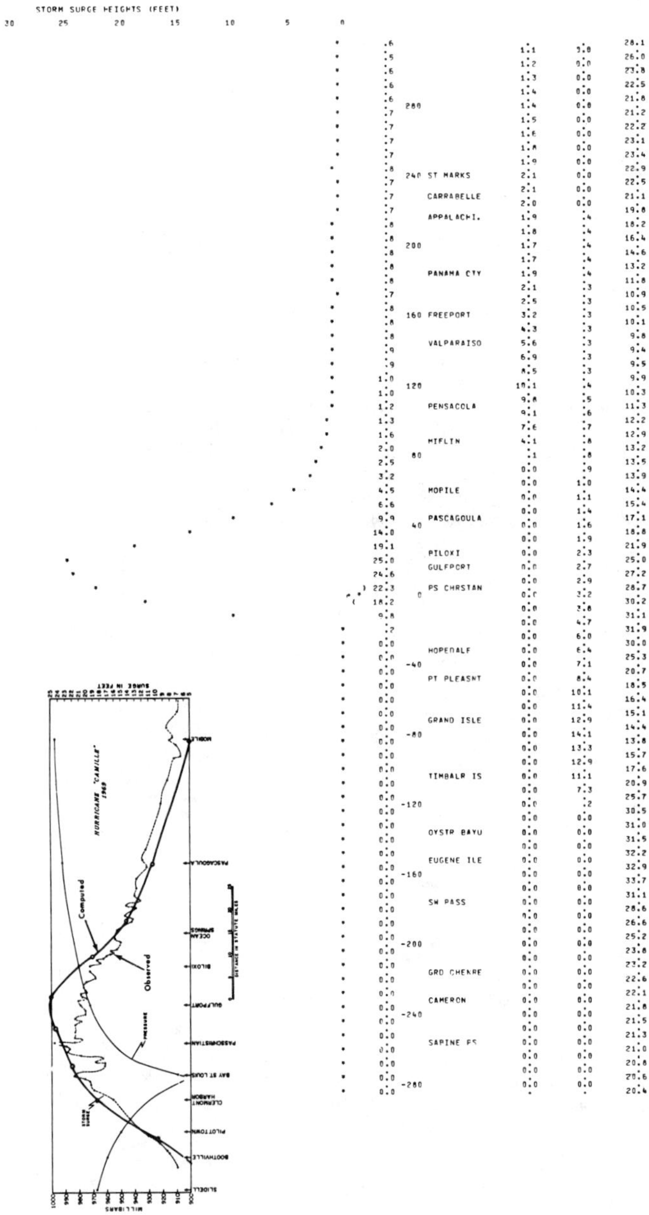

Figure 1 continued.

The vertical scale of the graph is 20 feet. If the peak surge exceeds this, then the ordinate is rescaled to 40 feet.

The horizontal scale of the graph is 600 statute miles long; it represents the curvilinear coast shifted onto a straight line. The dots on the scale are 4 miles apart. On alternate dots, highest local surge heights are printed in feet. Just above, and near the middle of the scale, is a symbol * ; this represents the landfall position on the coast. Left and right of landfall, and below the dots, are linear distances in statute miles; below this are selected cities as they lie along the coast. The printed distances away from the basin center are smaller than the actual curvilinear distances along the coast.

The graph is portrayed with asterisks. Only positive surges are printed. Negative surges do occur on the coast; if such information is desired, it is displayed on a separate output of snapshots of the surge profile against time (i.e., a time-history of the coastal surge). For storms reaching land, the time surge profiles are less interesting operationally than the envelope of high waters. Heights computed along the trailing ends of the envelope are slightly smaller than those obtained with an infinite real-time run, especially so for very slow storms traveling nearly parallel to the coast.

For operational planning, without resort to extensive recomputations, it is desirable to form an estimate of the surge if landfall occurs at a point other than the one originally forecast. To do this, we assume the original storm can strike anywhere on the coast, with the same constraints as for the original landfall. We only need to compare the relative shoaling factors (Jelesnianski, 1972) and restructure the original envelope with these relative shoaling factors.

The three horizontal scales that appear below the column of cities all display approximate surge heights. The first two give envelope surge heights for the storm reaching land 100 miles to the right and left of the original landfall. The last scale gives potential peak surge heights everywhere on the coast (i.e., a worst case estimate for each point on the coast). Note that when using the last scale, the peak surge is at about distance 'R', the radius of maximum winds, to the right of assumed landfall; exceptions would be coastal areas where there are severe changes in horizontal depth contours on the continental shelf.

In Fig. 1, for the particular basin used, the peak surge varies by more than a factor of three on the last horizontal scale; this indicates the significance of local bathymetry in surge generation and hence the importance of landfall position.

In principle, we could combine astronomical and meteorological tide for a total tide envelope. Although this appears to be a sensible procedure, such an envelope would be misleading. A display of this nature is meaningful only if we have an accurate knowledge of landfall in space and time and an accurate knowledge of the meteorological parameters; we just cannot forecast meteorological occurrences to the accuracy required. Neither would it be meaningful to display envelopes including highest and lowest expected values for a range of meteorological accuracy with time; the range between the two envelopes could be so large that the information would be almost useless.

The separate displays of surge envelope and tide forecasts allows the surge forecaster to correct for revisions in meteorology (i.e., he can update the surge prediction as his input data improves with time without constant rerunning of the program).

OUTPUT OF THE SLOSH MODEL FOR OPERATIONAL FORECASTING

The mechanics of running the SLOSH model are identical to SPLASH. The input meteorological data is the same. The initial and quiescent height of inland water bodies, and the sea, are set from pre-storm readings of tidal gages; the model does not use a pre-arranged, dynamic, initial set-up; it only requires the datum of the various surfaces before the storm affects them. The user must select a basin, from a library of basins, with an input data card.

The surge output of the SLOSH model is entirely different from SPLASH, Fig. 2. The main output is a surface envelope of highest waters, regardless of time, for the storm's existence. The output is abbreviated since the actual basin used extends further to the east and south into deep waters.

Fig. 2 is for hurricane Betsy, 1965, a large and severe storm. Available tide gage readings and measured high water marks are inserted for comparison. The model, with the New Orleans basin, has been run with meteorological data for four other historical storms, with similar results.

The '$' signs on the figure represent the storm track. The coastlines, islands, levees, roads, railroads, etc., are discreetly outlined by a line printer. The line printer cannot do a complete job of outlining the geographical features. To bring into focus the various geographical features, a transparent plastic overlay is supplied to users, for each basin, as an aid. The overlay outlines, in different colors, the various geographic reatures including topographic contours of terrain heights and vertical elevations of levees, roads, and other barriers.

For quick reference, only a portion of the overlay features are traced onto Fig. 2. The grid points are 4 miles apart. The origin of the grid is chosen for a particular station (New Orleans). The heights of the land surges, from inland inundation, are above MSL and not above terrain; the topographic contours on a plastic overlay allow the user to mentally subtract the contoured height elevations from the computed surge values to arrive at height of water above terrain.

Additional outputs are snapshots of surge heights at selected time intervals on the entire basin, and continuous marigrams at several selected grid points.

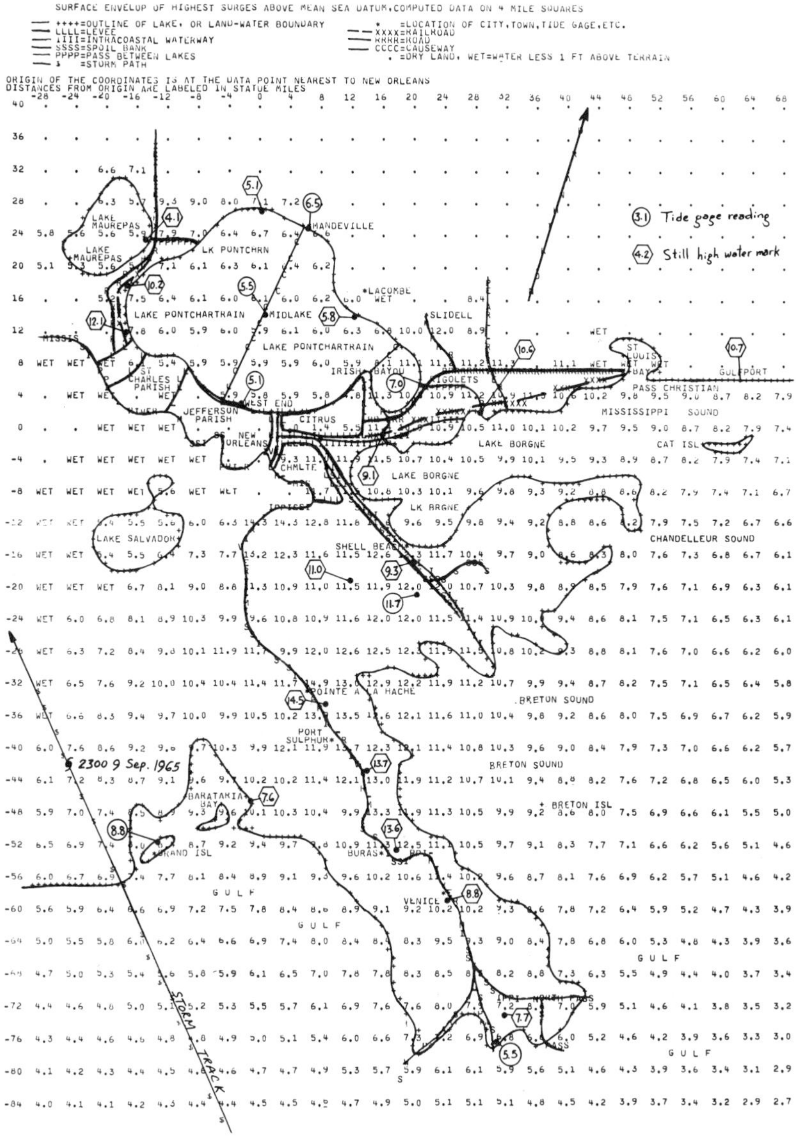

Figure 2. An example of the computer output by SLOSH. Values are computed surge heights for a surface envelope of high waters generated by hurricane Betsy, 1965.

The dynamics of surge generation on inland bodies of water is much more complicated compared to coastal surges from a landfall storm. Unlike SPLASH, there are no simple procedures to manually update the computed surge by SLOSH if the user wishes to amend his meteorological input data.

FUTURE OUTLOOK

The present NWS surge models use coarse meshes for computations and the grid lengths are invariant in space. The numerical methods are explicit and marching in time. The time step between computations varies directly as the grid spacing and inversely as the water depths. Hence for economy of computer use with large basins, the present models require a coarse mesh. It is possible, however, to have a fine mesh in the shallow waters of a large basin without unduly sacrificing economy; this can be done with use of variable grid sizes, providing the mesh is coarse in deep waters of the basin. The NWS models are now being updated for fine mesh in shallow regions, but with a different approach for each model.

A stretched coordinate system is now being implemented on the sheared coordinate system of the SPLASH model to compress the grid in shallow waters near the shore, and expanding to a coarse mesh in deep waters; the stretching transformation is applied in one-dimension only, is continuous seaward, and is a function of depth for a continuous, smoothed, depth-profile offshore. This simple method takes advantage of near-orderly, depth variations seaward from a coast. The result is better resolution in shallow waters offshore, especially so when the depths descend rapidly from the coast. The system should be operational for all of the Eastern Seaboard and Gulf coasts for the 1979 hurricane season; the basin data, for all of the 35 basins, is now being prepared.

A polar coordinate system has been developed for the SLOSH model. The grid varies continuously in space and is fine over inland bodies of water and coarse in deep waters. Unlike SPLASH, the grid variation is not a function of depth; we cannot take advantage of any orderly depth changes because the inland terrain and water bodies vary chaotically. The system is not fully operational because complete basin data has not yet been prepared for any basin.

Some benefits with variable grid sizes are:

1. Good resolution in shallow waters where surge dynamics are most pronounced.

2. A large basin can be encompassed with not many more grid points than those of a coarse mesh.

3. Only one computer run is required for operational forecasting with a large basin. This is a replacement for a standard procedure to run with a coarse mesh in a large basin, save input boundaries for a limited area fine mesh model, and then re-run with the limited area basin.

4. The time step between computations is not much different than required by a coarse mesh model.

REFERENCES

Jelesnianski, C. P., (1972), SPLASH (Special Program to List Amplitudes of Surges from Hurricanes); I. Landfall Storms, NOAA Tech. Memo. NWS TDL-46, 52 pp.

Jelesnianski, C. P., (1974), SPLASH (Special Program to List Amplitudes of Surges from Hurricanes); II. General track and variant storm conditions, NOAA Tech. Memo. NWS TDL-52, 55 pp.

Jelesnianski, C. P., (1976), A sheared coordinate system for storm surge equations of motion with a mildly curved coast, NOAA Tech. Memo. NWS TDL-61, 52 pp.

Jelesnianski, C. P., and Chen, J. (1979), SLOSH (Sea, Lake and Overland Surges from Hurricanes). To be published.

Jelesnianski, C. P., and Taylor, Albion D., 1973, A preliminary view of storm surges before and after storm modifications, NOAA Tech. Memo. ERL WMPO-3, Weather Modification Program Office, ERL, 33 pp.

DECISION MAKING AND THE COST OF IGNORANCE

Jack R. Benjamin[1]

1. INTRODUCTION

The purpose of this paper is to present a discussion of decision theory with a brief introduction to the associated information theory. No large water-related engineering project (or any other major engineering project, for that matter) has had a complete development or analysis consistent with decision theory. This does not mean that the decisions in such projects were wrong, only that the decision-making phenomenon in these large engineering projects differed from the methodology of the science of decision theory. Reasons for such differences are partly the unresolved issues of the value of human life, legal liability, the natures of the legal and political processes, and the like. Thus this paper is concerned with a methodology that may or may not directly apply to the economic aspects of hydrologic forecasting.

The decision tree of Figure 1 illustrates the components of decision theory and their relationships. First, there is assumed to be a decision maker labeled DM. This DM must choose between possible actions or alternatives labeled AA, BB, CC. Having chosen one of the possible actions, the DM faces the uncertain future, SS, MM, FF. The DM will receive value depending on the action chosen and the future that is received. This value is labeled U for utility.

The problem of the DM is to make the optimum decision (choice of actions) considering that the future is uncertain with the goal of receiving the greatest value or utility. The identification of the optimal action is made by using a decision rule.

Thus there is a DM who must identify the optimum action using the decision rule considering that the future is uncertain and value will be received by the DM depending on the action taken and the future received.

Before examining difficulties in the application of this model to issues in hydrologic forecasting, two simple examples will be considered.

1. President, Engineering Decision Analysis Company, Inc., 480 California Avenue, Suite 301, Palo Alto, California 94306

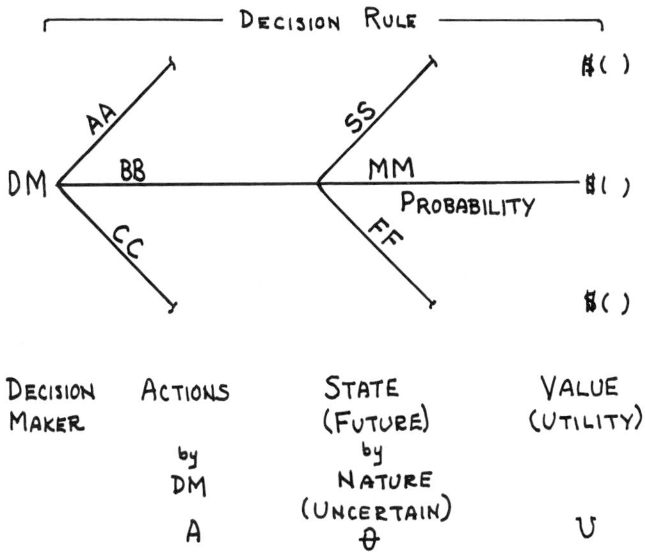

FIGURE 1

2. EXAMPLE 1

The following is a hypothetical problem. A very large dam-reservoir complex is being considered for possible construction. The site for the dam is not ideal since a very large city exists downstream of the dam; the geology of the damsite is complex and involves major uncertainties which are coupled with hydrologic uncertainties so that there exists some uncertainty about whether or not the dam should be constructed with respect to safety.

For simplicity, the DM is identified as an engineer. The decision tree is shown in Figure 2. The DM has a life expectancy of about 25 years so that failure of the dam-reservoir system from any cause during the next 25 years is extremely serious. The decision tree has two action branches, construct or not construct. The future state is failure or no failure in 25 years. The values associated with each action-state are as shown, with no failure having a very small positive value compared to the extremely large loss if the dam fails. If the decision is to not construct the dam, there is no gain or loss.

If the DM is risk averse (as is most humanity), the DM may use the minimax decision rule. That is, he will take the action which minimizes the maximum possible loss. The very large possible loss associ-

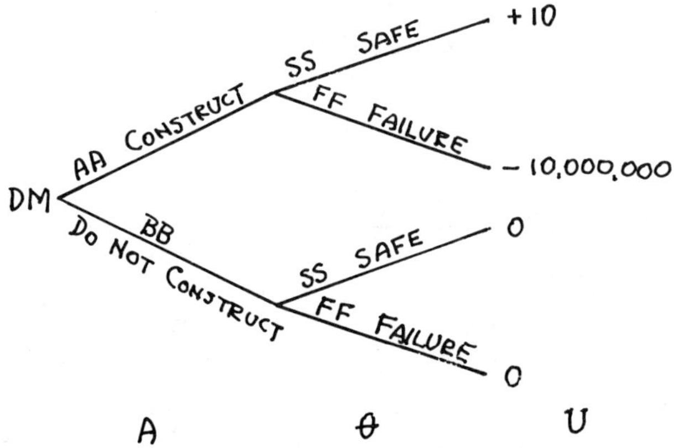

FIGURE 2

ated with constructing the project is much larger than the largest loss with no construction, so that the optimum decision is to not construct. Note that this decision has absolutely nothing to do with the design of the facility as long as it is conceivably possible for a failure to take place. If the probability of failure in 25 years is anything but absolutely zero, the optimum decision for the DM remains the same.

The contrast with conventional practice is apparent since this decision rule might appear to preclude any construction anywhere. This problem can be avoided by assuming by definition that bad outcomes can not occur. Reputable engineers have stated that they do not look downstream of a dam. The dam is by definition "safe." Also, the policy for the ranking for renovating old hydrologic systems has been stated to be the relative hydrologic hazard to the dam spillway. With such a rule, there is no automatic consideration of the cost of failure and downstream consequences.

The minimax decision rule is seen to produce problems since decisions are made based on only part of the information. That is, the conservatism in the design is not considered. If failure is possible, effectively, failure is assumed to be certain.

To illustrate another decision rule and a situation closer to the subject of concern, assume that the decision maker with the dam-reservoir system just discussed is a hydrologist whose concern is not failure of the system directly but rather the specification of the design spillway capacity. In simplest form, the DM again has a 25-year period of responsibility so that the actions are various spillway design capacities. The state or uncertain future for the next 25 years is the actual maximum flow through the spillway in that time. With such a

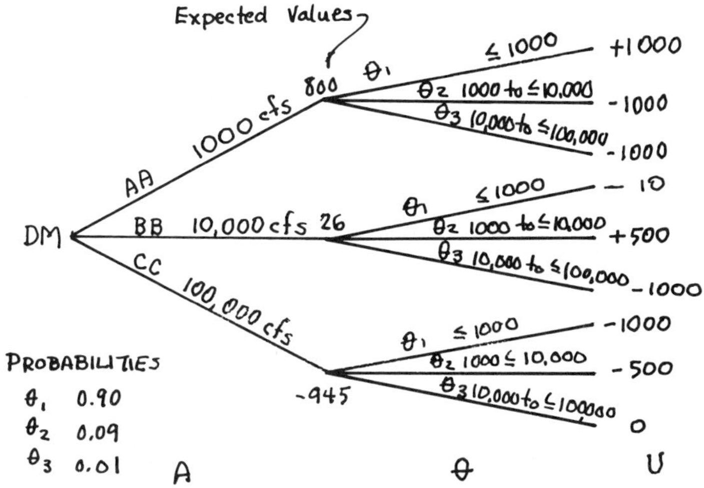

FIGURE 3

lottery, value is received if the design capacity is exceeded or if the design flow selected is absurdly large. Note that there is no mention of dam failure. Figure 3 shows this decision tree.

The actions are consistent with much of conventional engineering practice. The unknown future and values appear reasonable as well. Exceedance of design capacity involves a loss while the cost of construction of the spillway increases with design capacity. The hydrologist, DM, can no longer be involved with true dollar values, so that value is now stated in utiles, measures of utility or preference to the DM.

The minimax decision rule does not work well with the problem now considered since extreme states are not defined. The alternative expected value rule is the most generally applied rule in such studies. An expected value is defined as the product of a value by a probability measure. Thus, if there is a one percent probability of receiving one dollar in a gamble, the expected value is one cent. If two alternative actions are to be compared, the comparison is on the basis of expected value. That is, Lottery A is preferred to Lottery B if the expected value of A is larger than the expected value of B. Conversely, if a condition of indifference exists between Lotteries A and B, the expected values are equal by definition. If the values associated with A and B are known and one of the probability measures as well is known, the other probability measure can be calculated. Such indifference studies form the basis of utility theory.

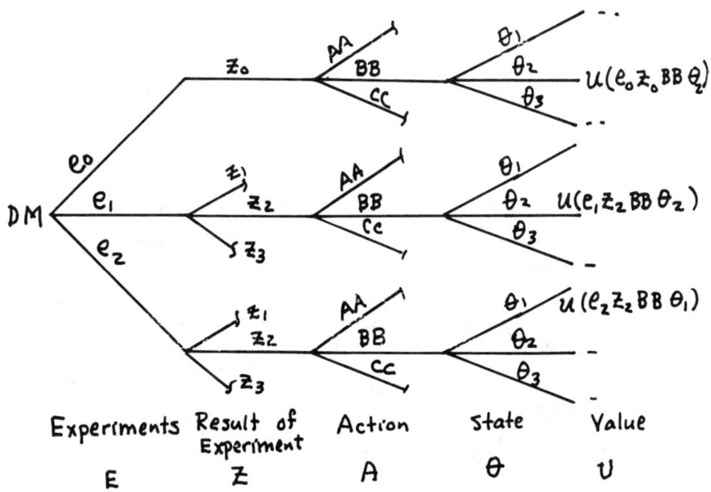

FIGURE 4

Now, considering the decision tree of Figure 3, the analysis using the expected value decision rule consists of calculating the expected values for each action branch and comparing the expected values to identify the optimum action. The calculation of expected values begins with the calculation of the products of values by the probabilities associated with that state and then summing these products to obtain the expected value of each action. If the probabilities and values are correct in Figure 3, the optimum action is to design for the smallest flow rate.

The expected value decision rule has been used in economic decisions of all types. The primary difficulty with its application in hydrologic-type decisions is that values are very difficult to identify and quantify. For example, failure of the system can involve both economic and human losses. Is there a dollar value to human life and injury? Despite many papers and meetings on this subject, the issue is unresolved. It is possible that engineers could make such an evaluation from an actuarial viewpoint. The issue, however, is one of law and politics, not one of engineering or mathematics, so that it remains unresolved.

When we consider that there are three distinct types of value (economic, human, and impact) and each can be very important in the decision process, the analysis of such situations is not trivial.

3. EXPERIMENTS

The two simple decision situations discussed so far are classed as decisions made under uncertainty. The future and likely the values in

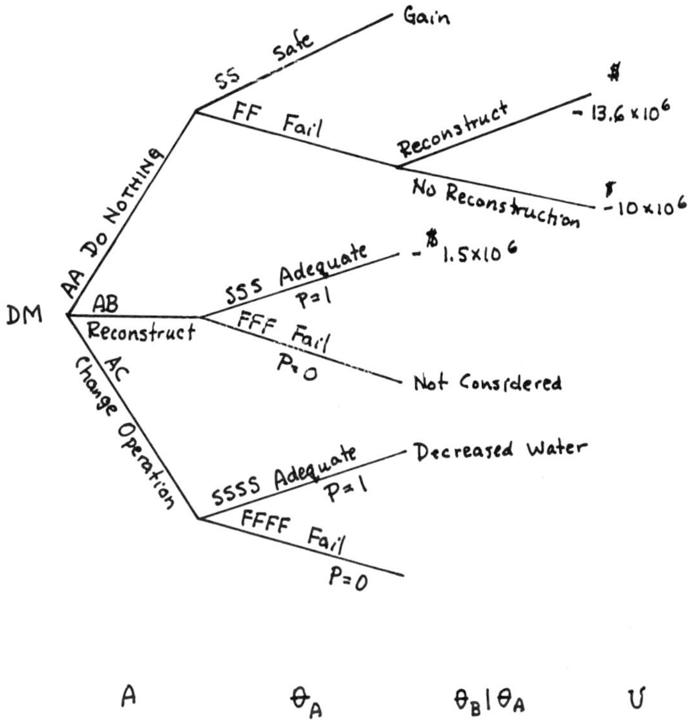

FIGURE 5

the decision tree are uncertain. If it is possible to obtain more information at some cost, the decision tree is expanded by adding an information branch (Fig. 4). The information branch contains several levels of experiments, each with its own cost and different ability to define the uncertain future. The analysis is termed "preposterior analysis" (Ref. 1) and its purpose is to define the optimum experiment prior to making any experiment. The details of such an analysis are beyond the scope of this paper. Numerous references exist on such Bayesian procedures.

4. BAYESIAN FORECASTING

Bayesian forecasting techniques differ materially from those of conventional statistics using actuarial flow data and plotting procedures. The mathematics are directly derived from preposterior analysis.

To illustrate the concepts, consider the spillway remodeling decision situation with McKay Dam (Fig. 5), based on the papers by Austin (Ref. 2), Kirkpatrick (Ref. 3), and Benjamin (Ref. 4). The dam was constructed in the period of 1923 to 1927 by the Bureau of Reclamation. The outlet works has a capacity of 1,400 cfs and the total design spillway capacity was 10,000 cfs. The Bureau report indicated that the spillway capacity should be increased to 27,000 cfs based on the PMF. The existing spillway has an estimated overtopping capacity of 19,000 cfs. The dam has performed adequately for 46 years and we assume, for purposes of illustration, that it has not been necessary to release more than 1,000 cfs at any time in 46 years. Data on the latter assumption are not available but it is not unreasonable since more than a 1,000 cfs release causes flooding. We assume that no other data are available and the issue is what is the probability of a spillway flow in excess of 17,000 cfs such as to produce overtopping of the spillway. The total release would then be of the order of 18,400 cfs.

The 46-year record of no release larger than 1,000 cfs translates into a binomial model from classical statistics, but the probability of an exceedance is unknown since none has occurred. A Bayesian simple approach is to assume that the peak flow only occurs one time a year, the probability of a series of 46 such events is $F(X \leq 10,000)^{46}$. The probability model for F(X) might be of the gamma family, for example, containing two parameters, both of which are uncertain. If $F(X \leq 10,000)^{46}$ is formulated, a joint distribution on the two unknown parameters is found if the expression is normalized. This may have to be done by computer but it is a simple operation. Then a data-based forecast of future events in terms of exceedance of 1,000 cfs or <u>any other level</u>, X_0, can be made by integration of the product of the conditional distribution of $F(X_0)$ given the parameters by the joint distribution of the parameters given the data over all values of the parameters.

With the McKay Dam situation, the PMF is found to be to have a return period of the order of millions of years as is the return period of spillway overtopping. The reliability of such a forecast is near to zero. It can only be said that if the past is assumed to be a reliable estimate of the future, future extreme events such as those of interest are so remote that the expected value of loss from overtopping is very close to zero. That being the case, the spillway remodeling to 27,000 cfs can only be optimum if the minimax decision rule is employed.

The important point in this discussion is that statistical decision methodology can be used as an additional tool in forecasting problems. Good Bayesian forecasters could even make a rational approach to the return period of the PMF since subjective probabilities are an integral part of the methodology.

5. THE COST OF IGNORANCE

What is the cost of not knowing that which we do not know? From a decision theory point of view, the question resolves into what is known as "opportunity loss." Computations using opportunity loss are made similarly to those with utility theory. Opportunity loss is the difference in value between that of the action-state combination being evaluated and the largest of all values. In effect, if we knew the future, we would choose the action with the largest certain value. If an action is chosen which is less than optimal, there is a loss.

Information theory considers the cost of ignorance in terms of the value of perfect information, VPI, The theory is well developed and includes such possibly confusing statements as the "prior expectation of the utility of terminal action posterior to receipt of perfect information." Such discussions are beyond the scope of this paper.

CONCLUSION

An extensive literature exists on value theory and the associated concepts of statistical decision theory. There are also some conflicting methodologies in classical probability and statistical theory that also imply that they yield better decisions. Thus the concepts presented in this paper should be viewed as representing a brief survey of the state of one art.

Finally, it is necessary to accept the truth that the very best of decisions can have an undesirable outcome.

REFERENCES

1. Benjamin, J. R. and Cornell, C. A., "Probability, Statistics, and Decision for Civil Engineers," McGraw-Hill, 1970.
2. "Inspection, Maintenance, and Rehabilitation of Old Dams," ASCE, 1974.
3. "The Evaluation of Dam Safety," ASCE, 1977.
4. Benjamin, J. R. "Statistical Decision Theory Applied to Dams," Stanford Conference, New Perspectives on the Safety of Dams, 1978.

HYDROLOGIC UNCERTAINTY AND PROJECT BENEFITS

By B. E. Martin and R. W. Julian*

INTRODUCTION

Water resource agencies have long been concerned with the impact of hydrologic uncertainty on project operations and on design. Most of the improvement in technique, however, has been in the domain of short-term project operations. Although many papers have been written advocating various stochastic modeling and optimization approaches to project design, there have been relatively few reports of actual applications. This paper utilizes a stochastic analysis to examine the impact of hydrologic uncertainty on lifetime project benefits. Two Bureau of Reclamation case examples are used: (1) an existing reservoir, Lake Berryessa, and (2) a possible future reservoir, an enlarged Shasta.

Two main aspects of future project life concern economists. First, the benefits per unit of project output can vary as future economic events and market forces unfold throughout the life of a project. Second, the quantity of project output is functionally related to future hydrologic events. This paper deals only with the latter aspect.

Forecasting hydrologic events over a project life of 50 or 100 years is, unfortunately, an impossibility. The traditional approach for dealing with this problem has been to assume that the historical streamflow record will be repeated during the project life, computing benefits accordingly and ignoring the hydrologic uncertainty. Stochastic modeling provides a means of dealing with the uncertainty, at least in a probabilistic sense.

Before proceeding with the technical discussion, some background information is provided on the two reservoir cases. Both involve Bureau of Reclamation reservoirs in the vicinity of the Sacramento Valley in northern California.

The Lake Berryessa data resulted from an analysis of the risks associated with the assumed water yield of the reservoir. This analysis was performed as part of the Bureau's reevaluation of the Solano Project for the Solano County Water Project Feasibility Study. Putah Creek, the source of water for the Solano Project, lies to the northwest of Solano County on the eastern slope of the Coast Ranges in Napa and Lake Counties. The average annual runoff of Putah Creek at Monticello Dam, which forms Lake Berryessa, for the period 1906-1974 is 375,000

---

*Regional Director and Civil Engineer, respectively, Mid-Pacific Region, U.S. Bureau of Reclamation, Sacramento, California acre-feet. The maximum annual runoff occurring in 1941 exceeded 1 million acre-feet. A minimum annual runoff of 35,000 acre-feet occurred in 1931. Lake Berryessa, with a storage capacity of 1,602,300 acre-feet, regulates flows along the lower reaches of Putah Creek, storing surface water for irrigation and municipal and industrial purposes.

The idea of an enlarged Shasta Dam and Reservoir has recently received attention, motivated in part by the 2-year drought of 1976 and 1977 which again focused attention on the long-range water needs of California. Both State and Federal studies point to the need for a major increase in water supply to maintain the California economy. Enlarging Shasta Lake of the Central Valley Project (CVP) is one of a limited number of alternatives for increasing the future water supply of the Central Valley Basin and other areas in California. This paper considers an enlarged size of 14 million acre-feet or an increase in storage capacity of about 9.5 million acre-feet. The existing Shasta Reservoir, completed in 1945, has a capacity of 4,552,000 acre-feet. Located at the northern end of the Sacramento Valley in Shasta County, it is fed by waters of the Sacramento, McCloud, and Pit Rivers. The mean annual inflow for the period 1922-1978 is 5,544,000 acre-feet. The lowest annual inflow of 2,479,000 acre-feet occurred in the year 1924. The highest inflow of 10,796,000 acre-feet occurred in 1974.

METHODOLOGY

As indicated previously, traditional methodology for analyzing reservoir yield uses the historical inflow to hypothetically operate the reservoir, thereby obtaining an estimate of the sustainable yield based on the period of record. Other accomplishments such as power, flood control, and water quality are similarly obtained. This procedure gives an indication of the possible future accomplishments of the project. Actually, the future reservoir inflow pattern and the corresponding accomplishments are unknown. The historic flow record gives just one realization of the occurrence pattern; the probability of the historical sequence repeating itself is infinitesimal. In other words, the historical record is merely a sample of what is possible. Stochastic hydrology addresses the inability to predict future inflows by examining the probabilistic aspects of the future as derived from the statistical properties of the historical sample.

The basic procedure can be summarized as follows: First, studies are made to determine the underlying statistical properties of the historical inflow. Based on this analysis, a mathematical model is constructed which preserves the historical statistical properties and generates synthetic inflows of any specified duration. By generating many sets or traces of inflow, the reservoir can be operated through a variety of possible inflow patterns, each of which gives a different estimate of project yield and accomplishments. Comparing these yields and accomplishments with the frequency of their occurrence enables one to estimate the risk or probability associated with a specified level of output. In the Berryessa analysis, 100 traces of monthly inflow, each having a duration of 70 years, were generated. In the case of enlarged Shasta, a larger sample containing 1,000 traces of annual inflow was generated, each trace having a duration of 50 years. For the benefit of those familiar with the mathematics of stochastic processes, a lag-one Markov model was used for both cases.

ANALYSIS AND RESULTS

Of the two cases, Berryessa, a single-purpose (water supply) reservoir operationally and financially independent of the CVP, is far simpler to analyze. Using an existing operation program, which assumes about 12,000 acre-feet of current upper basin use and about 20,000 acre-feet for downstream water rights, the reservoir was operated using monthly synthetic flows. The yield obtained reflects the firm marketable water supply. This yield times a weighted average of the existing water rates is used to represent annual lifetime project benefits. Figures 1 and 2 summarize the results of the Berryessa analysis by plotting the yield and benefits obtained from each 70-year trace against their probability of occurrence. For reference, the comparable results from the historic record plot at the 7 percent point. In other words, there is a 7 percent chance that in any future 70-year period the annual benefits will be equal to or less than the $945,000 estimated from the historic inflow. Figure 2, then, provides an indication of the impact of hydrologic uncertainty on lifetime project benefits. As can be observed, those benefits range from a low of $796,000 to a high of $1,368,000.

In contrast to Berryessa, a number of simplifying assumptions were needed to expedite analysis of the enlarged Shasta case. The more important ones are enumerated below:

1. Shasta was decoupled from the rest of the CVP when in fact standard practice is to operate the system as a whole, referencing water yield to the Sacramento-San Joaquin Delta and power accomplishments to the load center at Tracy.

2. Both inflow and reservoir operations were modeled on an annual basis, starting with a full reservoir.

3. Power and flood control operations were simplified.

4. Prior rights releases, Delta water quality releases, and dry year deficiencies were not dealt with in the operation program. These involve complex considerations such as annual and monthly antecedent conditions and inflow hydrology to other reservoirs and the Delta.

5. Rather than sizing the enlarged Shasta power facilities, firm capacity and energy were based on physical potential at the site.

The yields and dollar accomplishments associated with enlarged Shasta were derived by subtracting existing Shasta Reservoir results from the enlarged Shasta results, giving the annual yield and annual benefits of enlarging the reservoir for each 50-year trace. As similar simplifying assumptions were made in both cases, some of the departures from existing and future operating realities are minimized. Figures 3

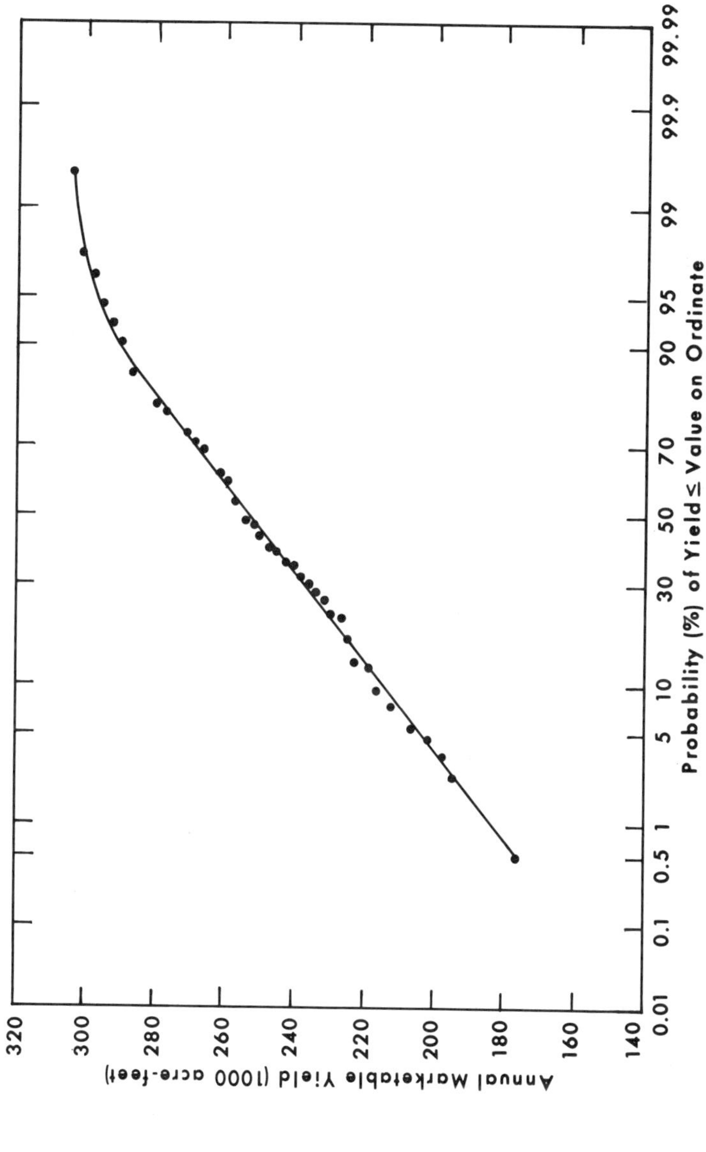

Figure 1. VARIATION OF YIELD FOR LAKE BERRYESSA

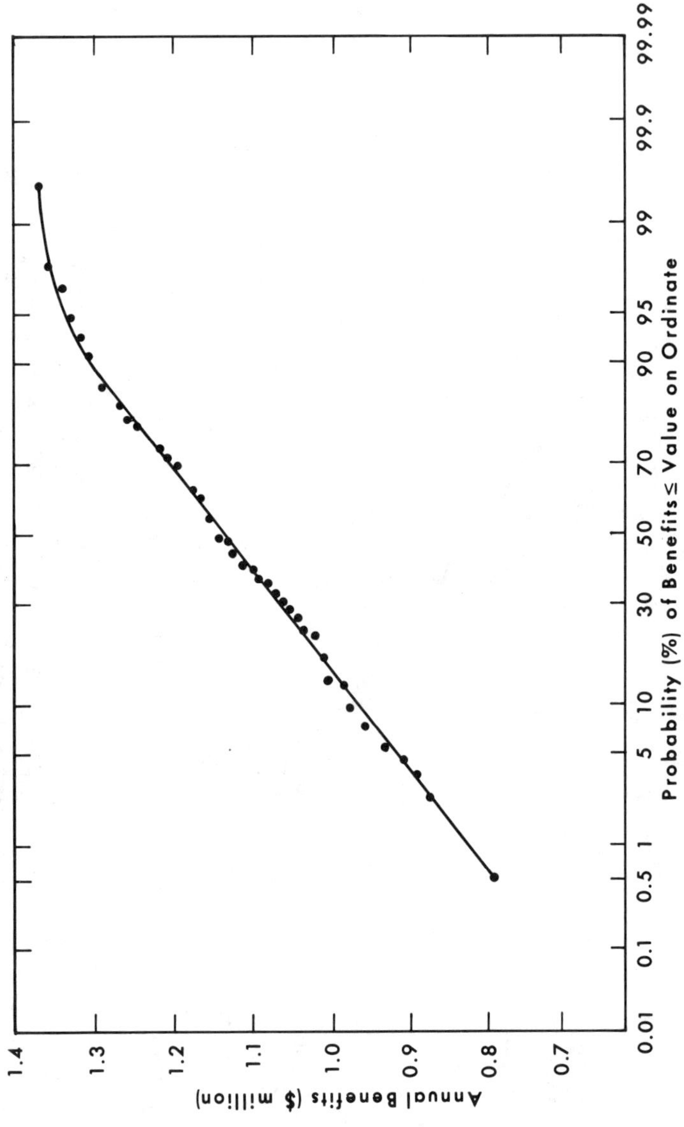

Figure 2. VARIATION OF BENEFITS FOR LAKE BERRYESSA

and 4 summarize results in a manner similar to the Berryessa analysis. However, the benefits in figure 4 include both water supply and power. Water benefits are based on an assumed incremental worth of $100 per acre-foot. Power benefits assume a 20 percent annual capacity factor for enlarged Shasta and a 30 percent factor for the existing Shasta. The difference in the two capacity factors reflects the greater opportunity provided by an enlarged Shasta to operate the powerplant in a peaking mode. Both dependable capacity and energy components were estimated for each reservoir size. Aggregate annual benefits for water and power, based on historical inflow are $140 million. The stochastically generated benefits range from a low of $72 million to a high of $306 million.

OTHER APPLICATIONS

The foregoing analysis has related hydrologic uncertainty to project benefits in terms of the underlying probabilities. But, what does one do about this uncertainty when planning projects? The traditional approach of basing evaluations on a repetition of the historic flow record has been a reasonable procedure for western projects designed since the drought of the late 1920's and early 1930's. The relatively low risk of a sequence of events worse than historical is borne out by this and other analyses.

One way of dealing with hydrologic uncertainty is to seek an operation rule curve which will take advantage of a more favorable sequence of flows should they occur in the future, while protecting against a reservoir overdraft should future flows be worse than those of historical record or any other selected trace. While various strategies can be formulated using historical data, only stochastic hydrology can provide numerous equally likely traces to test alternative strategies. As an example, a possible alternative to the existing fixed water supply operation of Berryessa would be a 2-tier water marketing program consisting of a low-risk firm yield coupled with a high-risk variable yield component. Such a program presupposes that a variable water supply can be tolerated either by deficiencies or by conjunctive ground-water utilization. Certainly such an assumption is not always possible or acceptable.

While a simple case like Berryessa can be optimized on the basis of water yield, analysis of the more common multipurpose reservoir such as enlarged Shasta would more logically focus on maximizing benefits. Methodology here would be to select a rule curve which maximizes the lifetime project benefits. The rule curve could be developed by trial on the basis of the historical or similar low-risk trace and then refined and tested on other traces, or by a more sophisticated approach using operation research and system analysis techniques. Whatever approach is selected, one should keep in mind that the process of planning new projects or modifictaions to existing ones, is long and complex, at least at the Federal level. The decisionmaking process which leads to a selected design does not lend itself to optimization but rather to accommodation and compromise. It must be recognized that some reservoir outputs and user deficiencies may be negotiable, while

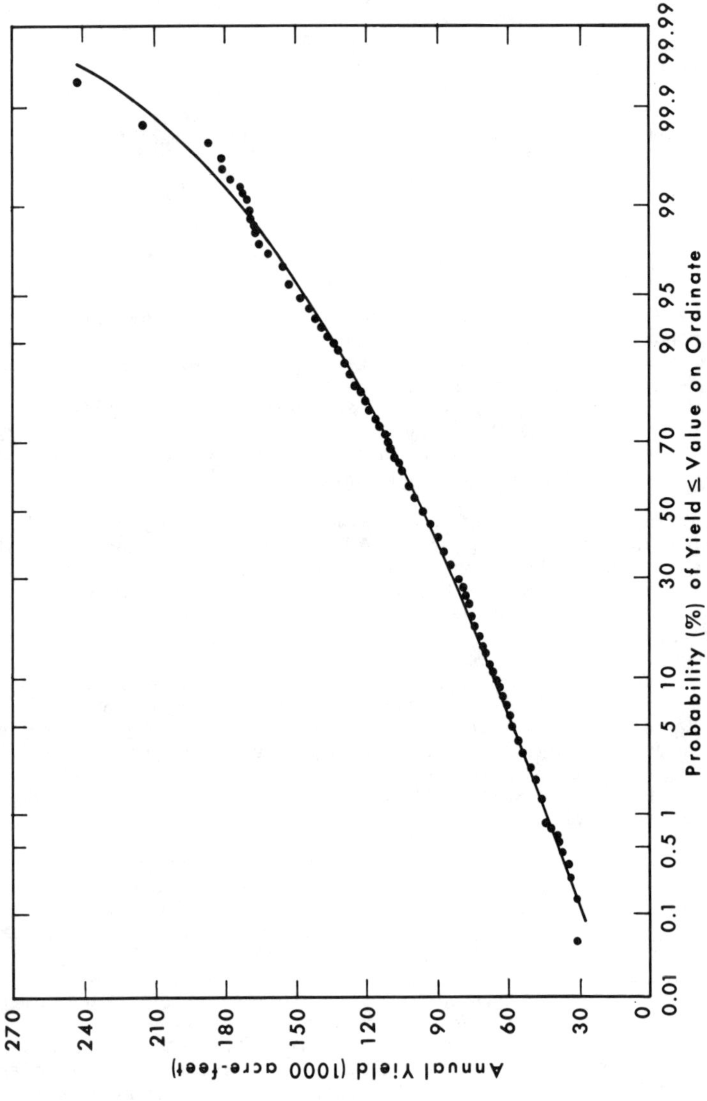

Figure 3. VARIATION OF YIELD ATTRIBUTABLE TO ENLARGEMENT OF SHASTA RESERVOIR

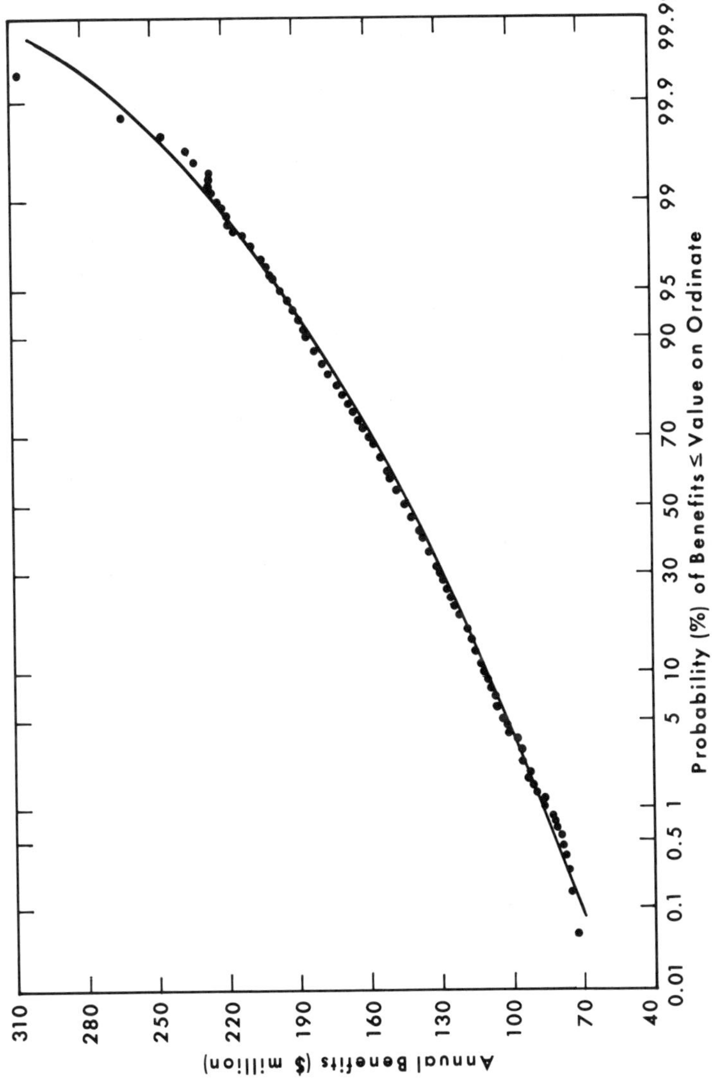

Figure 4. VARIATION OF BENEFITS ATTRIBUTABLE TO ENLARGEMENT OF SHASTA RESERVOIR

others may be more subject to government control and therefore more amenable to optimization. Equally important is the need to utilize planning techniques which are comprehensible among a variety of professionals and the public at large. This is not meant to imply that some of the new analysis techniques which have evolved in the past decade are of no value, but that the increased complexities of both the planning process and water resource systems must be recognized.

CONCLUSIONS

Stochastic analysis can provide useful probabilistic and operational information, although it does not remove hydrologic and related economic uncertainty. In planning a project, assumptions regarding future inflow obviously influence the assumed project benefits. As demonstrated earlier, a wide range of benefits is possible depending on whether the historical or some other trace is assumed as being the best representation of the future. However, inability to forecast the occurrence and magnitude of future droughts dictates a conservative posture with regard to assumed project capabilities and benefits.

THE NEW OBERS PROJECTIONS

Henry L. DeGraff*

The OBERS Projections of regional economic activity and population are being updated. For the benefit of those not familiar with the OBERS Projections, they resulted from a program to prepare long-range regional projections which was initiated in 1964 through an agreement among Federal water resources planning agencies, the Office of Business Economics (OBE) in the Commerce Department, and the Economic Research Service (ERS) in the Department of Agriculture. The acronym OBERS came from the combination of OBE and ERS. The names of both agencies have since been changed but the name OBERS will continue to be used because it has become very well established. The Bureau of Economic Analysis (BEA) is the successor to OBE and what was ERS is now part of the Economics, Statistics and Cooperatives Service. In 1965 the U.S. Water Resources Council (WRC) was established and the projections program became an integral part of the comprehensive water resources planning program and the periodic national assessment of water and related land resources. The projection effort is now a regular part of the BEA and Economics, Statistics and Cooperatives Service work programs.

The objective of the OBERS program was the preparation of a consistent set of long-range projections of economic activity and population to serve as a basis for planning for water resource development and management. Since their publication, the projections have been widely used for other purposes both public and private.

The 1972 projections were the first set formally published. This set was based on the Bureau of the Census series C national population projections. This series assumes a continuation of the 1960's high birth rates. In 1974 the most recent set of projections was published based on an approximate replacement level birth rate (Census-series E). An interim set of State level projections was subsequently prepared under special arrangement with the Environmental Protection Agency (EPA) in 1977. However, these projections were not intended as part of the "official" OBERS Projections.

The categories and detail of data being provided in the new set of projections have both been expanded. Both employment and earnings are being projected to the year 2030 with industrial detail provided at approximately the 2-digit standard industrial classification (SIC) code level (some combinations of 2-digit industries were dictated by data or other constraints). As before, population is being projected as a function of the economic base but will be shown by five-year age groups and by sex as well as in total.

*Assistant, Chief, Regional Economic Analysis Division, BEA, U.S. Department of Commerce, Washington, D. C.

The geographic areas for which the 1979 projections are being prepared have also changed. In the 1974 publication, data were presented for the BEA Economic Areas, States, Water Resource Regions and Subareas, and Standard Metropolitan Statistical Areas (SMSA's). The first change occurred in the BEA economic areas. The BEA economic areas are nodal functional areas delineated to facilitate regional economic analysis and, as will be described later, are a basic geographic unit for analysis in the projection system. Each area consists of an economic node--an SMSA or similar area that serves as a center of economic activity--and the surrounding counties that are economically related to the center. To the extent possible, each area includes the place-of-work and place-of-residence of its labor force. The areas cover the entire United States.

BEA economic areas were first delineated in 1969, based on data from the early 1960's. During the past decade, the regional distribution of economic activity changed. Some new centers developed and at the same time, some centers--particularly those in areas where agriculture was the basic industry--declined and were absorbed into other areas. Expanded transportation networks, by affecting patterns of travel, also led to changes in area boundaries. The Interstate Highway System was a major factor of this kind.

The 1977 delineation resulted in 183 areas--ten more than in 1969. Forty-nine or 27 percent of the first delineation remain unchanged and another 48 were changed by only one or two peripheral counties. The new delineation was based primarily on three sets of data: (1) journey-to-work data from the 1970 Census of Population; (2) newspaper circulation data for 1972; and (3) 1975 county commuting data developed from Social Security Administration and Internal Revenue Service records. Copies of the new BEA Economic Area Map and a book containing listings of the counties in each economic area are available for distribution.

The other geographic changes have to do with the types of areas for which projections will be prepared. Subsequent to the publication of the earlier reports, comments originating primarily with the State and local users indicated that (1) the data for the Water Resources Areas were of value to a relatively small group, and (2) since neither the water areas nor the economic areas were coterminous with State boundaries, the data were of limited value to State and local planners. Data for the water areas will not be presented in the upcoming revisions. Instead, data for States, Economic Areas, SMSA's, "State pieces" of economic areas, and "State pieces" of SMSA's will be presented. In other words, in cases where an economic area or SMSA lies in two or more States, the amount constituted by each State will be presented along with the total. Data for a total of 585 subnational regions will be presented.

The methodology behind the projections, as in the past, involves the extrapolation of historic trends in national-regional relationships. However, some changes have been made in the means by which these trends are extrapolated. A general overview of the system follows.

As before, a set of population projections for the Nation, prepared by the Bureau of the Census, served as a starting point. They provide the national framework for the BEA regional population projections and they also provide the national population scenario utilized by the Bureau of Labor Statistics (BLS) in preparing national projections of total output and employment. BLS employed an economic model to provide industrial detail for output and employment. BEA has utilized these BLS national projections. Historical earnings per worker for each industry have been projected at the national level by BEA and applied to the BLS employment base to provide the national earnings estimates by industry. The BEA regional projections are generated within this detailed national framework of future earnings, output, employment and population.

One distinction between the 1974 and 1979 procedures should be noted. In 1974, projections for the BEA economic areas were developed, reviewed, and (relatively) finalized prior to proceeding to the other geographic configurations involved. From a theoretical standpoint, this is a preferable approach as these areas represent integrated economic units. However, at the time the mathematical relationships were being established for the 1979 update, the regional data base was undergoing a major benchmark revision. A time series for States from 1958 forward had been finalized, but the data necessary for a time series for the economic areas were still under revision. Also there is a two-year lag in economic area data while the State data series lags by only one year. This lag can be reduced by another one-half year by using the State Quarterly series. Therefore, the decision was made to develop the primary projections at the State level for this update. While a review and revision is being conducted at the State level, none of the results will be considered as final until the projections have been disaggregated to the State pieces and the reaggregation of these State pieces to the economic area level is completed, reviewed, and finalized. Thus, as in 1974, the economic area is considered to be the primary economic unit. The economies of these areas display a variety of industrial interrelationships which may not prevail in the other geographic units being projected.

The distribution of the national projections to the regions is accomplished by a more complex routine than was used in 1974. There are four components to be disaggregated: earnings, employment, population, and the non-earnings components of personal income. Earnings and employment are, of course, disaggregated by industry. The earnings for most industries are distributed by an economic model which involves (1) the projections of basic industry earnings via shift-share analysis, (2) the projection of final demand for each industry via the application of the State's share of total U.S. earnings to the national projections of final demand, and (3) the disaggregation of total earnings to individual industries via the application of a regionalized input-output structure. The last step incorporates the projected national input-output results, the projected State production/demand ratios, and the projected final demands for the State. The system is iterated annually.

Earnings for agriculture and mining are projected exogenously by estimating value of production and deriving earnings from that. Military earnings are based on the assumption that military manpower will remain constant.

To obtain the important measure of total personal income, earnings of residents are increased by property-type income (rents, dividends, etc.) and transfer payments, and diminished by contributions for social insurance. While property-type income is estimated directly from the level of earnings, three categories of transfer payments and four categories of contributions are estimated via regression relationships with other elements of the model.

Projected employment is obtained by first projecting trends in State earnings per worker levels by industry and applying the results to projected earnings.

The process by which State projections of population by age and sex are developed from the projected employment base involves several steps. It is a standard cohort component population projection except for the means by which migration is established. First, an "expected" (no migration) population is projected, by age, via the application of age/sex specific death and birth rates. These rates, for each State, are projected as a function of the birth and death rates incorporated in the Bureau of the Census national population projections. Historical birth and death relationships in State/national relatives for each State are trended into the future. The actual population is projected via the addition or subtraction of net migration projections to the expected population in each age group. Different approaches to establishing migration levels are applied to three broad age groups--pre-work force, work force, and post-work force.

The ratio of working age population to employment for each State is projected via regression relationships with a variety of other regional indicators. Total population age 15-64 ("work force" ages) is then projected as a direct function of total (all industry) employment.

The difference between this projected total and the "expected" total for ages 15-64 is taken to be due to net migration for the group as a whole. It is necessary to disaggregate this total into five-year age groups. To do so, a model is employed to (1) estimate the gross migration flows associated with the net flow established above, (2) insert age/sex detail on the <u>gross</u> flows and (3) compute the implied net flows for each age/sex group. This completes the population projection for ages 15-64.

Projections of net migration for the pre-work force age (0-14) are developed in a straightforward fashion from the migration patterns previously established for the "parent" ages. Gross in and outmigration for ages 5-9 and 10-14 are established via a direct link to the flows for ages 30-49. (A variety of "parent" age groupings was tried before selecting this one.) For example, a State's share of total national movers aged 5-9 is the same as its share of total national movers aged 30-49. Net migrants aged 0-4 were projected via the application of birth rates to female net migrants aged 15-49.

Population 65 and over is not intimately linked to the economic base and there is dramatic variation among States. Several unsuccessful attempts were made to model the migration of this age group. Finally the fraction of retirement-aged population of each State's total population relative to the Nation was plotted on graph paper for the period of 1950-1978. These plots were then extended by the professional staff based on a consideration of the factors affecting the retirement age group in each State.[1]

What has been described might be referred to as the mechanical phase of the projections, work that can be duplicated with available models and data series. However, this is not the end of the projection routine. All of the results are being subjected to review by members of the staff of the Regional Economic Analysis Division and when trends and/or relationships appear out of line, adjustments are made to "correct" them. The original model is not being rerun but rather computer routines have been prepared which permit many types of adjustment. Computer runs can be made quickly so that the results of several alternative adjustments can be quickly evaluated before deciding on the appropriate adjustment.

As indicated earlier, the State projections are being disaggregated to the 585 geographical pieces which are necessary for reaggregation into the various levels at which the data will be published. The procedure for doing this is an extension of the share relationships between the pieces and the State total. The first reaggregation will be to the economic areas so that the relationships among industries and among earnings, employment, total personal income, and population may be reviewed and adjusted where necessary before finalizing the State projections.

The projections of agricultural production and land use are being prepared by the National Resources Economics Division of the Economics, Statistics and Cooperatives Service, U.S. Department of Agriculture. That information will be published separately.

As with previous OBERS Projections, these are considered to be "baseline" projections, with baseline denoting an initial statistical framework for use in the planning or evaluation of an action or operation. They are a casting into the future of recent historical trends in growth and economic relationships. They are in no sense a goal, an assigned share or a constraint on a region's economic activity. They carry no connotation as to desirability or undesirability. Especially they should not constrain a planner in considering alternative levels of growth which might be achieved by more or less resource development or management.

It is now planned that the 1979 OBERS projections will be ready to go to the printer in early summer with published results available as soon thereafter as they can be printed. It is expected that the reports

---

1/For a detailed explanation of the entire projection methodology, see a paper entitled, An Overview of the BEA Regional Projections Program-OBERS by Kenneth P. Johnson, Chief of the Projections Branch, Regional Economic Analysis Division, BEA.

will be available from the Superintendent of Documents. U.S. Government Printing Office.

The BEA Regional Industrial Multiplier System

A regional industrial multiplier system (RIMS) has been developed by the Regional Economic Analysis Division of BEA to be used in the analysis of the regional economic impact of any project or program which effects the industrial structure of the region. This system has been described in a WRC publication entitled, Industry Specific Gross Output Multipliers for BEA Economic Areas. It is WRC Guideline 5 and is available from the U.S. Government Printing Office. I wish to talk briefly about the system for several reasons: (1) I think it is not too well known, (2) a considerable amount of developmental work has been done on the system since the time of that publication and (3) BEA provides the service (for a fee) of producing the multipliers for specific industries in any specific group of counties which represent the area of impact.

The total regional effect of a project or program will involve regional industries other than those in which the initial change occurs. Other industries will realize changes in production , because they supply inputs directly to the initial-impact industry. The supplier industries, in turn, will require increased levels of inputs, some of which will be supplied by industries within the region. Additionally, in each industry experiencing a change in production, a change will occur in the earnings paid to those who provide labor inputs to the production process. With this increase in income to a region's households, corresponding increases in consumption expenditures will also occur. These increases, in turn, will influence the level of regional activity associated with the production of consumer goods, either directly, as in the case of retail and service establishments, or indirectly, as in the case of manufacturing and agriculture.

Regional impact analysis measures the additional rounds of regional economic effects that are triggered by an initial impact, by taking account of (1) interconnections between industries in the region, (2) the implications of increased regional household income, and (3) the resulting consumption expenditures of the region's households. Of course, regional differences in size and industrial structure can result in substantial differences in the intensity of the impact of a given initial change. The principal task of regional impact analysis is to take account of the regional characteristics which influence the intensity of the regional reaction to some initial impact.

One tool of impact analysis is the regional multiplier, which relates a total change in output, income, or employment, to some initial, exogenous change in regional economic activity.

One technique for deriving regional multipliers is by use of a regional input-output (I-O) model. An I-O model of an economy describes the flows of goods and services to markets and between industries and is an important tool for describing the structure of the economy. More importantly, however, it may also be used to analyze the implications of the changes in one portion of the economic system on levels of activity in other portions of the system. These interactions between industries represent, conceptually, endless chains of actions and reactions, and the I-O model serves as a convenient means of summing these effects.

This characteristic of the I-O model is useful in estimating the total economic effect of an exogenous change in an economy. The first step is to identify an initial change in the system. Usually, this means estimating a change in the final-demand sales of a particular industry. The I-O model then summarizes the many rounds of economic effects that are set off by the initial final-demand change.

This process is analogous to multiplying an exogenous change by a multiplier. The heart of the RIMS approach is the estimation of this multiplier value without the creation of a complete I-O table for the economy under study.

The source of the regional change to be analyzed can generally be represented by a change in output in a given industry or set of industries. For example, the economic effect of an increase in the availability of irrigation water might manifest itself in the form of production increases in cotton or potatoes, or in some other commodity. Similarly, an increase in recreational facilities may increase activity in certain trade of service industries.

RIMS produces gross output multipliers which can be used to estimate changes in regional total gross output resulting from changes in regional final demand in specific industries. In the RIMS model final demand includes specifically sales to Government, sales to other regions, and capital formation. A conversion of the gross-output change to changes in earnings and employment is also produced by RIMS.

Since RIMS provides multipliers which are specific to a particular industry in a particular region, it is possible to provide estimates of economic impact without building an I-O model for each region under study. Further, because of the relatively disaggregated sectoring plan, analysis may be performed at a detailed industry level, thus avoiding the errors which occur when differing industries are combined. The multipliers also offer a consistent set of assumptions across regions, making comparisons between regions more meaningful than would be the case if results were to be based upon different procedures and conventions. Even though the multipliers are based upon secondary data sources, with the basic information about the technical requirements of industries coming from the national I-O model, preliminary empirical tests of the accuracy of the multipliers indicate that they yield estimates which are not substantially different from those generated from regional I-O models based on survey data.

At the present time RIMS is based on the 1967 I-O table. A new national table for 1972 is due to be completed very soon. When available, that table will be utilized in the RIMS. It is also expected that when the new OBERS Projections are completed, the WRC will sponsor a revision of Guideline 5 which will produce multipliers for the revised 183 BEA economic areas.

Those who are interested in the application of the RIMS to a particular study area should contact BEA by phone or by letter. Several clarifying changes have been made in the system since Guideline 5 was published and assistance can be obtained in the procedures for estimating changes in final demand which is essential to deriving the multipliers.

PLANNING FOR USER ORIENTED HYDROLOGIC FORECASTS

by

L. Douglas James[1]

Introduction

This conference was called as a forum for the exchange of ideas on improving the forecasting of droughts and floods and on disseminating forecast information to potential users. The topic was posed with two questions: why? and how? Why do we seek improvements? How do we make improvements? The first question seeks goals to serve as a basis for improvement. The second seeks advances in scientific understanding that can be implemented so that the goals can be achieved. The understanding is needed in the areas of making more accurate forecasts, disseminating the information to users, and assisting users in making more constructive use of the information they obtain.

The questions of why and how have been posed in a logical order because goals furnish the basis for improving. Without criteria to judge progress, one cannot know whether conditions are getting better or worse. The logic has been followed in selecting the order for the sessions in this conference. The session this opening evening is looking at goals, and later sessions will explore how our understanding of hydrology and of information dissemination techniques can be better applied in achieving them. Of course, goals and methods are best formed iteratively because available means constrain the goals that can be achieved and feedback is required to establish goals that can operationally be achieved. Useful goals are those achievable by available means and the best means are those that accomplish most in achieving priority goals.

Beginning then with goals, what are we trying to achieve through improved hydrologic forecasting? The pure scientist might answer in terms of the intrinsic value of the greater understanding gained as we make better forecasts and disseminate them more effectively. The more applied orientation in the majority of us, however, is more comfortable with goals expressed in terms of the contributions advanced knowledge of drought and floods can make to the health, safety, and welfare of human beings and the vitality of the environment in which we live.

When one translates the lines of inquiry suggested by why and how into the terminology of the economist, the how of forecasting provides the supply. The answers to the why determine the demand. In other words, the demand to improve forecasting is found in the benefits people gain, individually and collectively. For the benefits to accrue, the forecasts have to be made, disseminated, and used. The improvement can come as forecasts are made more accurately, disseminated to deliver more of the information needed to those who can use it, and applied more efficiently by the user.

---

[1]Director, Utah Water Research Laboratory, Utah State University, Logan, Utah 84322

The supply and demand approach to defining the need for improved forecasting leads to a question that can be asked in a third three-letter word: who? Who will receive and use the forecasts? It is the uses they make that produce the results that in aggregate determine the goals the forecasts actually achieve and who a user may be is very important in determining what he responds to. It is the information needed by users that comprises the demand for forecasting.

Satisfying user demand requires a product designed to maximize user satisfaction. Users, however, vary. The ideal forecast for one may be totally useless for another. This fact makes it very important to determine who 1) are using forecasts and how they are using them to advantage, 2) are using forecasts inefficiently in terms of the value they could receive, or 3) could potentially gain significant value from forecasts that they are not using.

The approach taken in this paper to planning user-oriented hydrologic forecasts begins by identifying user characteristics that determine forecasting needs. User characteristics are defined on the basis of perspective of problems and opportunities. The four dimensions of user perspective are taken as 1) range of actions, 2) scope of control, 3) favored values, and 4) time frame. Each perspective is then explored through definition of its distinctive characteristics, discussion of recent trends in the institutionalization of dealing with it, description of the sorts of forecasting information needed, and evaluation of the implications of the above for making forecasting improvements. At the end, conclusions will be drawn in terms of guidelines on how to improve and how not to improve forecasting.

Perspective as a Basis for User Classification

Users of hydrologic forecasts vary from the federal agency operating a major system of reservoirs and possibly considering developing new ones to a farmer determining whether there will be enough irrigation water to mature a crop he is considering planting or a hiker worried about being caught in a flash flood. The effort to improve hydrologic forecasts needs to begin by recognizing the diversity of uses and users. The logical starting point in dealing with this diversity is by classifying uses and users according to a taxonomy that puts those best served by a common forecasting method in a common group.

Baier (1965:91) observes that perspective refers to "the demands, goals, or aims of persons holding certain special positions or jobs or functions in a society." The special position of interest here is that of using hydrologic forecasts in decision making. In other words, demands and positions interact in forming the goals, without which rational choice is impossible. As described by Moline (1968:195), consequences of taking a perspective include 1) limiting goals to some specific set, 2) relying in making subsequent decisions primarily on evaluative criteria tied to achieving those aims, 3) regarding other interests as less relevant, 4) biasing interpretation of incoming information to favor selection of an alternative achieving the goals, and 5) forming alliances with others of common perspective. A perspective thus serves as a perceptual filter that separates relevant from irrelevant information.

Since hydrologic forecasts are designed to be relevant information by those who formulate and distribute them, the application of the concept of perspective important for our purposes is in its use to identify goals associated with perspective selection that may be causing potential users to judge hydrologic forecasts to be irrelevant. Futhermore, the form of hydrologic information considered most relevant by the user depends on his perspective. Each user has his own perspective of how forecasting could help him and of how to obtain and apply hydrologic information.

Perspective has four principal dimensions (Larson, et al., 1979): range of actions, scope of control, favored values, and time frame. The range of action dimension recognizes that some users of hydrologic forecasts are interested in designing or operating large systems of reservoirs; others are interested in formulating nonstructural approaches to water resources management; and many individuals are faced with day-to-day water or land use decisions. The scope of control dimension recognizes that users vary in area of jurisdiction or concern from river basins to units as small as a city lot and vary in function among such water uses as irrigation, municipal water supply, and flood control. The values dimension recognizes that while the federal government has specifically stated, if not operationally defined, planning objectives (WRC, 1973), other users have objectives that are more volatile because they are less clearly defined and often conflict with one another. The time frame dimension varies from those interested in flood or water supply probabilities over the long run for water or land use planning to those provided warnings of immediate danger in order to protect their property or persons.

Structure of Forecast Design

The prevailing framework for evaluating hydrologic forecasts is to visualize the information provided as a common or public good which serves all recipients and has benefits equal to the sum of the benefits to the respective users. The sum, however, can be much larger if, instead of providing a common product to all users, the forecasts can be prepared and broadcast in a variety of ways, each individually tailored for a user group.

While the structure and dissemination of hydrologic forecasts according to this design could greatly increase the benefit to users, three issues need to be carefully considered in justifying the effort. The first is the technical feasibility of putting together such a system. Can a practical system be designed that will identify user groups by the kind of hydrologic forecast that would fit their perceptual preferences for better coping with drought or floods and deliver such forecasts to those who need it? The second is the economic feasibility of the effort. Would the benefits resulting from such a more effective system (measured in greater national income through better water management or reduced flood damages) exceed its costs? The third is financial feasibility. Where are forecasting agencies to get the necessary funds? What are the possibilities for collecting some of the costs from users? The scope of this paper is to explore these issues conceptually. Their resolution would require research and quantitative analysis far beyond the scope of effort available for preparing this paper.

The conceptual design of a user-oriented hydrologic forecast system thus requires 1) identification of user groups with similar perceptions of needs for hydrologic forecasts, 2) determination of how each group can profit from better forecasts and the form of information dissemination that will be best received, 3) abstraction from the total hydrologic forecast of the specific information needed by each group, and 4) communication of the results to the users. A complete system should also include monitoring of how the forecasts are being used and corrective feedback systems as difficulties are observed.

Range of Actions

Range of actions was named above as one of the four dimensions of perception because goals become closely linked in one's mind to the alternatives visualized for achieving them. In the context of application to hydrologic forecasts, a potential user who perceives his choice as to which alternative is most advantageous as depending on the information he receives from the forecasts is going to be much more receptive than will a user who sees little that he could do different no matter what information he obtains. Furthermore, the user seeking information before choosing from a range of actions probably has in mind what he wants to know and will find the forecast most useful if it is presented in those terms. In order to begin probing the constraints perceived actions place on how to tailor forecasts to meet needs, we can profitably begin by reviewing the range of actions being taken in water resources management and how this range has been changing in recent years.

Definition: The measures one can use in water resources management can be classified two ways: 1) individual or group (community) and 2) structural or nonstructural. The four possibilities resulting from these two two-way divisions can be illustrated in the context of flood control. Structural measures prevent water from inundating and damaging property; nonstructural measures reduce the amount or damageability of property in areas subject to inundation. An individual may employ the nonstructural measure of not locating in a flood hazard area or the structural measure of flood proofing to protect building contents when flood water rises around the outside. A community may employ such structural measures as channelization or detention storage or may instigate a nonstructural program employing various means to encourage more widespread or more efficient individual efforts.

A second set of illustrations may also be helpful. With respect to water conservation during droughts, an individual may employ the nonstructural measure of planning cropping patterns to reduce water requirements or the structural measure of drilling his own well. A community may augment water supply structurally by building a new reservoir. It may undertake a program to encourage water conservation and thereby nonstructurally reduce the need for expanding the water supply.

The actions one can take in implementing measures in any of these four categories depends on his role. For community measures, the desired use of hydrologic forecasts varies among technical personnel charged with project management, governmental leaders responsible for oversight of the technical management, and the public served. For individual measures, use depends on technical and financial constraints to what one is able to do and what one learns or sees being accepted by the community from the actions of his neighbors.

For the purpose of tailoring hydrologic forecasts, it is important to remember that each of these four classifications contains a multitude of possible actions. Some actions within a given classification require quite different information than others, and these sorts of differences, too, are reflected in differences in user perceptions.

Institutionalization: Over generations, the emphasis in water resources management has shifted from individual nonstructural to (with development of functional technology) individual structural to (with economies of scale favoring large over small systems) group structural to (with the environmental and social disruptions caused by very large systems) group nonstructural. Agricultural water supply has, for example, moved through 1) concentration in humid climates or along well watered valleys, 2) growth into more arid areas as farmers supplied themselves from wells or small diversions, 3) dependence on very large projects taking water up to hundreds of miles to major cities or vast agricultural developments, to 4) programs of water conservation to get more use from a fixed amount of water delivered. The trend within all water resources management purposes in recent years is from the third (large structural projects) to the fourth (nonstructural means) group of measures.

This trend toward government taking an increasing role in encouraging individuals to adopt water management policies deemed in the public interest places a much greater burden on hydrologic forecasting because of the key role reliable information credably presented has in eliciting a desired response. People are going to be very slow to conserve water or avoid flood plains unless they really understand why these actions are advisable, and hydrologic forecasts need to be carefully formulated and presented to achieve this acceptance. Information should be just one of the set of means carefully integrated into a governmental nonstructural program, also including incentives, expert help, and regulation (James, 1975), for effective results. So far, nonstructural programs have not achieved this sort of integration, and the gap between institutionally set ideals and what is being achieved is one of the major issues of our day. Collective decisions to protect environments do not stop individuals from harming them; collective decisions to conserve do not stop individuals from being wasteful.

Information Requirements: Six kinds of changes are needed in the formulation and dissemination of hydrologic forecasts as the emphasis shifts from structural to nonstructural programs (James, 1973). These are with respect to:

1. Geographical coverage. The structural designer only requires information on hydrologic probabilities at a selected facility site plus average probabilities over the flood plain or service area to estimate benefits for economic justification. The planner of a nonstructural program requires information on flood probabilities or water supply conditions at every point (flood plain property or water user location) where individuals need to make decisions. The design or operation of community projects largely depends on information oriented to larger streams where hydrologic records are generally more extensive and larger tributary areas facilitate very short-run forecasting because of the longer time lags for flow through the basin. Individual efforts, however, often require forecasts from very small drainage basins where forecasting methodology is much less precise. For example, a difference between the information requirements for structural versus nonstructural flood control programs is that the former can be satisfied by curves of flow and damage versus frequency referenced to one point along the stream whereas the latter requires maps indicating stages (and possibly velocity or other indices of flood severity) as a function of frequency at every point on the flood plain.

2. Repeated Presentation. The presentation to the public of hydrologic information on community structural projects only requires that sufficient technical detail to convince them that the project is

worthwhile. The greater detail only need be prepared for design engineers and can be delivered at one time soon after the study is completed. The hydrologic information for individual and nonstructural programs becomes completely ineffective if it is presented and then forgotten. The results must be kept continually on hand for whenever interested individuals inquire; and when they ask, it must be presented in a form that communicates with those who need help.

3. Convincing promotion. Once a structural project is complete, hydrologic forecasting remains important for optimal operation but the range of management action is greatly reduced by the cost sunk in construction. Nonstructural measures are much less capital and hence much more maintenance intensive and consequently are easily reversed. In other words, the commitment from constructing a structural project places much greater constraint on future action than does the commitment through legislative action to a nonstructural program of flood plain management or water conservation. A short-term lapse in enforcement at any time later can lead to extensive flood plain development or adoption of a high water use technology that can invalidate all that was accomplished in years of previous effort. On a more basic level, legislative bodies are constitutionally unable to make commitments that will prevent future sessions from changing the policy. Consequently, while hydrologic forecasting for structural faciltiy design or operation is presented to professionals who feel it part of their job to use it, the information for nonstructural programs must always be presented with the promotive emphasis of convincing the multitudes of individual decision makers (users and legislators) that taking a continuing action in response is worthwhile.

4. Universal response. The hydrologic information provided for community decision making need only convince a majority who can achieve implementation over any minority opposition. The information provided for nonstructural or individual decision making must reach and convince a multiplicity of decision makers. Every individual who does not receive and respond to a flood warning and every individual who wastes water while the hydrologic forecasting warns of impending drought represents a failure of the program.

5. Technical sophistication. The hydrologic information provided for structural design can be presented in technical terminology for use by engineers. This keeps forecasting dissemination and use in the realm of the physical sciences where hydrologists feel comfortable. The information for nonstructural and individual use must be presented in a manner that will be sufficiently understood to effectively counter whatever reasons people may have for disregarding it. The forecaster who develops an effective program must enter the relatively unfamiliar world of understanding how different sorts of people respond to social forces and how to reach different groups of people who respond in quite diverse ways (Hough, 1975).

6. Setting for interpretation. Information provided for the design or operation of structural measures can be provided in terms of expected or average values because the hydrologists can correctly assume that the engineer using it will operate against an information base that is relatively independent of recent events. This assumption is not at all true of forecasts delivered for individual or nonstructural response. Laymen respond much better to warnings of floods or droughts in the first few years after one has occurred than they do years later when their memories of past hardships have faded.

Implications for improvement:

There are obviously a great many possible water resources management alternatives, and each has special forecasting needs. The information disseminated should vary with whether the user is considering building a dam, evacuating his home during a storm, or letting his lawn turn brown during a summer drought. Obviously, the flow-frequency curves so useful for dam design are of no help to a flood plain occupant fearing rising waters during a stormy night, and the flows caused by a given summer thunderstorm are nearly as useless for dam design. These sorts of differences in need are generally recognized, and forecasters are experienced in dealing with them

The area where hydrologic forecasting needs to concentrate in order to improve is in providing a more credible information base for effecting nonstructural measures ranging from flood plain management to water conservation. Practitioners are not on top of what information in hydrologic forecasts best fits user perceptual preferences for coping with either floods or droughts. More important, they understand very little about how to vary forecast content with the intended audience.

The void can best be understood by exploring the variables conceptually. For a given use situation, one can well visualize the existence of some optimal course of action and some set of hydrologic information that maximize the managers ability to pursue that course. The information needs to be provided in two steps: technical development and targeted dissemination. The issues in technical development are the focus of other sessions of this conference and need be emphasized here only to the degree necessary to make sure that the techniques used in forecasting are producing what people perceive they need to know or are convincing when they seek to present what is delivered as even better.

Targeted dissemination should consider physical characteristics of the geographical area which contribute to local awareness and understanding of the need for information. Examples may include basin size, recent flood or drought experience, and ease with which laymen can identify with the problem. The dissemination should consider characteristics of the local community which affect capability to establish and maintain an effective information dissemination network. Examples here include the percentage of the residents who relate to the problem, the hydrologic expertise of available technical staff, and the acceptance of nonstructural methods by local officials. It should consider such characteristics of the individuals whose response in using the forecasts is needed as time and inclination to respond, ability to comprehend technical concepts, and perception of personal benefits. A model for using all three sorts of information in flood plain management program planning is available (James, 1975).

Planning the targeting of an information dissemination program will reveal that the physical and community factors are uniform for a given situation but that the individuals in a target area vary considerably in actual or perceived information needs. One sort of dissemination may be just right in eliciting the desired response from some but be ignored or cause adverse responses from others. Information dissemination design can be conceived in terms of finding a method that will achieve widespread response, observing who is not responding, finding a second method for residual nonresponders, and continuing adding methods until the desired coverage is obtained (James, 1976). Each addition should, of course, be considered in terms of whether the marginal improvement in response justifies the effort required to achieve it.

Scope of Control

Scope of control is an important dimension of perception because government and individuals are not going to be receptive to forecasts to which they perceive no options where their choices will make any difference. To proceed, we need to explore the range and trend in institutionalization of control.

Definition: Scope of control has three important aspects 1) geographical jurisdiction, 2) functional jurisdiction, and 3) powers of execution. Communities as well as individuals have geographical boundaries to their control and influence. Information telling how others could help by doing better is not very productive. A given agency may be responsible for flood control but not water supply and thus be only interested in forecasts relating to its functional jurisdiction. Greater functional jurisdiction problems, however, exist between land and water planning because people harmed by floods or water shortages are not responsive to forecasts that would help them when they perceive their problem as caused by people upstream who have increased flood flows or diverted low flows (James, 1971:135-138).

The distinctions in powers of execution encompass 1) differences in technical and financial resources ranging from large governments at one extreme and poor or uneducated individuals at the other, 2) differences in required interaction among control centers ranging from actions that one control center can implement on its own to actions that can only be implemented after prolonged negotiation among affected parties, and 3) measures which can be directly implemented by the control center (construction of a dam) and those for which a control center only has power to exert influence on others (flood plain occupancy or water conservation).

Institutionalization: The shifting emphasis from the structural to the nonstructural in water resources management has important implications for all three control aspects. Since the major impetus for the shift is coming from the federal level but the constitutional division of power in the United States places nonstructural programs within state or local responsibility, the scope of control is shifting toward more decentralized decision making. This requires greater coordination among the decentralized units (what good is it to get water users supplied by the community to conserve when violaters can obtain all the water they want from some other source?) and generates conflict among the jurisdictional levels (communities with the federal government).

The shift is also going from water projects directly implemented by the planning federal agency to nonstructural programs conceived as desirable from the federal viewpoint and then imposed on communities that may not be receptive to in turn impose on individuals who may be even less receptive. It is folly to fail to recognize that the indirectness in control resulting from acting through this double filter does not greatly increase the difficulty in achieving desired ends. The alternative of removing these filters and giving the federal government direct control over individual water use decisions, however, would be an unacceptable infringement on our democratic ideals. The key to the success of the nonstructural movement lies in finding means of getting large numbers of people to make choices in tune with nationwide goals while maintaining our constitutional separation of powers and democratic ideals.

The shift toward nonstructural measures is also pushing water resources planning into greater interaction with planning for land use, transportation systems, urban development, and many other areas that have historically largely been performed by other groups with radically different training, experience, clientel, and terminology. This aspect of the institutional trend has some important implications for hydrologic forecasting. The forecasts that for years have been designed to communicate with engineers and economists performing water resources planning now must be structured to meet the needs of landscape architects performing land and urban planning, environmentalists trying to protect natural processes, lawyers trying to devise and enforce workable regulations, and many other groups (Mulder, et al., 1979). The problems in meeting these needs are in part caused by differences in training and professional objectives but are greatly compounded by the degree to which many of these potential users are locked into decision-making processes that are not tuned toward hydrologic input.

Finally, the shifting emphasis from structural to nonstructural is greatly reducing the decision-maker's power to implement decisions. The process of designing and then contracting for construction of a dam is much more straightforward than using carrots and sticks to encourage local government to implement means for encouraging individuals to conserve water or restrict their use of flood plains.

Information requirements: The basic change in delivering hydrologic forecasts required by the institutional trend in scope of control is from forecasting for a few users with precise technical conceptions of their desired information to forecasting for multiplied users, many of whom may not even perceive that they can benefit. The changes in users from agencies managing large structural systems to many local governments and individual persons requires that the forecasts be delivered in technically much less sophisticated terminology. The change to users whose functional responsibilities are not water oriented requires expression in the terminology of new disciplines with the force necessary to be convincing. The reduction in power of execution accents the importance of delivery in a manner that will catch the attention of potential users and motivate them to effective use of the information. In summary, the perception dimension of "scope of control" adds to the dimension of "range of actions" in indicating changes in hydrologic forecasting that can improve effectiveness in terms of geographical coverage, repeated presentations, convincing promotion, universal response, technical sophistication, and setting for interpretation.

In an effort to probe more deeply into what needs to be done by communities to get flood plain occupants to protect their property from flood damage by flood proofing, Dexter, et al. (1978) canvassed 157 adopters in Atlanta, Georgia, and Charlotte, North Carolina. They sought views on the acceptability of flood proofing, information on factors leading to the choice and adoption of specific measures, expressions of degrees of satisfaction with choices made, and guidance from the results for promoting more effective flood plain management. Several findings should be particularly helpful for hydrologic forecasting.

1. Social acceptability is very important in determining the choices made in response to information. People do not want to take measures that are obviously at odds with peer mores.

2. People who have entered the community within the last four

years are much more receptive to new sorts of inputs into their decision making than are old timers. The relative newcomers are looking for better ways to accommodate to their new environment while old timers are slow to change accommodations that they have used for years. The principle also has important implications for those seeking to promote water conservation. Water saving practices can probably be fostered better in new residential areas or by working with people who have recently entered the community rather than by trying to convert people whose water using habits have been fixed for many years. While the newcomers may have been well set in their water using habits where they came from, the many changes in habits required by the experience of moving improves the probability of getting them to change water using habits as well.

3. Many people need technical assistance to handle unfamiliar sorts of information. General education programs and the availability of qualified expert help to deal with particular problems are both important.

Implications for Improvement: The scope of control of the targeted users is thus a second important dimension to perception that needs to be considered in formulating systems for delivering hydrologic forecasts. The key element needing to be added is obviously that of convincing the recipient that he has the power to use the information to his advantage. Perceptions have a very important role in the convincing as an appearance of real choices, or social acceptance of available choices, may well be far more convincing than a sophisticated presentation of how one can use information beneficially. Study is needed to determine the variation in perception of control among potential beneficiaries and what is important to people in various groups with common outlooks.

Values

The aims or values characteristic of an individual or a community are described by Albert (1956) as 1) the normative elements in the definition of a situation by which its desirability is judged, 2) either explicit criteria defined before making the judgment or implicit derivations from judgments made, 3) relatively persistent through time, and 4) interrelated among individuals and groups in culturally distinctive patterns. Specific values relate to perspective in an interactive way that makes it impossible to say whether the values result from perceptions or vice versa.

Definition: Values can be expressed on a community or national scale in terms of societal goals whose theoretical foundation was formulated within the discipline of welfare economics (Bergson, 1966) and extended more recently by the definition of goals related to environmental (Corwin, *et al*., 1975) and social (Finsterbusch and Wolf, 1977) impact assessment. Societal goals are usually defined with such concepts as income, environmental quality, or social well-being. On an individual scale, the theoretical foundation has largely been developed within the discipline of psychology and defined in a hierarchy of levels ranging from material comfort (food and shelter) to freedom to choose (Lasswell, 1971). Efforts have been made to work from this hierarchy of individual values in nonstructural design (Mack, 1977) but this promising effort has a long way to go.

The evidence as a whole indicates that key decisions are not made on the basis of reasoned maximizations of selected human goals but rather on the basis of perceived intrinsic merit (or fault) associated with such structural programs as irrigation of deserts, erosion and sedimentation control, or water quality enhancement (James and Rogers,1976) or such nonstructural means as zoning, water quality standards, or effluent changes. Collective decision making adds the elements of strategy or game playing in a polyarchial political process (Hart, 1974), and the interactions add a dynamic factor to expressed value preferences if not the values themselves.

Institutionalization: Collective values for planning structural projects emphasized net benefit maximization until recent Water Resources Council efforts produced the "Principles and Standards" that added environmental quality as a second goal (WRC, 1973). Procedures for applying these principles and standards provide for determination and presentation of an optimal plan with respect to each goal and of intermediate tradeoffs for decision making in a process involving public participation as well as review through official channels. Additional displays are also required on the impacts of the alternatives on social well-being and regional development. The ideal is to institutionalize a management scheme that will promote the quality of life by reflecting society's preferences for attaining these objectives.

A number of problems related to perceptions of values surface in this process. Several of these are enumerated as follows:

1. The societal goals used to guide formulation of nonstructural programs are not logically aggregated from the values that govern the actions of individuals. In fact, they may often be in strong conflict. Conflict is unavoidable when values are in direct conflict, but the worst aspect of the situation from the planning viewpoint is that the consequences of nonstructural alternatives are displayed with respect to societal values rather than with respect to the values that concern the other choices of the individuals making the decisions.

2. Available procedures are not adequate for predicting the consequences of defined alternatives in terms of the institutional goals and don't even attempt to predict them in terms of individual goals. We are not able to predict what nonstructural programs do to improve or harm the environment and are making no effort to predict what response to hydrologic forecasts does for people in terms of the values they hold dear.

3. Different participants in the decision making process vary in the importance that they attach to various objectives. This causes disagreement and negotiation, and the outcome depends on who is doing the negotiation, how much weight the negotiators have politically, and many other factors not related to project merit.

4. The major new federal management thrusts are not justified on the basis of achievement of institutionalized objectives but rather on their own intrinsic merit. Currently water quality control and flood plain management are the two most recent major programs, and neither is planned on the basis of the principles and standards ostensibly chosen for guiding all planning for water and related land resources.

The institutionalized trend toward multiobjective planning has been supported by advances in theory and techniques (Major, 1977). At the end of his presentation of the state-of-the-art of multiobjective planning, Major presented six suggestions for solving these and related problems:

1. Extend application of the Principles and Standards to encompass the activities of the federal water agencies not now covered. Decision

makers at other levels need to develop explicit statements of their own values to facilitate more objective resolution of value conflicts.

2. Extend the methodology of the Principles and Standards to encompass such concepts basic to the theory of multiple objective planning as separate discount rates by objectives and explicit weighting among objectives.

3. Extend use of the methodology from project design to include decision making on cost sharing and repayment issues. One of the major problems with the current approach is that it uses objectives other than net benefit maximization in project design decisions without specifying financial responsibility for their costs.

4. Strengthen mathematical modeling capability and train planners in their use so that available methods can be more readily applied at the field level.

5. Develop the institutional arrangements for water resources decision making so that the multiple objectives are not lost in the communications processes among those involved in planning, the communications between planners and the public in the formation of public sentiment, and the presentations of planning agencies to political decision makers within congressional committees.

6. Establish postplanning review and assessment as a mechanism for developing feedback for process improvement.

The movements toward formally defined objectives at the federal level and toward optimization techniques in multiple objective planning are widening the differences between the large agencies who have the resources to formalize criteria and optimize mathematically and those who do not. Levels of government below the federal could potentially formalize some set of multiobjectives criteria; but the results would no doubt be different than those expressed at the federal level. This conflict would in itself be healthy for water planning by disclosing issues for explicit discussion.

A much greater problem exists at the next step where individual response to forecasts are largely based on personal values, and these are largely unknown to the planners. Even were these values derived implicitly from observed decision, one would find considerable variance between the values people use in governing personal actions and the values people collectively establish for communities or nations. The water conservation program, for example, that a community establishes from consensus norms will vary considerably from the progream individuals select for themselves. The conflict exposes some basic issues of human rights versus governmentally imposed values.

One additional important aspect of the way values have been institutionalized can be found in the observation by Smith (1974) that "economic language has been used extensively ... to rationalize... decisions dictated by other than formal considerations." He found that the choices have been made through political processes and then defended to the public on the basis of economic merit even when they vary substantially from the true economic optimum. The corollary problem in multiple objective planning is that the techniques won't be used to optimize as described by Majors but rather that the additional objectives will give proponents of the alternatives selected additional ammunition for convincing the public of the wisdom of the selections. Already our experience is that the environmental quality objective has been used much more to persuade the public than a source of data for impartial analysis.

For the specific problem of promoting more effective application of hydrologic forecasts, one can see definite possibilities in multiple objective persuasion; and at this point, we return to ethical problems in government persuading people to change their values. To illustrate a situation, a survey of flood plain residents in Atlanta, Georgia, showed that many could be convinced to employ flood plain management practices contributing to a better environment even though they didn't see an economic advantage in doing so (James, 1972). If those promoting flood plain management cannot identify an important environmental advantage, are they justified in using this argument to encourage their desired response to hydrologic information?

Information requirements: If hydrologic forecasts are to be formulated to meet the needs of potential users, it is important to understand the personal values underlying those needs. The information can then be structured for each potential user group in the form of what each one wants to know. Hydrologic forecasts can be translated to the terminology of economic loss, risk to life, environmental damage, etc. An individual needs to know what is in responding for him as measured by values he believes important (as apposed to values that society as a whole believes should be considered important). A lot could be done to improve response to hydrologic forecasts by identifying user values and how hydrologic forecasts can be used to improve the probability of their realization.

Implications for improvement: The values of targeted users are thus an important dimension to perception. Their identification can play an important role in convincing potential users of how they can gain in their own eyes. Research is needed to identify the values people hold important in the sorts of decision making where hydrologic forecasts can provide significant input. Failure to consider values explicitly is to bias use of hydrologic forecasts towards individuals whose values match those of the forecaster. The result is to favor economic objectives and agency needs over the values of individuals and thus be counter productive with respect to adoption of nonstructural measures.

Time Frame

The time frame dimension of perspective is important because of the influence it has on all three other dimensions. A shorter time frame reduces the range of actions and degree of control and confines values considerd to ones that come immediately to mind as important.

Definition: Time frames are usefully distinguished in economics as very short run, short run, and long run according to key distinctives in the character of the alternatives available. In the very short run, the user can only work with the water already stored in the reservoir in deciding on a conservation policy or with the flood hydrograph already coming downstream in choosing between protection and evaluation. Between the two examples, the flood response is obviously the much shorter time frame.

In the short run, the conservation reservoir, the flood wall, or the zoning laws exist, and the user must choose how to operate these tools efficiently. In the long run, one is comparing alternatives for adding structural facilities or nonstructural capabilities.

Institutionalization: The institutionalized procedures for water resources decision making emphasize long run comparisons, give some consideration to the short run, and are very seldom applied in the very short run. Time seldom permits very short run analysis in flood control situations, and values which surface in the very short run are quite different than those used in long-run planning for water conservation. The implications for hydrologic forecasting can be seen when one realizes that forecasts largely provide inputs to very short run decision making. For short or long run decisions, one thinks in terms of probabilities instead of immediate consequences of a known event, the difference in concept and between physical consequences, and probabilities is an important barrier to understanding by many of the less sophisticated potential users.

Information requirements and needed improvements: The time frame within which hydrologic forecasts are to be used obviously should be considered in determining the detail and sophistication of the information provided and the methods of dissemination. Existing programs are structured to get needed information to potential users quickly enough for them to have time to respond, but response efficiency could probably be improved by explicit evaluation of the potential actions, scope of control, and predominate values operating during the available time frame and determination of how best to provide inputs to decision making as it really occurs within that time frame.

Optimization and Actuals

By way of recapitulation, improved forecasting implies change from actual present practice toward some new policy closer toward optimal. One model for promoting improvement would thus be to determine optimality, record current practice, identify discrepancies, evaluate alternatives for bringing practice closer to optimum, adopt those that promise success, monitor the results, and modify the practices that work poorly. More realistically, true optimality is too elusive to define with confidence; and one must often operate by proposing practices, predicting consequences, and adopting options that accomplish more good than harm. The greatest contribution to improved forecasting can no doubt be achieved by considering both models and applying each as possible.

Conclusion

The above analysis of why forecasts need to be improved by targeting to meet the needs potential users perceive can be best summarized in two recommendations each on what practices would seem likely to improve response and, perhaps more useful, what practices that ostensibly would seem to promote improvement should be used with caution because they are likely to be counter productive. The two recommendations on how not to improve are:

1. Religiously avoid the conceptual trap of thinking in terms of establishing a uniform system of forecasting and information dissemination. Deal with each situation as unique because it is.

2. Do not assume that individuals are going to necessarily be eager to follow community values. They need to be motivated in terms of their own values.

The two recommendations on how to foster improvement are:

1. Develop a strong base of data on how people with the various dimensions of perception respond to alternative hydrologic information and delivery systems. Such data provide important information for research and evaluation.

2. Work with potential users in expanding their range of actions, increasing their scope of control, and showing them how hydrologic forecasts can be used to further their personal values.

REFERENCES

Albert, Ethel M., The Classification of Values, American Anthropologist, v. 58, pp. 221-248, 1956.

Baier, Kurt, The Moral Point of View, New York, Random House, 1965.

Bergson, Abram, Essays in Normative Economics, Cambridge: Harvard University Press, 1966.

Corwin, Ruthann, _et al._, Environmental Impact Assessment, San Francisco: Freeman, Cooper, & Co., 1975.

Dexter, James R., _et al._, The Social Aspects of Flood Proofing: The Atlanta and Charlotte Experiences, Presented to ASCE Specialty Conference, Blacksburg, Vir., 1978.

Finsterbusch, Kurt, and C. P. Wolf, eds., Methodology of Social Impact Assessment, Stroudsburg, Penn., Dowden, Hutchinson & Ross, Inc., 1977.

Hart, Henry, Toward a Political Science of Water Resources Decisions, in Man and Water: The Social Sciences in Management of Water Resources, L. D. James ed., Lexington: Univ. of Kentucky Press, 1974.

Hough, G. W., Technology Diffusion: Federal Programs and Procedures, Mt. Airy, Md.: Lomond Books, 1975.

James, L. D., Surveys Required to Design Nonstructural Measures, Proceedings ASCE, v. 99, n. HY 10, pp. 1823-1836, Oct. 1973.

James, L. D., Formulation of Nonstructural Flood Control Programs, Water Resources Bulletin, v. 11, n. 4, pp. 688-705, Aug. 1975.

James, L. D., A Decision-Sequences Model of Residential Occupancy of Urban Flood Plains, Proceedings of International Symposium on Mathematical Modelling Techniques in Water Resources Systems, A. K. Biswas, ed., Ottawa: Environment Canada, 1972.

James, L. D., _et al._, The Flood Plain as a Residential Choice: Resident Attitudes and Perceptions and their Implications to Flood Plain Management Policy, Atlanta: Georgia Institute of Technology, Environmental Resources Center, ERC-0671, 1971.

James, L. D., Implementing Nonstructural Flood Control Measures in the Local Community, Proceedings of a Seminar on Nonstructural Flood Plain Management Measures, Davis, Calif.: U. S. Army Corps of Engineers, Hydrologic Engineering Center, 1976.

James. L. D., and J. R. Rogers, Economics and Water Resources Planning in America: 1776-1976, Water Resources Planning in America - 1776 to 1976, Preprint 2754, ASCE Annual Convention and Exposition, Philadelphia, 1976.

Larson, D. T., _et al._, Multilevel Modeling in Water Resources Planning, Presented at Water Systems 79, ASCE Specialty Conference, Houston, Texas, 1979.

Lasswell, Harold, Preview of Policy Science, New York: American Elsevier, 1971.

Mack, R. P., Comprehensive Values and Public Policy, in Methodology of Social Impact Assessment, K. Finsterbusch and C. P. Wolf, eds., Stroudsburg, Penn,, Dowden, Hutchinson & Ross, Inc., 1977.

Major, D. C., Multiobjective Water Resources Planning, Washington: American Geophysical Union, Water Resources Monograph 4, 1977.

Moline, Jon, On Points of View, American Philosophical Quarterly, v. 5, pp. 191-198, 1968.

Mulder, Jim, et al., Integrating Water Resources and Land Use Planning, Logan, Utah Water Research Laboratory, UWRL/P-79/01, 1979.

Smith, S. C., Economics and Economists in Water Resources Development, in Man and Water: The Social Sciences in Management of Water Resources, L. D. James ed., Lexington: Univ. of Kentucky Press, 1974.

U. S. Water Resources Council, Principles and Standards for Planning Water and Related Land Resources, Federal Register, v. 38, n. 174, pp. 24778-24869, Sept. 10, 1973.

MYTHS AND USES OF HYDROMETEOROLOGY IN FORECASTING

Gene E. Willeke[1]

INTRODUCTION

The practice of hydrometeorology, for just over the past half century, is a direct application of Murphy's Law: "If anything can go wrong, it will." Some of the elements of modern hydrometeorological practice were used by the Miami Conservancy District in its 1917 analysis of the flooding potential of the Great Miami River (Miami Conservancy District). The next major advance came in the 1930s, coinciding with the construction of a number of major dams. Although the issues posed by the Miami Conservancy District analysis (damages from flooding) and those of the Corps of Engineers, Bureau of Reclamation, Tennessee Valley Authority, and Weather Bureau in the 1930's (dam failure) were significantly different, in all these cases there was a common question: "How bad could things be if the most severe observed or calculated conditions should occur simultaneously," a question whose content is remarkably similar to Murphy's Law.

It is important to distinguish between two situations in which hydrometeorological analysis may be used. The first is as an aid to decision making, the usual decision pertaining to the sizing of spillways, outlet works, downstream channel capacity, reservoir freeboard, etc. The other is as a tool in representing reality. There is significant overlap between the two situations. However, there is difference in that a decision maker may accept the reality of an event while denying its relevance or significane to the decision.

In representing reality, we may well be unduly influenced by our approach to making decisions. For example, we routinely do frequency analysis of precipitation, streamflow, temperature, etc. For small recurrence intervals, i.e., large probability of occurrence, these methods appear to give us excellent information for decisions and appear to represent reality well. When annual maxima fit a probability distribution with little deviation, we tend to assume it fits throughout. However, this may be quite illusory. If there really are physical maxima in natural processes, and we know there are, they may well be encountered rather frequently. The speed of light, the wingspread of butterflies, and the heights of trees, relative humidity, temperature, and precipitation in a given time period at a given location may all have physically limited values. Though not necessarily common (say, 0.01 or less), they are not necessarily infinitesimal (say, 10^{-5} or less) either.

[1]Director, Institute of Environmental Sciences, and Professor of Geography, Miami University, Oxford, Ohio 45056.

In this paper, I wish to examine a few of the myths surrounding the practice of hydrometeorology, and to indicate some potential uses that are not part of conventional practice now. Myth is taken to mean one of the fictions or half-truths forming part of the ideology of civil engineering design practice, a notion based more on tradition and convenience than on fact. There may be a large amount of truth in the myth, but there is neither certainty nor ample evidence to support it.

THE MYTHS

The Myth of the Tails

"Statistical distributions applied to hydrometeorological events that fit through the range of observed data are applicable in the tails."

A statistical distribution is a mathematical expression. It may be shown to fit a body of observed data quite well. The adequacy of the fit in the tails is dependent upon the number of observations available to the analyst, the representativeness of the observations, and the appropriateness of the mathematical distribution. In a large, finite population where observations may be taken of each element of the population, a given distribution may often be shown to have an adequate fit not only in the central portion of the distribution but also in one or both tails. If the particular distribution chosen initially doesn't work well, another may be found that does fit will or at least better than the initial choice.

A statistical distribution is usually an "as-if" representation of reality. There is, for example, no theoretical reason why natural hydrologic processes should be random occurrences (a view Albert Einstein espoused). Rather, we find that, if random occurrence is assumed and a given distribution is chosen, observed data may be represented well and one might say the natural system functions "as-if" it were the mathematical function.

The most serious problems with "as-if" representations tend to come with extreme cases, in the tails of a distribution. Suppose a real distribution is not asymptotic in the tails, but rather reaches a true boundary. Precipitation, of course, always does this in the left tail, having a lower bound of zero (leaving aside the arguable proposition that evaporation is negative precipitation). It is usually assumed, however, that precipitation does not have an upper limit, but is asymptotic and ever increasing, though with extremely low probability of occurrence. If we posit a case in which there is a true upper limit, the distribution may fit the data very well except in that region.

An observed time series of annual maximum precipitation for a given time period, such as 6 or 24 hours, is a portion of a very long time series of such events. Considering climatic shifts, the very long time series is not infinitely long. In terms of the life span of a person, observed data series for precipitation are relatively long. For most records and for most of us, the length is of the order of our ages, 30-60 years. Such record lengths are, however, relatively short, and do not meet the criteria cited above for a good fit in the tails.

A further complication in extreme events work is the inherent inaccuracy of measurement. The largest extreme events are often measured in non-standard containers, forcing us to rely on bucket surveys. Such surveys are very useful, and may be reasonably accurate, but are not especially precise.

What then can we say about the applicability of statistical distributions to precipitation, streamflow, or other data relevant to a hydrometeorological analysis? We can and should regard such distributions as applicable throughout much of the range of the distribution. We should regard both tails as very uncertain and not seek to use them for long recurrence intervals. For U. S. data, I think the upper limit should not exceed 50-100 years. When used even at such levels, it should be used in the context of decision analysis and sensitivity tests, rather than as good representations of the natural situation.

The Myth of Infinitesimal Probability

"The probability of occurrence of Probable Maximum (or Minimum) events is infinitesimal."

The basic origin of this myth is in the set of theoretical distributions used to describe precipitation or streamflow frequency. By way of illustration, we may consider the Chow frequency factor K (Chow, 1951). The K value for the 100 year event is about 4. The mean of the annual maxima is added to the product of K and the standard deviation of the annual maxima series in order to obtain the magnitude of an event of a given probability. Hershfield (1961) examined maximum observed rainfalls and found that K factors of 14 to 19 were required to reach these maximum values. While such K values imply an infinitesimal probability, the relatively large number of occurrences of such events lead us to different conclusions. It should also be noted that PMP calculations are usually higher than would be indicated by the use of Hershfield K factors (Hansen, et al., 1977).

I think we must first either reject the validity of the theoretical distributions in the tails or accept the proposition that our data base does not permit an adequate fit in the tails and is perhaps too short to define the moments of the data series adequately. Second, we should be willing to accept a distribution with a fairly high probability for the probable maximum event. If we do this, we can recognize that the maximum event may occur on the average of once in several hundred years, rather than once in 10,000, 100,000, or 1,000,000 years.

For appropriate comparisons, we should alter our data series for frequency analysis in such a manner as to coincide with the data used in PMP analysis. This would require using the entire storm as the unit of analysis. It would require altering the geographic area from a point (as is now done in frequency analysis) to a meteorologically homogeneous region. Then, either annual maximum events (the more likely choice) or all storms would comprise the data series. The data points would be depth over a given area for a given duration.

At present, comparing rainfall frequency to PMP is like comparing Winesaps and Rome Beautys. They're both apples, but of a different kind. The work required to use such a data series would be somewhat larger than for point analyses. Satellite and radar data plus denser networks make the task easier than it has been in the past. (Of course, point analyses are in no way precluded by such a shift in focus.)

With such a frequency analysis, moisture, seasonal, orographic, and other adjustments could be made if desired in the same manner as for a PMP estimate. In the absence of such analyses at this time, we can only conjecture as to the altered picture of relative frequency. However, it appears we would find small indicated K factors as well as probability statements of the following kind: "In meteorologically homogeneous area A, there is a 0.01 probability of all season precipitation of 8" in 24 hours for areas of 1000 square miles at the index location."

The meteorologically homogeneous area is, in some respects, a more usable unit of area than the watershed. In the general case, a given storm will cover many watersheds. A special case is that of the nested watershed. When we are interested in flood flow on a watershed and its subwatersheds, it may be handled more readily in terms of the storm than the streamgage. Management of water, including coping with flood damages, is typically done on a political subdivision basis (state, group of states, county, city, etc.) rather than the river basin. There is little evidence that water resource management is moving in the direction of basin management.

The Myth of Impossibility

"Hydrometeorological estimates of extreme events are so large that they cannot or will not occur."

Returning to an earlier issue, this statement can be analyzed from either a decision or reality standpoint. In making a decision, the consequences of experiencing an extreme event may be such that the extreme event may be disregarded as either irrelevant or inconsequential. This is an economic and social judgment. If the consequences would be catastrophic, however that word is defined, it is foolhardy to reject the possibility of occurrence.

From a reality standpoint, the situation is otherwise. For a variety of reasons (data, model inadequacy, record length, etc.), the magnitude of extreme events is uncertain. This may give rise to the experience of an actual event exceeding an extreme event estimate. The August 1978 storm in Texas provides one of our most recent examples. A bucket survey report is being investigated by the National Weather Service of a 48 hour rainfall of 48 inches, about 2 inches larger than the 48 hour all-season PMP for 10 square miles (Miller, 1979). There are already several verified records in excess of 30 inches from the same storm. This is not an isolated case. A number of storms have either been near or have exceeded PMP estimates. Hydrometeorological Report 51, PMP Estimates East of the 105th Meridian, notes 30 storms in the last century with moisture maximized precipitation magnitudes within 10 percent of PMP (Schreiner, 1978). Storms after 1972 are not in that list.

The profession has not as yet undertaken the estimation of Probable Maximum Drought, as I advocated 4 years ago in a presentation to the ASCE annual convention in Denver (Willeke, 1975). We know, however, from both precipitation records and tree ring records that some truly extraordinary drought events have occurred, including such modern events as the mid-1970's California dought, the Northeast drought of the early 1960's, and the Dust Bowl drought of the 1930's.

For those inclined to think in such terms, this myth will be observed as a direct contradiction of Murphy's law.

The Myth of Stationarity

"The distribution of events is not changing in a trend, cyclic, persistent, or catastrophic fashion."

In discussing this issue, we are confronted with the necessity of being careful in our choice of words.

It appears stationarity is assumed for several reasons:

-- The records do not convincingly demonstrate trends, cycles, persistence, or climatic shifts.

-- For many design situations, the economic life of the project is short enough to justify ignoring long-term trends, cycles, and shifts.

-- Computational simplicity.

Each of these reasons is valid for most cases. Diurnal and seasonal variations are well demonstrated and accepted, and used in analysis (for the most part). Cycles and trends in annual series cannot yet be demonstrated convincingly on U. S. precipitation or streamflow data. Persistence appears to have some support in the records. In large part, this may be a result of two factors, short record length (less than 50 years for most data) and little analysis of data with such goals in mind.

For design situations utilizing average recurrence intervals of less than 50 years or so, stationarity would have to be preferred over the alternative possibilities, since we don't find clearcut cycles, trends, or consistent persistence in the short records available to us. For larger events, with long average recurrence intervals, the possibility of their existence may not be so easily dismissed. The most prudent approach to including such factors is probably sensitivity analysis.

THE POTENTIAL

The potential uses of hydrometeorology are more numerous than for simply determining the potential magnitude of the rarest floods. The most likely new use is at the opposite end of the spectrum, the extreme drought (Willeke, 1975). Essentially the same principles used in flood estimation can be used in droughts: transposition, envelopment, moisture optimization (in this case, minimization), etc.

To date, the agencies involved in planning for drought conditions have not engaged in such analyses. A reasonable argument can be made that the drought events of interest are those with relatively short average recurrence intervals, say, 10-25 years. A serious drought could, however, occur at any time and the economic and social impacts could be devastating both in arid and humid regions. Some notion of drought potential seems to be available to use as a result of dendrochronology studies. If these studies are applicable, and I believe they are (with or without the stationarity assumption), then there is good reason to anticipate some extremely severe drought conditions.

Turning from drought and floods, the civil engineer may well give attention to other geophysical events of consequence. We may consider the hydrometeorological corollaries of the human adjustments to natural hazards. Snowfall, for example, has been considered almost entirely in the context of runoff potential. The magnitude of the snow event itself is, however, of consequence, especially in urban areas. Ice storms have similar consequences. This is an especially promising area of investigation because the conditions leading to rain, freezing rain, and snow are so similar. Minor temperature shifts can completely alter the form of precipitation. Work on hurricanes has been done from a hydrometeorological standpoint and appears to need, primarily, continual updating and refinement.

In each of the above phenomena, the general hydrometeorological principles can be applied. Of course, the relative importance of each parameter varies among the phenomena. White and Haas (1975, p. 22) estimate the potential damages from a repeat of the 1911 Mississippi Valley earthquake at over $100 billion. If similar estimates were made for extreme urban snow, rural drought, and ice storms, they would be staggering. The importance of devising suitable human adjustments, especially if coupled with the possibility that the real frequency of such events may be in the order of once in several hundred years, would be more evident than it is now.

Forecasting of this type is rather different from most operational forecasting. It states that an event of a given magnitude may (will) occur at some time in the future, but without specifying when. The forecast will usually be narrowed somewhat to a particular season, but little more may be possible.

The situations where long average recurrence intervals are considered justified in planning and design are those where such an event would have catastrophic consequences. Generally, such consequences are social, economic, political, or environmental in character. Loss of life is the one most often used, in the design of hydraulic structures. As is stated above, other events such as droughts, urban snow, ice storms, and hurricanes also merit such consideration.

Since we are considering hydrologic forecasting at this conference, it is desirable to consider what should be forecast. I have already addressed general kinds of things that should be of interest, such as drought, urban snow, ice storms, and hurricanes. However, such things as storm direction, orientation and shape need to be addressed more thoroughly than they have been. These aspects of a storm are even more important than the exact quantitative precipitation forecast for urban flooding situations. If we know a storm's areal extent, we can direct our efforts toward the exact areas subject to flooding and can increase the credibility of forecasts in the general public. Radar and satellite imagery are helping us move in that direction.

Observational networks are quite deficient for dealing with hydrology as we now use it. Built on a base of dedicated, helpful observers, and operating on a meager budget, it has performed reasonably well for the old hydrology. For the hydrology of conceptual and stochastic simulation models, it needs a change. In particular, the relationship of precipitation and streamflow gages needs to be more balanced. The separation of the responsible agencies, their different budgeting structures, and the general meagerness of budget for both agencies have precluded coordinated network design. The parameters of direction, orientation, and shape cannot be addressed by our ground observational networks. Radar and satellites can't do the whole job, either. Gage representativeness has nearly always taken a back seat to ease of observation. The antiquated Thiessen weighting procedures should be abandoned in favor of zonal index stations, with the index station locations being established from isohyetal patterns of observed storms. The zonal index approach has been used in a limited range of cases, but in general the inferior Thiessen method is employed.

REFERENCES

Chow, V. T., "A General Formula for Hydrologic Frequency Analysis," Transactions of the American Geophysical Union, vol. 32, pp. 231-237, 1951.

Hanson, E. M., F. K. Schwarz, and J. T. Riedel, "Probable Maximum Precipitation Estimates, Colorado River and Great Basin Drainages," Hydrometeorological Report No. 49, Superintendent of Documents, U. S. Government Printing Office, Washington, D. C., 1977.

Hershfield, D. M., "Estimating the Probable Maximum Precipitation," Proceedings, ASCE Journal of the Hydraulics Division, vol. 87, pp. 99-106, 1961.

Miami Conservancy District, Storm Rainfall of Eastern United States, Technical Reports, Part V, Dayton, 1917.

Miller, J. F., Personal Communication, February 28, 1979.

Schreiner, L. C., and J. T. Riedel, "Probable Maximum Precipitation Estimates, United States East of the 105th Meridian," Hydrometeorological Report No. 51, Superintendent of Documents, U. S. Government Printing Office, Washington, D. C., 1978.

White, G. F., and J. E. Haas, Assessment of Research on Natural Hazards, The MIT Press, Cambridge, 1975.

Willeke, G. E., "Hydrometeorology is More than PMF," Paper presented at the Annual Convention of the American Society of Civil Engineers, Denver, October, 1975.

APPLICABILITY OF SPACE ACQUIRED DATA FOR WATER RESOURCES MANAGEMENT

by

Vincent V. Salomonson*, V. Ralph Algazi* and Donald R. Wiesnet*

I. Introduction

In many parts of the United States and other parts of the world, there is a need for more complete and more timely information that can be used to improve the management of water resources. This need for more and better information arises because of expanding populations and commensurate activities in agriculture, energy production, and industry. As the need for more water grows, the margin for inefficient management decreases and everyone becomes more sensitive to variations in water supply and quality.

Satellites and attendant sensors have the inherent capability to provide many observations over large areas on a repetitive basis. Until recently the character of these observations and the ability to extract information from them was not very compatible with typical water resources management problems such as seasonal or daily snowpack runoff estimation for individual regional or local watersheds, surface water inventories, and the gathering of land use information useful for updating storm water management plans in or near rapidly expanding and changing metropolitan areas.

In 1972, two spacecrafts were launched that carried remote sensors that appeared to have application for water resources management purposes. The National Aeronautics and Space Administration (NASA) launched the Landsat satellite carrying the Multispectral Scanner Subsystem (MSS). The national Oceanic and Atmospheric Administration (NOAA) included on the polar-orbiting operational environmental satellite called NOAA-2 a Very High Resolution Radiometer (VHRR). Since that time the utility of the observations taken by these systems and some others that came later (e.g. the Visible and Infrared Spin Scan Radiometer (VISSR) on NOAA geostationary satellites) have been applied and tested in many situations that demonstrated the utility of these data. This paper will describe those instances where these capabilities do clearly appear to be of practical value. In addition, some discussion of

---

*V. V. Salomonson, Hydrospheric Sciences Branch, Laboratory for Atmospheric Sciences, Goddard Space Flight Center, Greenbelt, Maryland.
*V. R. Algazi, Department of Electrical Engineering, University of California, Davis, California.
*D. R. Wiesnet, Environmental Sciences Group, National Environmental Satellite SErvice (NESS), National Oceanic and Atmospheric Administration, Washington, D.C.

improvements taht can be expected in the near future (less than 5 years) will be provided. Finally, some of the remote sensing research that is going on that should be of benefit in the longer term (within 10 years) will be briefly described.

II. Present Capabilities

Table 1 provides a brief description of the Landsat MSS, the NOAA VHRR, the VISSR on the NOAA geostationary satellites. As can be seen, the Landsat MSS has the advantages of having better spatial resolution and a greater number of narrow spectral bands. The VHRR and the VISSR, however, have the advantage in terms of providing more frequent coverage and they routinely gather data in the thermal infrared (10.5 - 12.5 μm). The thermal infrared capability enables one to estimate surface temperatures or the amounts of outgoing emitted long wave radiation. These systems when used together tend to complement each other. This approach has, for example, been used in the case of snow-mapping and will be described later.

The information that comes from the NOAA and LANDSAT systems is basically in digital form. This digital data is processed into imagery forms or placed on computer compatible tapes. Descriptions of the data formats and costs can be obtained by contactingthe EROS Data Center in Sioux Falls, South Dakota for information on the Landsat data or by contacting the Satellite Data Services Division of the National Environmental Satellite Center in Washington, D.C.

The imagery provided for the MSS is in a form (241 x 241 mm prints or transparencies) that can almost immediately be overlain on 1:1,000,000 scale maps for comparison purposes and for the rough updating of information on the maps (e.g. new roads, lakes, housing developments, etc.). The MSS digital data can be processed to provide information at other scales with greater cartographic accuracy although National Map Accuracy Standards will not generally be met at scales larger than 1:250,000. The NOAA VHRR data is not usually provided in a cartographically compatible formal although it can be processed so as to be compatible with maps at smaller scales (See Algazi, 1976; Ford, Algazi and Agee, 1978; Legeckis and Pritchard, 1976).

Landsat MSS and NOAA VHRR data are routinely and regularly gathered over the United States and most of North America. Tape recorders with limited storage capability relative to the total data that could be gathered in a given day are presently used to gather data in other parts of the world. As a result, the coverage is somewhat irregular. Some data from the MSS and the VHRR are available, however, for nearly all land areas of the world.

A) Landsat MSS Applications

With four spectral bands in the 0.5 - 1.1 μm region of the electromagnetic spectrum, the Landsat MSS basically is responding to radiation reflected from surface such as bare soil, water, snow, vegetated surfaces, man-made surfaces, or clouds. Taking advantage of even the relatively coarse spectral resolution of the MSS, these

Table 1. Basic Characteristics of the Landsat Multispectral Scanner (MSS) and the NOAA Very High Resolution Radiometer (VHRR)

Characteristic	MSS	VHRR
Spatial Resolution	79 m	1 km
Spectral Bands	0.5 - 0.6 μm	0.55 - 0.70 μm
	0.6 - 0.7 μm	10.5 - 12.5 μm
	0.7 - 0.8 μm	
	0.8 - 1.1 μm	
Temporal Coverage Capability	once every 18 days (some 9 day coverage available when two Landsat spacecraft were operating)	daily (twice daily in the infrared)

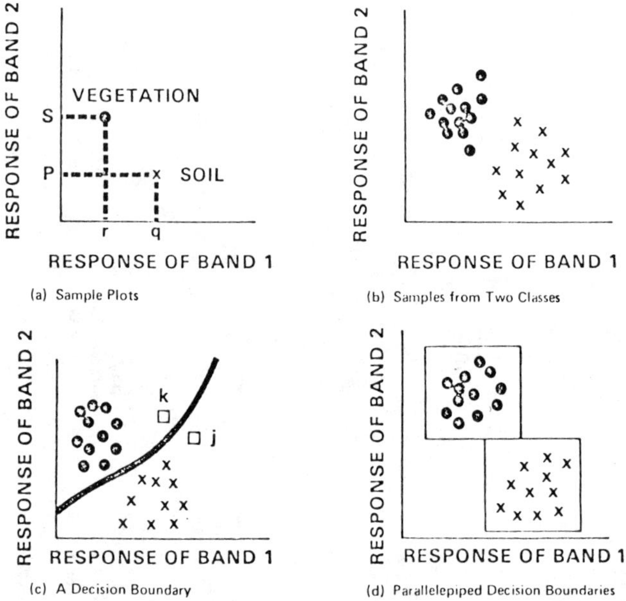

Figure 1: Sketches illustrating the approach used to classify digital data from Landsat MSS (Alford, 1978).

observations can be processed so as to objectively identify classes of surface cover both in terms of color expressions in imagery or digital expressions in computer printouts.

The success of these approaches basically rests on the fact that surface features have different spectral reflectance characteristics. Vegetation, for example, reflects highly only in the near infrared while water has low reflectance in all bands and snow generally reflects highly in all bands. Assigning shades of red to reflective radiation in in the near infrared (0.8 - 1.1 μm), blue to radiation in the 0.5 - 0.6 μm ("green" region) and yellow to the 0.6 - 0.7 μm ("red" region) allows the creation of false color composites where vegetation appears red, water is typically a dark blue, and the soils have a nore bluish cast to them than normally appear to the eye. Figure 1 (Alford, 1978) illustrated how the digital observations from the MSS can be statistically analyzed and classified so as to objectively identify classes of surface cover using training samples or unsupervised clustering approaches.

Using the principles very briefly indicated above there have been several classes of applications to water resources management that have been completed and analyzed. These analyses have compared inventories and feature identification case studies using satellite data with conventional approaches in terms of accuracy, time-savings or timeliness of data acquisition, and cost savings. An important advantage of the use of satellite data in digital form, is that, after full geometric correction, the processed data is available in a form of a grid cell data base. This information can be superimposed and combined with data available from other sources. In the broadest sense, it seems clear that the Landsat data do offer a relatively inexpensive data source that can complement conventional water resources management data gathering approaches. The number of cases in which the satellite data are competitive with conventional approaches in terms of accuracy, timeliness, time savings, and cost are fairly limited at present but there are some noteworthy examples. The following paragraphs will attempt to delineate those areas where satellite data have been successfully applied and the varying role they have played relative to conventional methodologies.

1. Watershed Physiography

The cartographic character of the Landsat imagery leads one to try and evaluate its utility for extracting watershed physiographic information that could be related to watershed yield and general runoff characteristics. This would involve measuring the extent of various surface cover types, (bare soil, vegetation, water, massed works of man), drainage density, basin shape, stream lengths, stream sinuosity, etc. Various reports of work along these lines have been reported including that provided by Rango et al. (1975). These results indicate that photointerpretation techniques can be applied to Landsat imagery that are consistently competitive with information provided on 1:250,000 scale maps. In well-dissected terrain with modest amounts of vegetation cover, the satellite derived information is competitive with information available with current maps at larger scales (e.g. up to 62,500). Obtaining Landsat imagery at special times such as when an area has been covered with light snow will enhance drainage features and allow more extraction of information than normally can be accomplished. In many areas of the world where maps are not available, or incomplete, or much out-of-date, the Landsat data are clearly advantageous.

2. Irrigated Land Surveys

For drainage basins in agricultural areas, the Landsat data can be used to estimate acreage covered by bare soil and various types of crops. The satellite-derived information is updated frequently in the United States because of the 18-day repeat coverage of the satellite and regular data acquisition. Even the limits imposed by cloud cover allow the acquisition of several observations during the crop growing season. For the average user of satellite data, the major constraint, in this instance, as well as many others related to water resources management, has been the slowness in delivery of the data. More will be said about this problem later in the paper.

As far as water resources management is concerned, information as to cropping practices and acreage devoted to various crops is needed for or related to accurate estimates of water demand. Water demand estimates are particularly critical in the Western United States because of the preponderant use of water by irrigation. The competition of irrigation needs with power, industry, and municipal or residential needs is continuing to grow as populations expand rapidly in these regions.

As a very exemplary case in point, satellite data appear to be quite useful in surveying total irrigated acreages in semi-arid and arid regions. Draeger (1977) and Estes et al. (1978) report studies in Oregon and California, respectively, that are illustrative. These studies show not only that the total irrigated acreage can be estimated with accuracies comparable to or exceeding that achievable with conventional approaches. More importantly, the estimates can be obtained more repetitively than with conventional approaches and at a fraction of the cost.

A more complete evaluation of the use of satellite data is being carried out in California by NASA, the University of California, and the California Department of Water Resources. In addition to the applicability of the satellite data for estimating total irrigated acreage, the ability of the satellite observations to provide crop-type information with sufficient accuracy within the irrigated land acreage is being evaluated. Preliminary results obtained to date look quite promising.[1]

It will, of course, take time for agencies responsible for making water demand estimates to completely evaluate satellite aided approaches and phase this information gathering approach into their procedures. But even in the most conservative sense, it seems safe to suggest that satellite information is a valuable and useful source of

---

[1]E. Bauer, NASA, Ames Research Center, Personal Communication.

data for irrigated land surveys that should be utilized by all parties having responsibility or need for such information.

3. Urban/Suburban Lands

A difficult problem in water resources planning and management is that associated with accounting for changes in runoff caused by changes in land use. A noteworthy case in point occurs in the rapidly changing and expanding urban fringe zone surrounding metropolitan areas. As residential housing, shopping centers, industry, etc., expands into what was once agricultural areas, the character of the water runoff and the associated water quality changes markedly. Often storm water management plans need to be updated frequently in order to reach equitable and reliable decisions.

Typically, investigations to update storm water management plans employ models that simulate the hydrology of the watersheds. These models translate basic data describing the hydrology of the watershed into useful information for planning. An essential subset of the basic data is the surface or land cover information which can be related to the fractions of impervious (high runoff) area and pervious (low runoff) area or runoff coefficients relating the precipitation input to the runoff output. Typical models of the type being referred to are the Storage, Treatment, Overflow and Runoff Model (STORM) (Corps of Engineers, 1976) or the procedures employed by the U.S. Department of Agriculture, Soil Conservation Service (Soil Conservation Service, 1972; Soil Conservation Service, 1975).

The acquisition of the land cover input for these planning models is often quite time consuming and the subsequent costs are appreciable, particularly for larger watersheds covering tens and hundreds of square kilometers. For these watersheds, the spatial resolution and other capabilities provided by the Landsat observations appear, based on existing results, to provide a time and cost saving method for acquiring land cover information that can be updated easily and made compatible with other planning data.

Initial studies of the utility of Landsat data first attempted to establish that the accuracy in the determination of land cover data were sufficient to pursue further evaluation of time and cost when employing a Landsat-aided approach. Examples of this type of work includes that of Ragan and Jackson (1975) and Algazi and Suk (1977) which basically showed that satisfactory agreement could be obtained between the Landsat-determined land cover determinations and conventionally derived estimates of land cover.

The studies referred to above showed results that indicated considerable cost and time savings accrued from using Landsat MSS data. Jackson, Ragan and Fitch (1977) conducted a more extensive comparison between the results obtained with conventional and Landsat based modeling. In this study the consulting engineering firm of Water Resource Engineers, Incorporated had been retained to conduct a watershed planning study on a highly urbanized 50.5 square kilometer watershed in the Virginia suburbs of Washington, D.C. In meeting their responsibilities to their client, the consulting firm used the best available conventional data to define parameters required by STORM which was operated to examine the impact of a number of flood alternatives. In order to evaluate the utility of Landsat MSS data as a tool for urban water resource planning, the consultant then ran STORM under the same conditions as those required to meet his obligations to his client, but with the input data based on Landsat estimated parameters. One phase of the study required that STORM be operated as a continuous streamflow generator using 52 years of hourly precipitation data as the input. The simulated discharges for each year were then used to develop a synthetic flood frequency curve. The conventional version of STORM estimated annual damages of $3,140,500, while the Landsat based STORM yielded annual average damages of $2,761,700, twelve percent lower than the conventional results. The decision on which flood control project to adopt was the same for each of the models. STORM was then used in the same way to estimate the impact of the project with the conventional version showing an annual damage of $89,000 compared to $86,500 using the Landsat version. The important result was that one hundred ten man-days at a cost of $14,000 were required to perform tasks using conventional methods in contrast to the 6.9 man days and $6,000 required using the Landsat aided approach. Costs diminished to near $2,000 for other watersheds in the same area because of not having to reorder satellite data, increased familiarity with the data, etc.

From the results of the studies previously described and others (See also, McKeon et al., 1978) one can conclude that by employing Landsat on urban/suburban watersheds one can obtain land cover information with accuracies comparable to that derived from conventional sources while realizing 50-80% savings in time and cost. This approach is being further evaluated by the U.S. Army Corps of Engineers aided by NASA, the University of California, and the University of Maryland in a quasi-operational setting. The Landsat-aided approach to gathering land-cover information is being tested on several watersheds across the United States (Algazi et al., 1979). This project has been in progress since 1976 and should be completed in 1979. The results to date seem to be corroborating earlier studies, but technology transfer and cost-effectiveness evaluations have not been completed (Salomonson et al., 1978). Because of the dramatic decrease in the cost of digital computing and storage, the combined usage of digital satellite data and computer technology is likely to decrease further in cost in the future.

4. Surface Water Observations

In several situations related to Water resources management, the area covered by surface water and its relationship to vegetation or its relative quality may be desirable. Landsat MSS observations can be used to survey large areas and assess changes in surface water area such as that associated with floods (Rango and Salomonson, 1974; Deutsch and Ruggles, 1978), the extent of vegetation cover and land use associated with flood plains (Sollers et al., 1978) marshes or wetland conditions (Carter, 1977), and the turbidity of the water which typically is associated with various concentrations of sediment or pollutants of some kind (Boland and Blackwell, 1975). These applications are limited by the spatial resolution and eighteen-day repeat cycle of Landsat. For example, the spatial resolution and related aspects of the Landsat spacecraft itself may not be sufficient to produce products meeting geodetic or map accuracy standards called for in legislated survey requirements. Nevertheless, the Landsat data do often add an additional data source that gives a meaningful overview or it provides a means of identifying anomalous areas, extrapolating conventional measurements to areas not normally covered, or directing ground-based efforts in ways that produce economy of effort.

One of the most exemplary examples of the use of Landsat data is that surrounding the U.S. Army Corps of Engineers responsibility for completing a dam inventory for all impoundments exceeding 50 acre feet in storage. The Corps is employing Landsat data as a means of locating these water bodies on a cost effective basis. They are using a computer processing approach applied to Landsat MSS data that was developed by Johnson Space Center, NASA personnel. The technique has been applied in several parts of the country and is now being employed in the Seattle District of the Corps of Engineers.[2]

5. Snow and Ice Monitoring

One of the land cover themes that is relatively easy to identify in Landsat observations is snowcover and ice cover on lakes. Difficulties do occur sometimes in separating snow or ice from clouds in imagery, but due to the different geometric patterns displayed in imagery associated with clouds versus snow, usually no major difficulty is experienced other than the fact that no snow or ice can be seen when complete cloud cover exists. As already indicated, both Landsat and NOAA VHRR observations are useful and complement each other in snow and ice mapping (Barnes and Bowley, 1974).

Insofar as water resources management is concerned the most apparent and useful application of snowcover mapping from satellites comes in mapping snowcover on watersheds in mountainous areas. (Rango and Itten, 1976). There has been some question, however, as to whether useful reductions in forecast error could result from use of satellite snowcover area measurements, and, if so, what the magnitude of the reduction might be. The principal difficulty revolved around the fact that very few years of satellite snowcovered area data actually exist. Recently, however, Rango et al. (1979) have shown that using a combination of aircraft and satellite snowcover area measurements do reduce forecast error. The largest reductions in error due to the use of satellite snowcovered area measurements comes in watersheds having an erratic snowpack accumulation pattern relative to elevation or poor coverage by precipitation stations or snow course measurements.

The most limiting factor in the use of satellite snowcovered area measurements now seems to be associated with procedural data acquisition and delivery problems. However, the evaluation of satellite data in a quasi-operational test is being conducted by six major federal agencies and three state agencies (Rango, 1975). This test is nearing its completion and will include reports benefit and cost estimates plus

---

[2]Personal Communication, L. P. Murphy, U.S. Army Corps of Engineers, Ft. Belvoir, Virginia.

descriptions of how and where to apply the satellite aided approaches. These reports should become available in late 1979. The next section laborates further on the use of the NOAA VHRR in snow and ice mapping.

6. NOAA Operational Environmental Satellite Applications

Research efforts in hydrology by the National Environmental Satellite Service (NESS) and others related to NOAA satellite sensors such as the VHRR have produced interesting and productive operational products. The results are commonly generated from research that appears to be promising and that has a definite user identification.

Research papers on the areal extent of snowcover appeared very early after the first weather satellites were launched as snow was readily discernible on these early satellite images. Papers by Barnes and Bowley (1968), Wiesnet and McGinnis (1973) and Wiesnet (1974) have stressed the value and accuracy of river basin snow extent mapping. The daily coverage of North American basins by NOAA/VHRR is of particular value to hydrologists. Subsequent studies of continental and hemispheric snowcover have led to a set of entirely new data for climatologists on snowcover (Matson, 1977) and a possible technique for forecasting winter snowcover on a continental basis (Wiesnet and Matson, 1976). Using the Eurasian snowcover data by Wiesnet and Matson (1976), Hahn and Skukla (1976) of the Geophysical Fluid Dynamics Laboratory at Princeton, demonstrated a previously postulated relationship between Eurasian snowcover and the development of the summer monsoon that followed.

The frequent coverage afforded by the NOAA satellites is also of significant value in tracking snowcover variations in the Western United States. Because of the value of these data the National Environmental Satellite Service (NESS) now has an observational river basin snow mapping program that provides satellite-observed snowcover estimates for 30 basins in the U.S. and Canada. These results are provided to the hydrologists in river forecast centers and other federal, state, and local offices within 30 hours of observation. The advantages of the large area synoptic overview of snowcover as provided by satellites sensors such as the NOAA VHRR are clearly exemplified by the informational impact provided by satellite images over the Sierra Nevadas and other mountainous areas during drought years versus normal or greater than normal snowcover years. Hydrologists and the public at large can easily and quickly grasp the extent the relative severity and extent of variations in snowcover as well as other hydrologic features such as surface water area, by examining satellite observations.

Wiesnet et al. (1974) were the first to apply NOAA/VHRR infrared observations to the problem of flood mapping. Thermal maps were prepared of the 1973 Mississippi River floods, successfully delineating the inundated areas and demonstrating that the 1-km spatial resolution of the NOAA thermal infrared could be used effectively and quickly for reconnaissance of large flooded river valleys. Disaster planning and relief-service mobilization would benefit from this important application of satellite information. The twice-daily IR overpasses of NOAA were timely and could be prepared quickly if needed.

Floods on much smaller rivers are not definable, however.

NOAA/VHRR and VISSR images have been successfully used to monitor the breakup of ice on large rivers in North America. The Great Lakes have been monitored for ice buildup, movement, and breakup. A recent paper by Quinn and others (1978) contains detailed descriptions of ice conditions and numerous NOAA satellite images of Great Lakes ice. The NOAA-USCG Ice Center in Cleveland uses both satellite images and Side-Looking Airborne Radar images from USCG aircraft to advise mariners transiting the lakes of current ice conditions. Early thermal maps of the Great Lakes have not been developed into operational weekly charts. Lastly, the NOAA satellites are especially well suited to monitor large ice sheets such as Greenland, Iceland and, the largest ice sheet of all, Antarctica.

As previously stated, many of these research activities have been transformed into operational products. The river basin snow-mapping doen at NOAA/NESS has been described by Schneider et al. (1976). Charts showing ice conditions in the Great Lakes based entirely on satellite data are produced weekly and are mailed directly to users. These ice charts are accurate to ± 5km. Great Lakes surface temperature charts are likewise prepared when observable (usually weekly) and are mailed to users. These charts have a relative accuracy of ± 0.5°C and an absolute accuracy of ± 1.5°C. A northern hemisphere weekly snow and ice chart is distributed to interested marine users and climatologists. A computer-generated product, the Composite Minimum Brightness Chart is prepared daily for both Northern and Southern Hemisphere. It basically "filters out" the clouds and produces a small scale chart of hemisphere snow and ice.

III. Planned Capabilities

On October 13, 1978, a polar-orbiting satellite named TIROS-N was successfully launched by NOAA. It represents the initial third generation of satellites to be operated by the National Environmental Satellite Service. This satellite will be followed by two others, one scheduled for launch in April, the other in December. These satellites will utilize the Advanced Very High Resolution Radiometer (AVHRR) which retains the capabilities of the previously described VHRR and includes new channels in the thermal infrared spectrum (3.55 - 3.93 μm) and one in the near infrared (0.72 - 1.1 μm). This new sensor is expected to provide additional information on snow and sea ice conditions, soil moisture related conditions, and improved surface temperature estimates.

Beyond the satellites and sensors previously described there are several research and technology efforts underway which are expected to provide more meaningful and useful data to hydrologists. Table 2 summarizes some of these future efforts. Landsat-D, in particular, should very significantly improve upon the capabilities of the Landsat 1-3 MSS series. Higher spatial resolution, better spectral resolution and observations in new spectral intervals are among the technological advances being incorporated in the Thematic Mapper on Landsat-D (Salomonson, 1978). Beyond Landsat-D there should be remote-sensing capabilities forthcoming that allow better estimates of soil moisture, snowpack properties, and precipitation distribution. Schmugge (1978), for example, summarizes the progress that has been made in applying remote sensing to the estimation of soil moisture; Algazi et al. (1977) also report on an attempt to use NOAA 3 and 4 data to determine soil moisture. For hydrology, the microwave portion of the electromagnetic spectrum has several fundamental characteristics that make it seem key to remote sensing of the parameters noted above. More temporally frequent observations are also key in observing dynamic hydrologic phenomena. This is a key facet of the pointable and geosynchronous sensor and spacecraft concepts noted in Table 2.

Another aspect involved in the utilization of satellite remote sensing observations is the speed of data delivery to users and the formating or packaging of the information contained in or derived from these observations. The NOAA direct-readout capability for the VHRR and AVHRR is a positive example of sppeding data delivery. This aspect has hindered the application of Landsat data in many instances.

There are significant efforts being made to develop improved computers for ground-based and spacecraft data processing. The Landsat-D data processing systems will be substantially advanced in the sense that they will not only be able to accomodate the higher data rates associated with the higher spatial and spectral resolution of the Thematic Mapper, but they will also improve upon the overall data throughtput and delivery by using communication satellite technology and high density digital tape reading and recording technology. It is envisioned, however, that researchers familiar with data processing, remote sensing, and water resources management needs must work closely together to define requirements and articulate clearly the need for improved data formatting and delivery if routine and wide spread application of these data for water resources management is to be achieved.

IV. Conclusions

Remote sensing observations from spacecraft such as the Landsat and NOAA series have been applied in several instances. These applications have shown in nearly every instance that the large area synoptic overview as provided by these Multispectral Scanner Subsystem (MSS) and Very High Resolution Radiometer (VHRR) is a valuable, ancillary source of information when used in combination with conventional data. In cases involving, for example, irrigated land surveys, land use and surface cover inventories on large watersheds in or near urban areas (greater than several square kilometers in size) surface water and snowcover inventories. The use of satellite data can result in very significant time and cost savings in watershed planning and runoff estimation efforts. Several technological advances, such as those associated with TIROS-N and Landsat-D should improve the utility of satellite remote sensing data for water resources management. More effort needs to be devoted to increasing reliability and speed in data delivery to operational hydrologists, engineers and planners as well improving the packaging and formatting of the data in order to reach the point where these data are widely and routinely applied or accepted into standard water resources management procedures and efforts.

Table 2. Planned Or Considered Spacecraft Missions That Would Have Features Relevant To Hydrology And Water Resources.

VEHICLE/SENSOR	SPECTRAL BANDS	SPATIAL RESOLUTION	IMAGE/SCENE AREAL COVERAGE	REVISIT INTERVAL	REMARKS
PLANNED					
LANDSAT D/ THEMATIC MAPPER (1981)	0.45– 0.52 μm 0.52– 0.60 μm 0.63– 0.69 μm 0.76– 0.90 μm 1.55– 1.75 μm 2.08– 2.35 μm 10.40–12.55 μm	30 m 30 m 30 m 30 m 30 m 30 m 120 m	185 x 185 km	16 DAYS	SECOND GENERATION EARTH RESOURCES SATELLITE SYSTEM
CONSIDERED					
IMAGING RADAR (1980's)	– L BAND ON EARLY SHUTTLE – 2 BANDS AND DUAL POLARIZATION ON LATER SORTIE FLIGHTS	25–100 m	100–200 km SWATH	TBD	– FLIGHT OF AN L-BAND SAR MAY OCCUR ON EARLY SHUTTLE FLIGHT IN EARLY 1980'S – LATER ADVANCED SAR MIGHT HAPPEN IN MID-1980'S TIME FRAME
MULTICHANNEL PASSIVE MICROWAVE (1980's)	K TO P-BAND	20-50 km	≈1000 km	≈3 DAYS	LARGE ANTENNA TECHNOLOGY INVOLVED
HIGH RESOLUTION & POINTABLE OBSERVING SYSTEMS	LANDSAT BANDS	10-30 m	VERY SELECTED AREAS	VARIABLE	SOLID STATE TECHNOLOGY MOST LIKELY INVOLVED
GEOSYNCHRONOUS SATELLITE	LANDSAT BANDS	100-500 m	LOCAL TO SUB-CONTINENT AREAS	FRACTION OF A DAY	NOT LIKELY UNTIL AT LEAST 1990's

V. References

Alford, W. L., 1978: Digital Image Analysis of Landsat Data. Proceedings of the Twelfth International Sympoisum on Remote Sensing of Environment, 20-26 April, Manila Philippines, Environmental Research Institute of Michigan, Ann Arbor, Michigan, pp. 329-354.

Algazi, V. R., 1976: Water Supply Studies by the Davis Campus Group, Chapter 2 in Annual Progress Report: An Integrated Study of Earth Resources in the State of California Using Remote Sensing Techniques (R. N. Colwell, Principal Investigator) NASA Grant NGL 05-003-404, University of California, Berkeley, California.

Algazi, V. R., and Minsoo Suk, 1977: Satellite Land Use Acquisition and Applications to Hydrologic Planning Models. Proceedings of the Eleventh International Symposium on Remote Sensing of Environment, Environmental Research Institute of Michigan, Ann Arbor, Michigan, pp. 1171-1181.

Algazi, V. R., et al., 1977: Remote Sensing Determination of Imperviousness and Soil Moisture for the Calibration of HEC Hydrologic models, Final Report on HEC Contract DACW05-76-P-2102.

Algazi, V. R., G. E. Ford and D. I. Meyer, 1979: A Non Interactive Approach to Land Use Determination. To be presented at the 5th Purdue Symposium on Machine Processing of Remotely Sensed Data, June 27-29, 1979.

Barnes, J. C., and C. J. Bowley, 1968: Snow Covered Distribution As Mapped from Satellite Photography, Water Resources Research, 4, 257-272.

Barnes, J. C., and C. J. Bowley, 1974: Handbook of Techniques for Satellite Snow Mapping, Final Report NAS5-21803, National Aeronautics and Space Administration, Goddard Space Flight Center, Greenbelt, MD

Boland, D. H. P. and Blackwell, R. J., 1975: The Landsat-1 Multispectral Scanner as a Tool in the Classification of Inland Lakes, Proceedings of the NASA Earth Resources Survey Symposium, Houston, Texas, Volume I-A, pp. 419-442.

Carter, V., 1977: Coastal Wetlands: The Present and Future Role of Remote Sensing. Proceedings of the Eleventh International Symposium on Remote Sensing of Environment, April 25-29, Ann Arbor, Michigan. Environmental Research Institute of Michigan Ann Arbor, Michigan, pp. 301-323.

Corps of Engineers, 1976: Storage, Treatment, Overflow, and Runoff Model (STORM), Users Manual, U.S. Army Corps of Engineers, Hydrologic Engineering Center (HEC), Davis, California.

Deutsch, M., and F. H. Ruggles, Jr., 1978: Hydrological Applications of Landsat Imagery Used in the Study of the 1973 Indus River Flood, Pakistan. Water Resources Bulletin 14(2), 261-274.

Draeger, W. C., 1977: Monitoring Irrigated Land Acreage Using Landsat Imagery: An Application Example. Proceedings of the Eleventh International Symposium on Remote Sensing of Environment, April 25-29, Environmental Research Institute of Michigan, Ann Arbor, Michigan, pp. 515-524.

Estes, J. E., J. R. Jensen, and L. R. Tinney, 1978: Remote Sensing of Agricultural Water Demand Information: A California Study. Water Resources Research, 14(2): 170-176.

Ford, G. E., V. R. Algazi and B. G. Agee, 1978: On the Correction of Geometric Distortion in Satellite Acquired Images. Proceedings of 1978 Asilomar Conference on Circuits, Systems and Computers.

Hahn, D. G., and J. Shukla, 1976: An Apparent Relationship Between Eurasian Snowcover and Indian Monsoon Rainfall, Journal of the Atmospheric Sciences, 33, 2461-2462.

Jackson, T. J., R. M. Ragan, and W. N. Fitch, 1977: Test of Landsat-Based Urban Hydrologic Modeling. Journal of the Water Resources Planning and Management Division ASCE, 103, 141-158.

Legeckis, R. V., and J. A. Pritchard, 1976: Algorithm for Correcting the VHRR Imagery for Geometric Distortions Due to the Earth's Curvature and Rotation, NOAA Technical Memorandum, NESS-77, [National Technical Information Service, Springfield, Virginia, 22151].

Matson, M., 1977: Winter Snow Cover Maps of North America and Eurasia from Satellite Records 1966-1976, NOAA Technical Memorandum, NESS, 84, National Oceanic and Atmospheric Administration, Washington, D.C.

McKeon, J. B., L. E. Reed, R. H. Rogers, R. M. Ragan and O. C. Wiegand, 1978: Landsat-Derived Land Cover and Imperviousness Categories for Metropolitan Washington: An Urban/Non-Urban Computer Approach. Proceedings of the American Society of Photogrammetry Annual Meeting, Washington, D.C., February 26-March 4.

Quinn, F. H., R. A. Assel, D. E. Boyce, G. A. Leshkevich, C. R. Snider, and D. Wiesnet, 1978. Summary of Great Lakes Weather and Ice Conditions, Winter 1976-77, NOAA Technical Memorandum, GLERL-20, Great Lakes Environmental Research Laboratory, Ann Arbor, Michigan, p. 141.

Ragan, R. M., and T. J. Jackson, 1975: Use of Satellite Data in Urban Hydrologic Models. Journal of the Hydraulics Division, ASCE, 101, 1469-1475.

Rango, A., and V. V. Salomonson, 1974: Regional Flood Mapping from Space. Water Resources Research 10 (3), 473-484.

Rango, A., and K. I. Itten, 1976: Satellite Potentials in Snowcover Monitoring and Runoff Prediction, Nordic Hydrology, 7, 209-230.

Rango, A., V. V. Salomonson, and J. Foster, 1977: Extraction and Utilization of Space-Acquired Physiographic Data for Water Resources Developments, Water Resources Bulletin, 11, 1245-1255.

Rango, A., J. Z. Hannaford, R. L. Hall, M. Rozenzweig, and A. J. Brown, 1979: The Use of Snowcovered Area in Runoff Forecasts, Journal of the Hydraulic Division, ASCE, (accepted for publication).

Salomonson, V. V., A. Rango, and W. Dallam, 1978: Applications Systems Verification and Transfer (ASVT): Water Management and Control Project, Interim Report: 1 January 1978 - 31 October 1978. Goddard Space Flight Center, Greenbelt, Maryland.

Salomonson, V., 1978: Landsat-D, A Systems Overview, Proceedings of the Twelfth International Symposium on Remote Sensing of Environment, April 20-26, Manila, Philippines, Environmental Research Institute of Michigan, Ann Arbor, Michigan, pp. 371-385.

Sollers, S. C., A. Rango, and D. L. Henninger, 1978: Selecting Reconnaissance Strategies for Floodplain Surveys. Water Resources Bulletin, 14 (2), 359-373.

Soil Conservation Service, 1972: National Engineering Handbook, Section 4: Hydrology. Soil Conservation Service, U.S. Department of Agriculture, Washington, D.C.

Soil Conservation Service, 1975: Urban Hydrology for Small Watersheds, Technical Release No. 55. Engineering Division, Soil Conservation Service, U.S. Department of Agriculture, Washington, D.C.

Wiesnet, D. R., and D. F. McGinnis, 1973: Hydrologic Applications of the NOAA-2 Very High Resolution Radiometer, Proceedings of the American Water Resources Association Symposium on Remote Sensing in Hydrology, Burlington, Ontario.

Wiesnet, D. R., D. F. McGinnis, and J. A. Pritchard, 1974: Mapping of the 1973 Mississippi River Floods by the NOAA-2 Satellite, Water Resources Bulletin, 10, 1040-1049.

Wiesnet, D. R., 1974: The Role of Satellites in Snow and Ice Measurements, Proceedings of the Symposium on Advanced Concepts and Techniques in the Study of Snow and Ice Resources, International Hydrological Decade, Monterey, California, December.

Wiesnet, D. R. and M. Matson, 1976: A Possible Technique for Winter Cover in the Northern Hemisphere and Eurasia, Monthly Weather Review, 104, 828-835.

THE BENEFITS OF UNCERTAINTY: PROBABILITY IN SHORT AND LONG TERM HYDROLOGIC FORECASTING

RAFAEL L. BRAS*

ABSTRACT

Engineers and scientists are commonly in the difficult situation of having to predict hydrologic events based on limited knowledge of reality, imperfect models and incomplete data sources. The theory of probability allows explicit consideration of the above uncertainties allowing improved hydrologic modeling. Examples are given where the joint use of probability and physical hydrology permit improved long term predictions and short term discharge forecasting. Long term predictions refer to statements of recurrence of future discharge events. Short term discharge forecasting deals with daily flood and river stage predictions done by the U.S. National Weather Service.

INTRODUCTION

Randomness of an event may be due to the analyst's lack of knowledge, the scale at which it is observed or to inherent stochasticity. Whatever the source of randomness, and in spite of possible philosophical arguments, probability theory offers a unified approach to handle uncertain events.

Statistics are a tool for data handling. Using them can lead to reasonable inferences about system behavior, even to model formulation. Nevertheless, they can only restate the information content of the data sources. To the physical scientist, for example, hydrologists, this may lead to frustrating experiences and a growing realization that his knowledge is being ignored. Reasonable use of probability theory is the cure to some of the hydrologist's problems.

The acknowledgement of hydrologic random processes permits the use of tools like derived distributions and optimal estimation theory. This paper will illustrate how the concept of derived distributions can incorporate the best deterministic hydrologic knowledge to obtain probabilistic statements of the recurrence of hydrologic events. In particular, the theoretical distributions of the volume of runoff above a given threshold and the volume of water infiltrated from any given storm will be presented. These are problems of long term prediction with solutions generally unattainable from available data.

Deterministic knowledge of the dynamic rainfall runoff process, represented by the National Weather Service River Forecasting Model, can be used with optimal estimation theory to greatly improve streamflow forecasts in an on-line adaptive framework. Results of this

*Assistant Professor of Civil Engineering, Massachusetts Institute of Technology, Cambridge, Massachusetts 02139 U.S.A.

short-term hydrologic forecasting model will be given.

DERIVED DISTRIBUTION TECHNIQUE (From Chan and Bras, 1979)

The mechanics of derived distributions are well established in probability theory [Benjamin and Cornell, 1970]. The conceptual framework is the following.

Assume that you have a variable, Q, functionally related to a vector of parameters θ by:

$$q = Q(\theta) \tag{1}$$

The elements of the vector θ are random variables with a given joint probability density function, $f_\theta(\theta)$, and corresponding cumulative distribution, $F_\theta(\theta)$.

Due to the randomness of θ, the variable Q is also a random variable with cumulative distribution,

$$F_Q(q) = \int_{R_q} f_\theta(\theta)\, d\theta \tag{2}$$

where R_q is the region within the possible values of θ for which $Q \leq q$.

THEORETICAL DISTRIBUTION OF FLOOD VOLUMES ABOVE A GIVEN THRESHOLD

Chan and Bras (1979) concentrated in obtaining the theoretical distribution of the volume of the hydrograph above a given threshold. The variable in question is illustrated in Figure 1. The volume above a threshold discharge is required for the design of flood storage devices in urban areas. For example, the threshold discharge shown in the figure may be the maximum capacity of treatment of water from a combined sewer system. The exceedance volume must be stored or spilled, possibly contaminating receiving water bodies.

Since hydrograph volume data is scarce or nonexistent, the approach taken was to relate the available probabilistic description of the rainfall process to a deterministic model of flood volumes in order to derive the distribution of the latter.

Integrating the kinematic wave equations for overland flow, Chan and Bras (1979) obtained the following deterministic expression for the volume above a given threshold:

$$V_{th} = t_r(L\, i_e - q_{th}) + \frac{q_{th}^{1+1/m}}{\alpha^{1/m}\, i_e} - \frac{L\, q_{th}^{1/m}}{\alpha^{1/m}} \tag{3}$$

where L is length of overland flow

α is a kinematic wave parameter function of slope and Manning's coefficient (Eagleson, 1970)

m is a kinematic wave parameter taking a value of 1.67 for overland flow

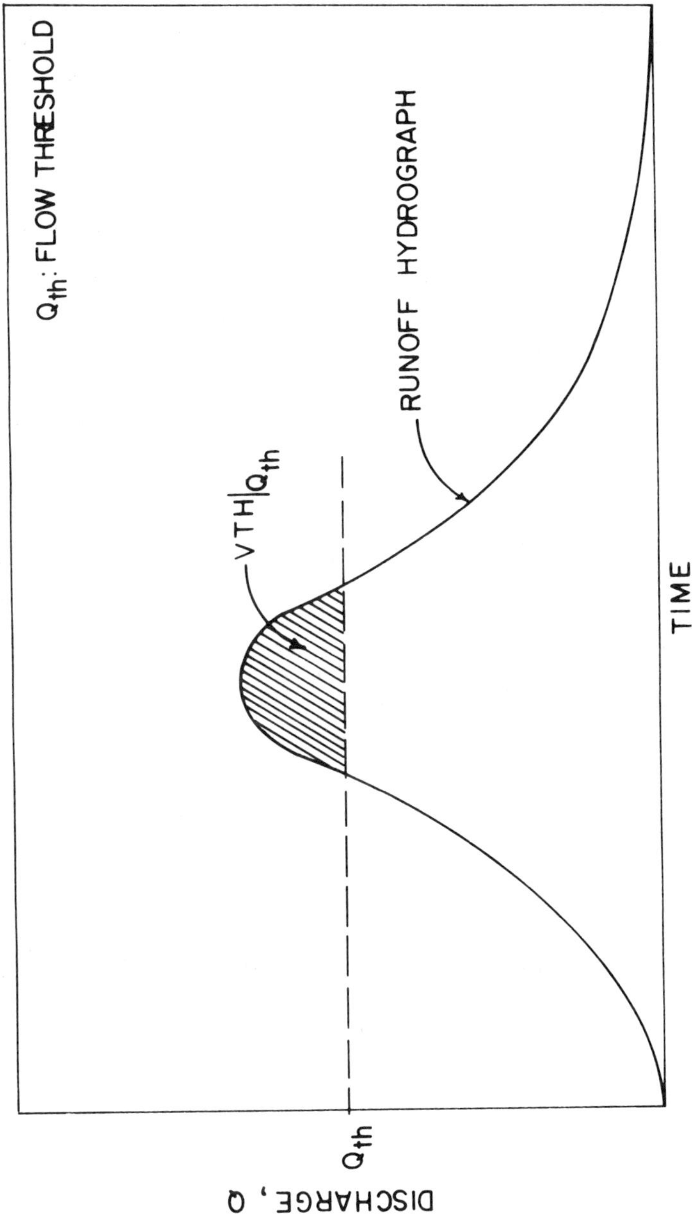

Figure 1 Volume above a Given Threshold Discharge, from Chan and Bras, 1979

q_{th} is the known threshold discharge as illustrated in Figure 1

t_r is the duration of effective rainfall

i_e is the effective rainfall intensity, assumed uniform in time

V_{th} is the volume of water above threshold q_{th}

Equation (3) can be easily solved for t_r and lines of equal volume, V_{th}, plotted as a function of i_e and t_r for constant basin parameters L, α, m and q_{th}. Figure 2 gives the lines of constant volume, the shaded region indicating values of i_e and t_r, yielding volumes less than V_{th1}. The kinematic wave equations determine the lines of zero volume, $V_{th} = 0$. The dashed line represents the condition of storm duration equal to the basin time of concentration, given by

$$t_r = t_c = \left(\frac{L}{\alpha}\right)^{1/m} i_e^{1/m-1} \tag{4}$$

For $t_r < t_c$, zero volumes occur at and to the left of the curved line given by

$$t_r = \left(\frac{q_{th}}{\alpha}\right)^{1/m} \frac{1}{i_e} \tag{5}$$

For $t_r > t_c$, zero volumes result whenever

$$i_e < \frac{q_{th}}{L} \tag{6}$$

The volume expression is parameterized by the storm properties i_e and t_r. These are random variables with given joint probability distribution. Following Eagleson (1972), the joint probability density function of the storm properties i_e and t_r is taken as

$$f(i_e, t_r) = \frac{\beta\lambda}{K} \exp(-\lambda t_r - \frac{\beta i_e}{K}) \tag{7}$$

where λ is the parameter of an exponential distribution for storm duration, so $1/\lambda$ is the mean duration

β is the parameter of an exponential distribution of point rainfall intensity, so $1/\beta$ is the mean of storm intensity

K is a factor reducing point rainfall depth to average depth over a given area, A_r

All the necessary terms for the integration of the derived distribution expression (Equation (2)) are then available. Region, R_q, is shaded in Figure 2. The distribution $f_\theta(\theta)$ is given in Equation (7). Essentially, the integration to be performed is,

$$F_V(V_{th})\Big|_{q_{th}} = \int\int_{R(i_e,t_r)} f(i_e, t_r)\, di_e\, dt_r \tag{8}$$

$$= \int_{i_{e_1}}^{\infty} \int_{t_{r_1}}^{t_{r_2}} f(i_e, t_r)\, dt_r\, di_e + F_V(0)$$

where

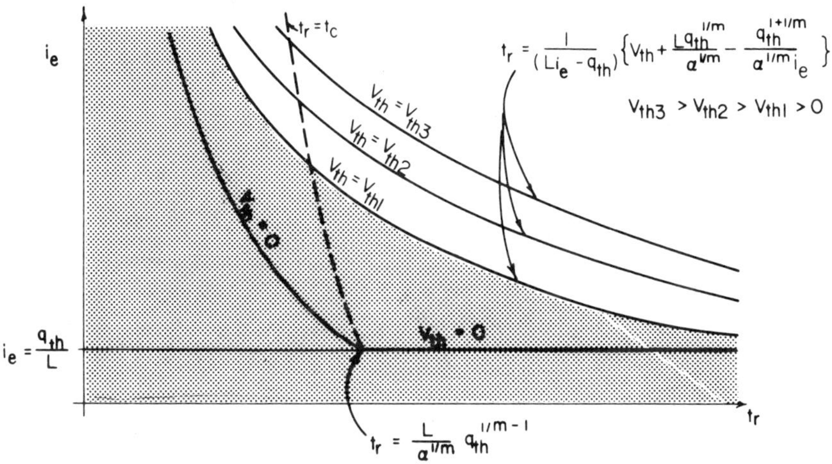

Figure 2 Lines of Constant Volume above a Given Threshold Discharge and Region of Integration, from Chan and Bras, 1979

$$i_{e_1} = q_{th}/L$$

$$t_{r_1} = \frac{q_{th}^{1/m}}{\alpha^{1/m}\, i_e}$$

$$t_{r_2} = \frac{1}{L\, i_e - q_{th}} \left\{ V_{th} + \frac{L\, q_{th}^{1/m}}{\alpha^{1/m}} - \frac{q_{th}^{1+1/m}}{\alpha^{1/m}\, i_e} \right\}$$

$F_V(0)$ is the finite probability of V_{th} taking the value zero which is given by the integral of $f(i_e, t_r)$ from the axes to the lines of $V_{th} = 0$ in Figure 2.

After considerable manipulations, Chan and Bras (1979) found that the previous integral has a very good approximate solution,

$$F_V(V_{th})\Big|_{q_{th}} = 1-a_4 e^{-a_4 a_2+\rho_2} \Big\{ 2\sqrt{(a_1 + a_3)/a_4}\, K_1[2\sqrt{(a_1 + a_3)a_4}\,] - \Delta_2 E_2[(a_1 + a_3)/\Delta_2] \Big\} \qquad (9)$$

Parameters a_1, a_2, a_3 and a_4 are known constant, function of known quantities λ, L, q_{th}, α, m, β and K and of the variable V_{th}. $K_1[.]$ is a first order modified Bessel function of the second kind. ρ_2 and Δ_2 are essentially known quantities generally function of parameters a_1, a_2, a_3 and a_4. $E_2(x)$ is a tabulated exponential integral function of the second order,

$$E_2(x) = \int_1^{\infty} \frac{e^{-xt}}{t^2}\, dt$$

The typical behavior of the cumulative density function of volume given by Equation (9) is shown in Figure 3. There, the parameter L is varied while the others remain constant.

The important features are that the probabilistic distribution is mixed. There is a finite probability of attaining zero volume above a given threshold and a continuous probability density for volumes greater than zero. As length decreases, the peak discharges relative to q_{th} decrease, leading to an increasing probability of zero volume above a fixed threshold discharge. Any parameter influencing peak discharge in a similar manner will have the same effect on the cumulative distribution function. For example, decreasing α or increasing q_{th} have the same qualitative effect on the distribution as decreasing length of overland flow.

The distribution in Equation (9) can be used to obtain recurrence of annual exceedance series. Eagleson (1972) has shown that the recurrence T_E of a flood volume event can be approximated by (using n as the mean number of events per year),

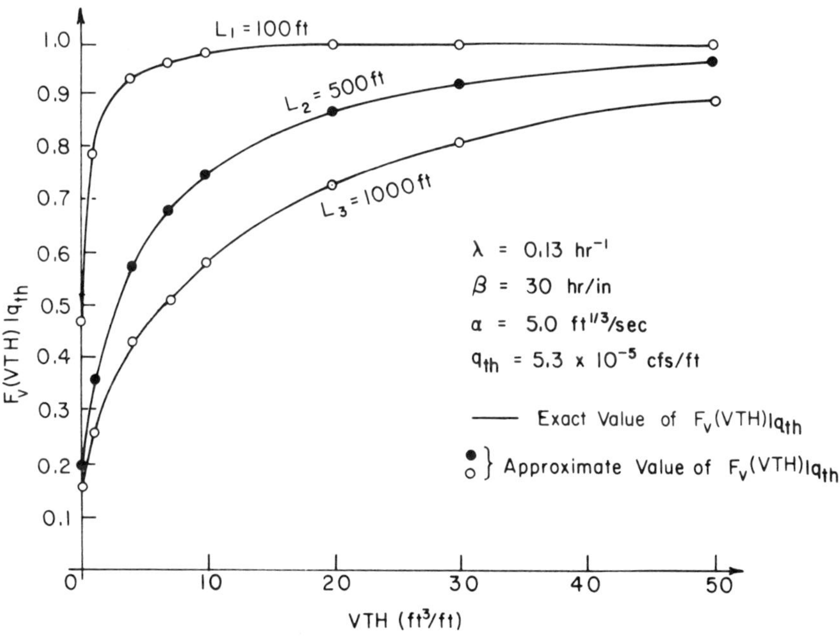

Figure 3 Cumulative Density Function of Volume above a Given Threshold Discharge, from Chan and Bras, 1979

$$\frac{1}{T_E} \cong n(1 - F_V(V_{th})) \tag{10}$$

Chan and Bras (1979) used Equations (9) and (10) together with data from a 23.29 acre catchment in Gray Haven, Baltimore, Maryland [Tucker, 1969; Leclerc and Schaake, 1973] to obtain the flood volumes frequency curves given in Figure 4. For example, if a recurrence interval of 25 years were chosen, the design of a water treatment plant with 5 cfs (3.23 MGD) would require a storage volume of 115,000 cubic feet.

THEORETICAL DISTRIBUTION OF INFILTRATED WATER FROM A GIVEN STORM

In studying the optimal scheduling of irrigation in rainfed agriculture, Cordova and Bras (1979) obtained the theoretical probabilistic distribution of water infiltrated into a homogeneous soil from any given storm.

Based on the equations obtained by Eagleson (1978), the cumulative infiltration from a storm of uniform intensity i and duration t_r is defined as

$$V(i, t_r) = \begin{cases} i\ t_r & \text{for } t_r \leq t_o \\ At_r + S(t_r/2)^{1/2} & \text{otherwise} \end{cases} \tag{11}$$

where A is the gravitational infiltration rate as modified by capillary rise,

$$A = \frac{1}{2} K(1)[1 + (\frac{\theta_o}{n})^c] - \omega \tag{12}$$

and S is infiltration sortivity

$$S = 2(1 - \frac{\theta_o}{n}) \left[\frac{5n\ K(1)\ \Psi(1)\ \phi(d, \theta_o)}{3\ m\ \pi}\right]^{1/2} \tag{13}$$

where n is porosity
$K(1)$ is saturated hydraulic conductivity
$\Psi(1)$ is saturated soil matric potential
$\phi(d,\theta_o)$ is infiltration diffusivity function given by Eagleson (1978)
ω is the capillary rise from the water table
m is the pore size distribution index
d is the diffusivity index
c is the pore connectivity index
θ_o is the initial moisture content

t_o is the time required for the soil infiltration capacity to match the average rainfall intensity. From Eagleson (1978),

$$t_o \approx \frac{S^2}{2(1 - A)^2} \tag{14}$$

Equation (11) defines curves of equal infiltrated volume as a

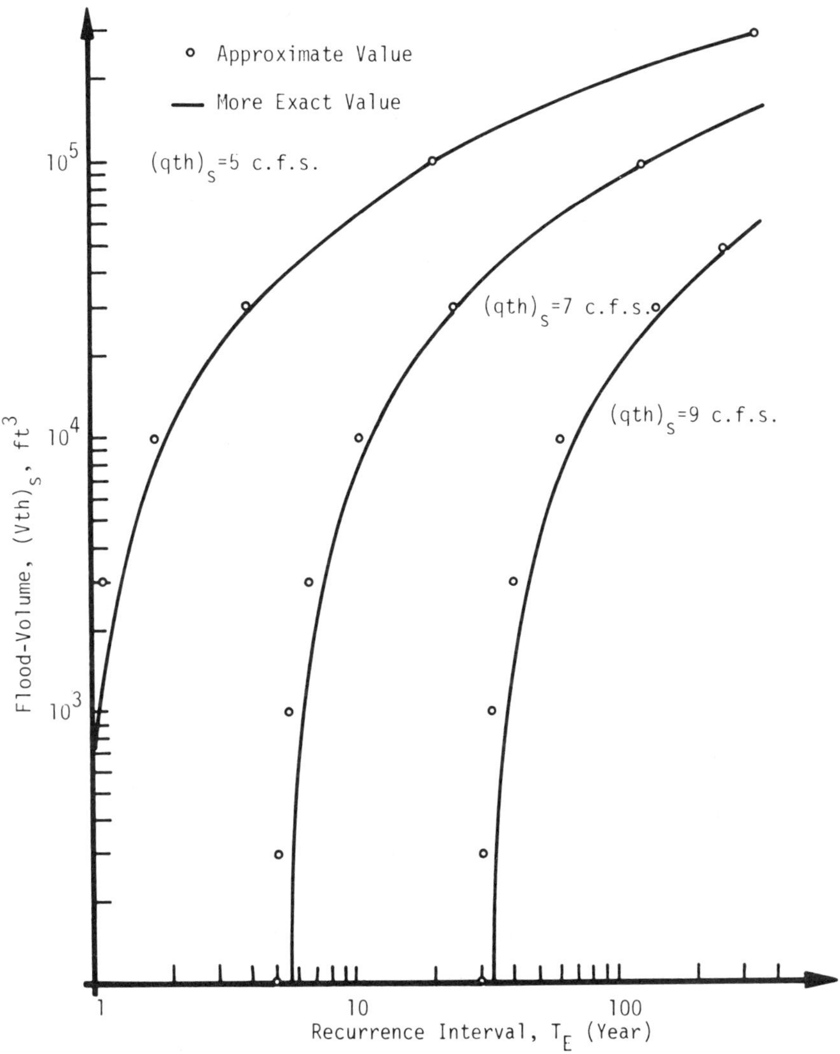

Figure 4 Flood-Volume Frequency Curves for the Gray Haven Catchment, from Chan and Bras, 1979

function of i and t_r. This is illustrated in Figure 5. The space within a curve of equal volume is divided in two regions:

$$\text{Region 1 where } V = V(i, t_r) \qquad \text{for } t_r \leq t_o$$

$$\text{Region 2 where } V = V(t_r) \qquad \text{for } t_r > t_o$$

The derived distribution problem can then be formulated as

$$F_V(V) = P[V \leq v] = \iint_R f(i, t_r)\, di\, dt_r \tag{15}$$

where R is shown shaded in Figure 5.

Equation (15) is equivalent to

$$P[V \leq v] = \int_0^{i^*}\int_0^{v/i} f(i, t_r)\, di\, dt_r + \int_{i^*}^{\infty}\int_0^{t_o^*} f(i, t_r)\, di\, dt_r \tag{16}$$

where for a given value of V, v, and $t_r = t_o^*$

$$v = i^*\, t_o^*$$

and

$$t_o^* = \frac{S^2}{2(i^* - A)^2}$$

which can be shown to lead to

$$i^* = \frac{4vA + S^2 + S\sqrt{8vA + S^2}}{4v}$$

Using the same joint distribution between i and t_r given in Equation (7) (without K) yields,

$$F_V(V) = P[V_I \leq v] = 1 - e^{-\beta i^* - \lambda t_o^*} - \beta \int_0^{i^*} e^{-\beta i - \frac{\lambda v}{i}}\, di \tag{17}$$

Equation (17) is easily integrated numerically. Its first two moments can also be explicitly obtained (Cordova and Bras, 1979). Figure 6 shows the excellent behavior of Equation (17) (continuous line) when compared to detailed simulations of a real case study with 27 years of historical hourly rainfall records. The precipitation periods refer to seasonal differences in rainfall behavior.

SHORT TERM DISCHARGE FORECASTING

The National Weather Service river forecasting model is an example of a conceptual formulation of the hydrologic cycle. It encompasses the best deterministic knowledge of the behavior of the different components of the rainfall-runoff process. Any conceptual model can be represented by a state-space equation,

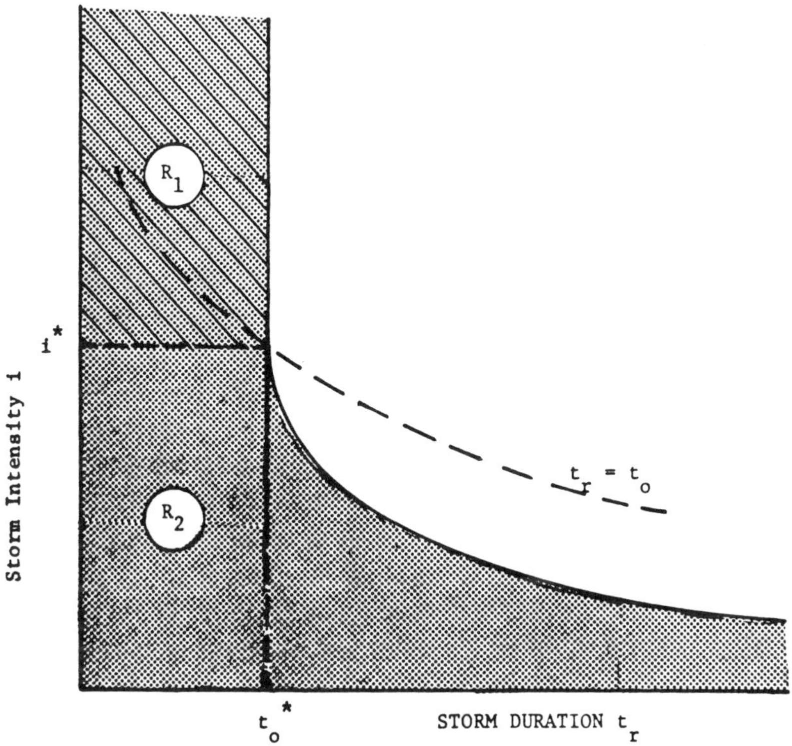

Figure 5 Integration Regions for Deriving the c.d.f. of the Cumulative Infiltration from a Given Rain Storm, from Cordova and Bras, 1979

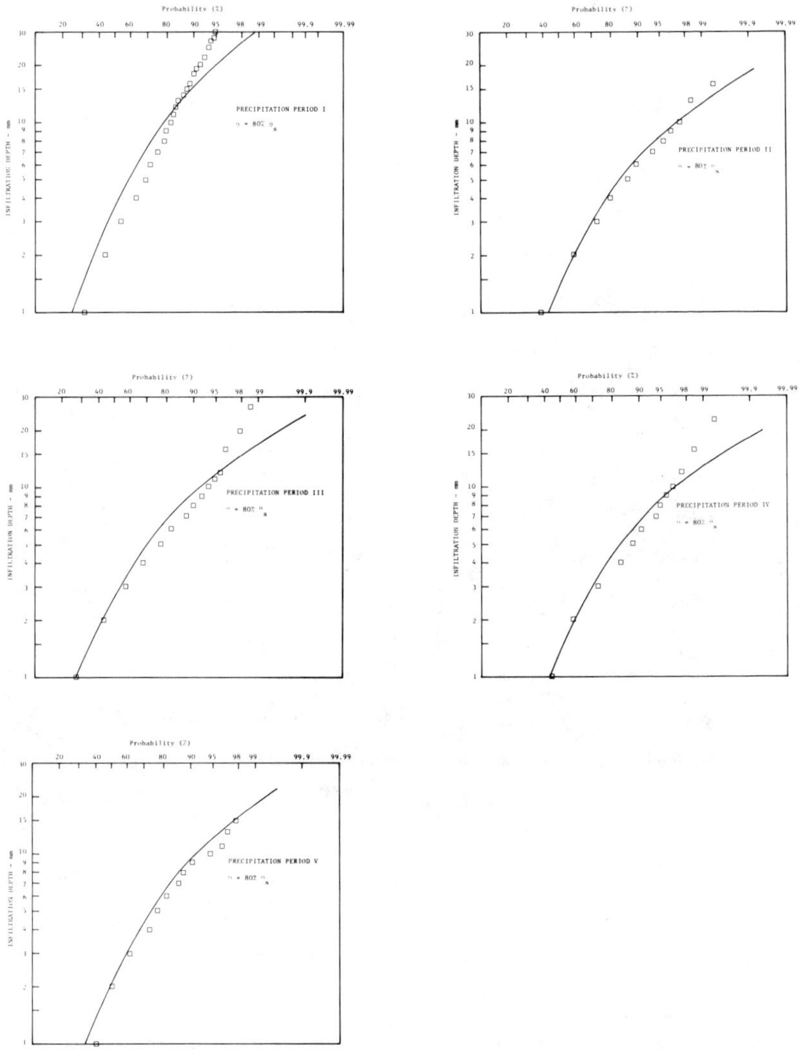

Figure 6 Theoretical Cumulative Distribution of Infiltrated Water (Continuous Line) Compared to Simulation Results (Squares) for Initial Moisture Conditions of $0.8\theta_s$ and Different Rainfall Periods, from Cordova and Bras, 1979

$$\underline{x}(t + \Delta t) = \underline{\Phi}(\underline{x}(t), \underline{u}(t)) \quad (18)$$

where $\underline{u}(t)$ is a vector of inputs, like rainfall, snowmelt and evapotranspiration

$\underline{x}(t)$ is a vector of state variables of time t. The states are by definition that set of variables which completely describes the system at any one moment and encompasses the effects of all past inputs into the system. In a conceptual hydrologic model, they are the water stored at various hydraulic elements within the soil and channels

$\underline{\Phi}(\cdot)$ is a function, possibly nonlinear, that operates on inputs and states to obtain new state variables at discrete times into the future.

Reality is that the model represented by Equation (18) contains many sources of uncertainty:

1. model error due to necessary simplifications of reality
2. input uncertainty since the hydrologist never perfectly knows the rainfall, snowmelt and evapotranspiration
3. parameter errors due to usually non-optimal calibrations of the model
4. the initial state of the system is rarely known with certainty and so introduces errors that propagate over finite lengths of time.

In order to acknowledge the above uncertainties, it is possible to convert the deterministic system of Equation (18) into a stochastic difference equation,

$$\underline{x}(t + \Delta t) = \underline{\Phi}(\underline{x}(t), \underline{u}(t)) + \underline{\Gamma}(t)\, \underline{w}(t) \quad (19)$$

where $\underline{x}(t)$ and $\underline{u}(t)$ are now considered random vectors and $\underline{w}(t)$ is a stochastic sequence representing the model or process noise.

It is possible to also acknowledge errors in the observations of the state vector $\underline{x}(t)$ by using a measurement equation of the form,

$$\underline{z}(t) = \underline{h}(\underline{x}(t)) + \underline{v}(t) \quad (20)$$

where $\underline{z}(t)$ is measured output and $\underline{v}(t)$ is a random vector representing measurement errors. Usually $\underline{w}(t)$ and $\underline{v}(t)$ are assumed to be jointly independent white noise sequences. They are also independent of the inputs $\underline{u}(t)$ and the initial condition $\underline{x}(0)$.

Probability theory gives the formulation for forecasting state vector $\underline{x}(t)$ given real time observations $\underline{z}(t)$. The following explanation comes from Kitanidis and Bras (1978).

Solving the probabilistic model, Equations (19) and (20), in its general form, is equivalent to finding the probability density function of the state of the system. Two problems are of primary interest.

1. Prediction. The problem is evaluating the pdf of $\underline{x}(t + \Delta t)$ given the pdf's of $\underline{x}(t)$, $\underline{u}(t)$ and $\underline{w}(t)$. In this case, by definition of marginal distributions:

$$p(\underline{x}(t + \Delta t)) = \iint p(\underline{x}(t + \Delta t), \underline{x}(t), \underline{u}(t))\, d\underline{x}(t)\, d\underline{u}(t) \tag{21}$$

where p(t) denotes probability density function. By assuming $\underline{x}(t)$ and $\underline{u}(t)$ jointly independent:

$$p(\underline{x}(t+\Delta t)) = \iint p(\underline{x}(t+\Delta t)|\underline{x}(t), \underline{u}(t))\, p(\underline{x}(t))\, p(\underline{u}(t))\, d\underline{x}(t)\, d\underline{u}(t) \tag{22}$$

The conditional probability density function $p(\underline{x}(t + \Delta t)|\underline{x}(t), \underline{u}(t))$ can be estimated from the system dynamics, Equation (19). Equation (22) is the recursive solution to the prediction problem. Since the probability density function, $p(\underline{x}(t))$, fully describes the physical system, it can be considered as the infinite dimensional state of the stochastic system. However, analytical solutions are not available for the general nonlinear problem, while the form of the equation is not suitable for direct implementation on a digital computer. For numerical solution, some type of approximate representation of the probability density function through a finite set of parameters is required. In practice, for the general case, such solutions are tractable only for small dimensional problems.

2. Filtering. The second problem of interest is estimation of the pdf of $\underline{x}(t + \Delta t)$ given measurement of the output $\underline{z}(t + \Delta t)$. By direct use of Bayes' theorem:

$$p(\underline{x}(t+\Delta t)|\underline{z}(t+\Delta t)) = \frac{p(\underline{z}(t + \Delta t)|\underline{x}(t + \Delta t))\, p(\underline{x}(t + \Delta t))}{\int p(\underline{z}(t+\Delta t)|\underline{x}(t+\Delta t))\, p(\underline{x}(t+\Delta t))\, d\underline{x}(t+\Delta t)} \tag{23}$$

where $p(\underline{x}(t + \Delta t)|\underline{z}(t + \Delta t))$ is the pdf of $\underline{x}(t + \Delta t)$ conditional on the result of the measurement $\underline{z}(t + \Delta t)$ while $p(\underline{x}(t + \Delta t))$ is the pdf of $\underline{x}(t + \Delta t)$ prior to a measurement. Consequently, for the solution of the filtering problem, it is first required to solve the prediction problem. With respect to the practical applicability of Equation (23), the same remarks are valid as in the case of Equation (22).

Mathematically, Equations (22) and (23) provide the solution to the forecasting and filtering problems. However, when the system equations (19) and (20) are nonlinear, analytical solutions are in general unavailable. On the other hand, for the large dimensionality of the conceptual hydrologic models and for the required accuracy, numerical solutions are not tractable. A special case that gives analytical solutions is when Equations (19) and (20) are linear and all variables are Gaussian. In this case:

$$\underline{x}(t + \Delta t) = \Phi(t)\, \underline{x}(t) + G(t)\, \underline{u}(t) + \Gamma(t)\, \underline{w}(t) \tag{24}$$

$$\underline{z}(t + \Delta t) = H(t + \Delta t)\, \underline{x}(t + \Delta t) + \underline{v}(t + \Delta t) \tag{25}$$

Now the problem of forecasting becomes equivalent to estimating the first and the second moment of the probability density function of

$\underline{x}(t + \Delta t)$ given measurement only up to time t, which, by employing the conditional density notation, are denoted by $\hat{x}(t + \Delta t|t)$ and $P(t + \Delta t|t)$, respectively. For $\underline{x}(t)$, $\underline{u}(t)$ and $\underline{w}(t)$ jointly Gaussian, $\underline{x}(t + \Delta t)$ is also Gaussian with first two moments:

$$\hat{\underline{x}}(t + \Delta t|t) = \Phi(t)\ \hat{\underline{x}}(t|t) + G(t)\ \hat{\underline{u}}(t) \tag{26}$$

$$P(t+\Delta t|t) = \Phi(t)\ P(t|t)\Phi^T(t) + G(t)\ U(t)\ G^T(t) + \Gamma(t)\ Q(t)\Gamma^T(t) \tag{27}$$

where $\hat{\underline{x}}(t|t)$ and $P(t|t)$ are the first two moments of $x(t)$, $\hat{u}(t)$ and $U(t)$ are the first two moments of the input $\underline{u}(t)$ and $\overline{Q}(t)$ is the covariance matrix of the zero mean model error $\underline{w}(t)$. The random vectors $\underline{x}(t)$, $\underline{u}(t)$ and $\underline{w}(t)$ are assumed independent between themselves, without loss of generality. Equations (26) and (27) are used to recursively estimate the first two moments of $x(t + N\Delta t)$, $N = 1, 2, \ldots$ which is the solution to the forecasting problem.

The linear-Gaussian case provides a tractable solution to the filtering problem. The result, known as the Kalman filter, is that the posterior probability density function of the state $\underline{x}(t + \Delta t)$ given the measurement $\underline{z}(t + \Delta t)$ is Gaussian with moments:

$$\hat{\underline{x}}(t + \Delta t|t + \Delta t) = \hat{\underline{x}}(t + \Delta t|t) + K(t + \Delta t)\ \underline{\nu}(t + \Delta t) \tag{28}$$

$$P(t + \Delta t|t + \Delta t) = [I - K(t + \Delta t)\ H(t + \Delta t)]\ P(t + \Delta t|t) \tag{29}$$

$$K(t+\Delta t) = P(t+\Delta t)\ H^T(t+\Delta t)[H(t+\Delta t)\ P(t+\Delta t|t)\ H^T(t+\Delta t) + R(t+\Delta t)]^{-1} \tag{30}$$

$$\underline{\nu}(t + \Delta t) = \underline{z}(t + \Delta t) - H(t + \Delta t)\ \hat{\underline{x}}(t + \Delta t|t) \tag{31}$$

where $\underline{x}(t + \Delta t|t + \Delta t)$ is the mean vector and $P(t + \Delta t|t + \Delta t)$ is the covariance matrix of $\underline{x}(t + \Delta t)$ given observations up to $t + \Delta t$ and $R(t + \Delta t)$ is the covariance matrix of the zero mean measurement error.

The form of the filter, Equations (28) and (30), is simple and intuitively appealing. The mean of the state posterior to the measurement is given by the mean of the state prior to the measurement plus a differential correction proportional to the step-1 prediction error $\underline{\nu}(t + \Delta t)$. The weight of the correction term $K(t + \Delta t)$, called the Kalman gain matrix, is a function of the prior covariance matrix of the state $\underline{x}(t + \Delta t)$ and the covariance of the measurement error. If it is desirable to relax the Gaussian assumption and the random vectors $\underline{x}(0)$, $\underline{w}(.)$ and $\underline{v}(.)$ are partially characterized by their first two moments, then Equations (26) and (28), respectively, give the minimum variance linear estimates of $\underline{x}(t + \Delta t)$ a priori and a posteriori to the measurement $\underline{z}(t + \Delta t)$, with corresponding variances given by Equations (27) and (29).

In order to employ the Kalman filter results, the system Equations (19) and (20) must be linearized. Although due to the linearization the solution may be suboptimal, in practice this technique provides a systematic and computationally efficient method to estimate the first two moments of the state and has proven its usefulness in a variety of applications.

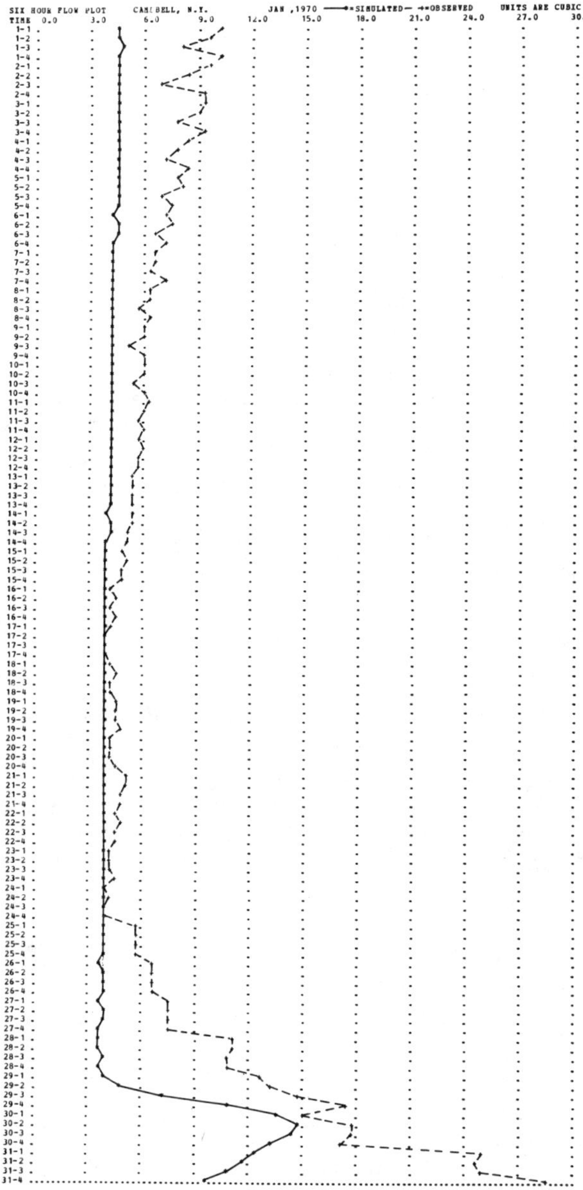

TIME	SIMULATED	OBSERVED
1-1	4.624	10.245
1-2	4.644	9.537
1-3	4.654	8.009
1-4	4.609	10.245
2-1	4.558	9.509
2-2	4.579	8.377
2-3	4.590	6.933
2-4	4.546	9.311
3-1	4.496	9.339
3-2	4.518	9.141
3-3	4.529	7.896
3-4	4.487	9.226
4-1	4.438	8.518
4-2	4.461	7.839
4-3	4.473	7.103
4-4	4.432	8.320
5-1	4.384	7.924
5-2	4.407	8.037
5-3	4.419	7.018
5-4	4.379	7.641
6-1	4.333	7.245
6-2	4.356	7.443
6-3	4.369	6.650
6-4	4.331	7.160
7-1	4.286	6.481
7-2	4.309	6.650
7-3	4.323	6.254
7-4	4.285	7.075
8-1	4.242	6.424
8-2	4.265	6.424
8-3	4.279	5.688
8-4	4.243	6.424
9-1	4.200	6.084
9-2	4.224	5.858
9-3	4.238	5.009
9-4	4.203	6.000
10-1	4.162	5.886
10-2	4.186	5.886
10-3	4.200	5.320
10-4	4.166	5.943
11-1	4.125	6.283
11-2	4.150	6.000
11-3	4.164	5.717
11-4	4.131	6.084
12-1	4.092	5.660
12-2	4.116	5.886
12-3	4.131	5.603
12-4	4.099	5.830
13-1	4.061	5.490
13-2	4.085	5.490
13-3	4.100	5.264
13-4	4.068	5.547
14-1	4.031	5.264
14-2	4.056	5.320
14-3	4.071	4.952
14-4	4.040	5.122
15-1	4.004	4.896
15-2	4.029	5.151
15-3	4.045	4.698
15-4	4.015	4.896
16-1	3.980	4.245
16-2	4.004	4.471
16-3	4.020	4.245
16-4	3.991	4.471
17-1	3.957	4.075
17-2	3.981	3.990
17-3	3.997	3.990
17-4	3.966	4.019
18-1	3.932	4.330
18-2	3.959	4.386
18-3	3.976	4.273
18-4	3.948	4.273
19-1	3.914	4.641
19-2	3.940	4.443
19-3	3.957	4.443
19-4	3.927	4.811
20-1	3.892	4.245
20-2	3.918	4.330
20-3	3.934	4.330
20-4	3.903	4.556
21-1	3.867	5.037
21-2	3.893	5.066
21-3	3.910	4.811
21-4	3.879	4.811
22-1	3.842	4.613
22-2	3.869	4.698
22-3	3.886	4.443
22-4	3.854	4.556
23-1	3.818	4.245
23-2	3.845	4.302
23-3	3.862	4.302
23-4	3.831	4.443
24-1	3.794	4.019
24-2	3.821	4.103
24-3	3.839	3.934
24-4	3.807	4.019
25-1	3.770	5.773
25-2	3.798	5.773
25-3	3.816	5.632
25-4	3.784	5.688
26-1	3.747	6.679
26-2	3.775	6.594
26-3	3.794	6.594
26-4	3.761	6.594
27-1	3.724	7.528
27-2	3.754	7.415
27-3	3.773	7.415
27-4	3.741	7.415
28-1	3.704	10.980
28-2	3.735	10.980
28-3	3.756	10.839
28-4	3.742	10.839
29-1	3.872	12.622
29-2	4.778	13.244
29-3	7.302	14.659
29-4	10.883	17.461
30-1	13.439	15.140
30-2	14.760	17.687
30-3	14.341	17.687
30-4	13.329	16.952
31-1	12.343	24.989
31-2	11.663	24.564
31-3	10.678	24.762
31-4	9.699	28.413

Figure 7 Six Hour Discharge Forecasting (Continuous Line) for the Cohocton River, Compared to Observations during January, 1970, Using the N.W.S. Off-line Rainfall-Runoff Model (from Kitanidis and Bras, 1978)

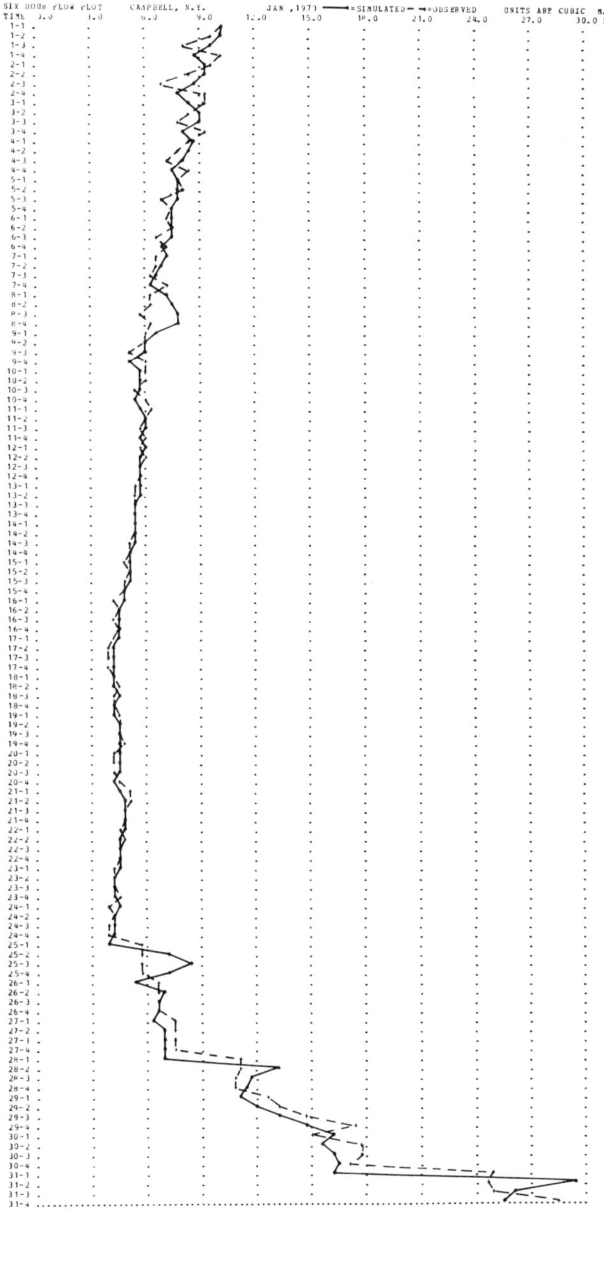

TIME	SIMULATED	OBSERVED
1-1	10.261	10.245
1-2	10.102	9.537
1-3	9.647	8.009
1-4	8.570	10.245
2-1	9.307	9.509
2-2	9.320	8.377
2-3	8.684	6.933
2-4	7.690	9.311
3-1	8.455	9.339
3-2	8.857	9.141
3-3	8.915	7.896
3-4	8.224	9.226
4-1	8.664	8.518
4-2	8.505	7.839
4-3	8.054	7.103
4-4	7.540	8.320
5-1	7.926	7.924
5-2	7.887	8.037
5-3	7.924	7.018
5-4	7.409	7.641
6-1	7.534	7.245
6-2	7.419	7.443
6-3	7.401	6.650
6-4	7.031	7.160
7-1	7.130	6.481
7-2	6.850	6.650
7-3	6.732	6.254
7-4	6.432	7.075
8-1	7.058	6.424
8-2	7.434	6.424
8-3	7.869	5.688
8-4	7.702	6.424
9-1	6.525	6.084
9-2	6.058	5.858
9-3	5.880	5.009
9-4	5.250	6.000
10-1	5.652	5.886
10-2	5.795	5.886
10-3	5.842	5.320
10-4	5.546	5.943
11-1	5.750	6.283
11-2	6.003	6.000
11-3	5.968	5.717
11-4	5.797	6.084
12-1	5.913	5.660
12-2	5.751	5.886
12-3	5.794	5.603
12-4	5.658	5.830
13-1	5.718	5.490
13-2	5.574	5.490
13-3	5.509	5.264
13-4	5.352	5.547
14-1	5.440	5.264
14-2	5.337	5.320
14-3	5.317	4.952
14-4	5.100	5.122
15-1	5.112	4.896
15-2	5.001	5.151
15-3	5.081	4.698
15-4	4.860	4.896
16-1	4.890	4.245
16-2	4.530	4.471
16-3	4.531	4.245
16-4	4.394	4.471
17-1	4.471	4.075
17-2	4.282	3.990
17-3	4.159	3.990
17-4	4.102	4.019
18-1	4.104	4.330
18-2	4.277	4.386
18-3	4.367	4.273
18-4	4.328	4.273
19-1	4.322	4.641
19-2	4.513	4.443
19-3	4.477	4.443
19-4	4.453	4.811
20-1	4.642	4.245
20-2	4.405	4.330
20-3	4.367	4.330
20-4	4.339	4.556
21-1	4.458	5.037
21-2	4.753	5.066
21-3	4.885	4.811
21-4	4.793	4.811
22-1	4.764	4.613
22-2	4.643	4.698
22-3	4.639	4.443
22-4	4.488	4.556
23-1	4.503	4.245
23-2	4.335	4.302
23-3	4.301	4.302
23-4	4.280	4.443
24-1	4.355	4.019
24-2	4.156	4.103
24-3	4.148	3.934
24-4	4.053	4.019
25-1	4.029	5.773
25-2	7.249	5.773
25-3	8.475	5.632
25-4	7.346	5.688
26-1	5.535	6.679
26-2	6.875	6.594
26-3	6.732	6.594
26-4	6.507	6.594
27-1	6.360	7.528
27-2	6.753	7.415
27-3	6.867	7.415
27-4	6.898	7.415
28-1	6.916	10.980
28-2	13.275	10.980
28-3	11.783	10.839
28-4	11.525	10.839
29-1	11.223	12.622
29-2	12.116	13.244
29-3	13.339	14.659
29-4	14.602	17.461
30-1	16.347	15.140
30-2	15.577	17.687
30-3	16.205	17.687
30-4	16.392	16.952
31-1	16.230	24.989
31-2	29.475	24.564
31-3	26.245	24.762
31-4	25.637	28.413

Figure 8 Six Hour Discharge Forecasting (Continuous Line) for the Cohocton River, Compared to Observations During January, 1970, Using the Adaptive, On-line Version of the N.W.S. Rainfall-Runoff Model (from Kitanidis and Bras, 1978)

Kitanidis and Bras (1978) formulated the National Weather Service river forecasting model in the framework of Equations (19) and (20). Some linearization was required in order to utilize an extended Kalman filter formulation [Schweppe, 1973] in a real time forecasting framework. It is impossible in the limited space available to discuss all the peculiarities of the approach. The interested reader is referred to the original reference. Nevertheless, Figures 7 and 8 illustrate the type of improved forecasting ability obtained. Figure 7 gives observed and 6 hour lead time forecasted streamflows for the Cohocton River basin (upstate New York) during the month of January 1970. The forecast is being made with the National Weather Service model operating in its normal deterministic, off-line, mode and making no use of real time observations. Figure 8 gives the forecasted flows obtained using an adaptive, on-line, operation which acknowledges uncertainties and uses real time streamflow observations in a feedback mode.

CONCLUSIONS

This paper has illustrated the solutions to three particular common hydrologic forecasting problems. The explicit acknowledgement of uncertainties permits the use of powerful probability theory concepts while at the same time preserving and using valuable deterministic hydrologic formulations. The techniques illustrated were derived distributions and optimal estimation using the Kalman-Bucy filter. It should be clear that the solutions presented, given in terms of physical hydrologic parameters, would have been impossible to obtain without exploiting the tools of probability theory.

ACKNOWLEDGEMENTS

The author wants to acknowledge Mr. Siu-On Chan, Dr. Peter K. Kitanidis and Dr. Jose R. Cordova, who are responsible for most of the work presented here. The reported work was part of the above individual theses. The derivation of the volume above a given threshold was sponsored by the National Science Foundation, Grant No. ENG770499. The distribution of the irrigation volume was obtained as part of the work performed for on-going N.S.F. Grant No. ENG 78-20245. The river forecasting problem solution was sponsored by the Department of Commerce, National Weather Service, Hydrologic Research Laboratory, Contract No. 7-35112.

REFERENCES

1. Benjamin, J. R. and Cornell, C. A., Probability, Statistics and Decision for Civil Engineers, McGraw-Hill Book Company, 1970.

2. Chan, S.-O. and Bras, R. L., "Urban Storm Water Management: Distribution of Flood Volumes," Water Resources Research, accepted for publication, 1979.

3. Cordova, J. R. and Bras, R. L., "Stochastic Control of Irrigation Systems," Technical Report No. 239, Ralph M. Parsons Laboratory for Water Resources and Hydrodynamics, Department of Civil Engineering, M.I.T., 1979.

4. Eagleson, P. S., Dynamic Hydrology, McGraw-Hill Book Company, 1970.

5. Eagleson, P. S., "Dynamics of Flood Frequency," Water Resources Research, 8(4), 1972.

6. Eagleson, P. S., "Climate, Soil and Vegetation 1. Introduction to Water Balance Dynamics," Water Resources Research, 14(5), 1978.

7. Eagleson, P. S., "Climate, Soil and Vegetation 2. The Distribution of Annual Precipitation Derived from Observed Storm Sequences," Water Resources Research, 14(5), 1978.

8. Eagleson, P. S., "Climate, Soil and Vegetation 3. A Simplified Model of Soil Moisture Movement in the Liquid Phase," Water Resources Research, 14(5), 1978.

9. Eagleson, P. S., "Climate, Soil and Vegetation 4. The Expected Value of Annual Evapotranspiration," Water Resources Research, 14(5), 1978.

10. Eagleson, P. S., "Climate, Soil and Vegetation 5. A Derived Distribution of Storm Surface Runoff," Water Resources Research, 14(5), 1978.

11. Eagleson, P. S., "Climate, Soil and Vegetation 6. Dynamics of the Annual Water Balance," Water Resources Research, 14(5), 1978.

12. Kitanidis, P. K. and R. L. Bras, "Real Time Forecasting of River Flows," Technical Report No. 235, Ralph M. Parsons Laboratory for Water Resources and Hydrodynamics, Department of Civil Engineering, M.I.T., 1978.

13. Leclerc, G. and Schaake, J. C., Jr., "Derivation of Hydrologic Frequency Curves," Technical Report No. 142, Ralph M. Parsons Laboratory for Water Resources and Hydrodynamics, Department of Civil Engineering, M.I.T., 1972.

14. Schweppe, F. C., Uncertain Dynamic Systems, Prentice-Hall, 1973.

15. Tucker, L. S., "Availability of Rainfall Runoff Data for Sewered Drainage Catchments," ASCE Urban Water Research Program, Technical Memorandum, No. 8, ASCE, 1969.

HYDROCLIMATOLOGY - A NEEDED CROSS-DISCIPLINE

Ronald F. Kilmartin [(1)]

International Engineering Company, Inc.
San Francisco, CA 94104

Abstract. Major irrigation projects are planned for long time spans compared to most human endeavors, with 50-year planning periods not unusual. Climatic fluctuations within the planning period are customarily considered in design, but climatic trends pervading an entire planning period have not been generally considered. Although the direction and magnitude of future climatic changes over the next 50-100 years may not be certain, there is agreement that the climate of the past 3 or so generations has been quite abnormal compared to the last millennium as a whole. If the 1000-year norms are robust and the last 100 or so an inertial or random aberration, the consequences of projecting statistics from the latter period forward over the 50-year period ahead may be quite adverse. A new cross-discipline - hydroclimatology - is suggested for developing quantitative techniques for considering the effects of long-term climatic fluctuations, i.e., the "Joseph events" in development planning. The potential for using worldwide pressure, tree-ring, and other data for back extension of local records and climatic fluctuation modeling is examined, and recent developments in hydrologic and climatologic record extension and modeling are noted. A structural notch for hyroclimatology in water resources project analysis is urged, together with global cataloging and coding of the most useful data sets.

1. INTRODUCTION

Mandelbrot and Wallis (1968) cited two events of Biblical history as useful analogies that have since become part of the hydrologist's language. By 'Joseph event', they meant that precipitation or streamflow could be either well above or well below normal for very long periods of time. By 'Noah event', they meant that extreme precipitation can at times be very extreme. In applied design practice the 'Noah event' corresponds to the spillway design or probable maximum flood problem, and the cross discipline of hydrometerology has evolved over the years to specialize in this unique area. In their quest for knowledge on super-floods or 'Noah events', hydrometeorologists developed an analytic approach that included a detailed consideration of the physics of the synopic scale atmospheric processes involved.

---

(1) Area Manager, Nicaraguan Projects, Managua, Nicaragua.

The 'Joseph event' corresponds to the reservoir storage design or critical-period problem. For the 'Joseph event', analytical techniques have thus far not progressed much beyond a consideration of basin hydrological processes. But the 'Joseph event' is not merely a within-basin process, it is a basin response to a major anomaly in the atmospheric circulation, an anomaly that is often near hemispherical in areal-magnitude.

The thrust of this paper is that increased attention to atmospheric processes is needed in modeling 'Joseph events' and for several cogent reasons; first, to extend the streamflow and precipitation database for better statistical sampling; second, to quantify the effects - if any - on the historic sample of the fairly sharp climatic abberation which we know occurred in many parts of the world over most of the data collection period; and third, to develop techniques for explicit consideration of climatic change in water resources project analysis over the 50-100 year planning period ahead.

2. MOTIVATION

Reconstructions of temperature histories of the last 1000 years and more (Bryson, 1974; National Academy of Sciences, 1975) indicate that mid-latitude Northern Hemisphere temperatures of the last 3 generations have been substantially higher than over most of the past millennium. Indeed, these years - the last 80-100, included a temperature maximum in the northern mid-latitudes which peaked in the 1940's. In Central England for example, these temperatures were higher than any experienced since about 1400 A.D., and mean rainfall was higher than at any time during the past 550 years ± (Lamb, 1972).

Considering the entire millennium back to 1000 A.D., Lamb (1972) showed that an even more pronounced warming occurred in England during the years 1100-1400 A.D., culminating in an earlier optimum in the 1200's. But in Iceland, the sharp climax of 1940 exceeded all temperatures of the previous 1000 years (Bryson, 1974).

Considering the 100 years ahead and the 1000 immediately behind, climate would appear to be a dynamic planning parameter, and to a certain extent it currently is. However there are obviously some problems in considering historic data samples and extracting information from them which deserve exposition, criticism, and better approaches than we now have.

The water resources planner who is looking ahead to the next 50 years would like to know what the local sequence of climate and streamflow will be like, but since a prediction like that is not operationally feasible, he must invoke decision analysis techniques. In this approach, he will estimate the probability distributions of climatic elements and associated streamflows that will prevail over the 50-100 year period ahead. With these distributions, he can develop models for Monte-Carlo generation of stochastic sequences of streamflow, precipitation and potential evapotransportation, and in turn use these sequences in simulation testing and economic analysis of project operations over the planning period ahead. The random or stochastic sequences are the foundation for system optimization studies. By simulation under a selected combination of stochastic sequences, the project's reliability under potential natural drought hazards can be assessed, and resource allocations adjusted to develop the desired overall performance (Kilmartin and Burges, 1976).

The probability distributions used for climatic elements and streamflows and the models for generating sequences from these distributions are thus key factors at the overall project planning stage. Considering the regional and sometimes national ramifications of such projects, it is vital that all the information and data that are available be utilized in deriving the various models required, and their parameters.

Turning first to the streamflow database, we find it is very sparce in much of the world between 30N and 30S, where the need is greatest for development planning purposes. Only in a few locations within this belt are streamflow records of more than 10 to 20 years available. Even when all the advanced countries are included, virtually no streamflow information antedates 1900 and most such records are less than 20 to 50 years in length.

It is sobering to consider that what little we do know of historical streamflow during these years may have been strongly affected by the rather sharp climatic aberration that occurred over the same period. We do know that the direction and local manifestation of this aberration were not globally or even latitudinally consistent within a hemisphere (e.g., Van Loon and Williams, 1976). Latitudinally, major circulation features tend to vary in a semi-harmonic fashion about the globe, but are constrained to some extent by physical features such as oceans and major orographic barriers; thus, trends in areas relatively close to each other may have been affected much differently by a circulation shift; precipitation and temperature may show an increase, or decrease, or no change at all over the period; changes in mean levels which did occur may or may not have paralleled the average trend to and away from the climatic 'optimum' that peaked about 1940. But let us cite a case where this apparently did occur, our own Colorado River Basin, and note the consequences:

> The 35-year mean annual virgin flow at Lees Ferry was estimated at 17 million acre-feet for the period 1896-1930, compared to 13 million for the period 1931-1965 (National Academy of Sciences, 1968). The former figure was the basis for a string of interstate compacts and treaties, and also the basis for design of major storage reservoirs comprising the river system and projects supplied by it. The four million acre-feet shortage has had costly remifications over much of the West ever since, and will continue to do so forever, because of the legal and economic basis that are now locked in.

In statistics, the number 30 is usually considered an adequate sample size for estimation of the mean; by that criteria the first 35 years of the Colorado River record should have been sufficient. However, in this case it was not; regional precipitation anomalies occurred over the drainage basin during these two periods (Bradley, 1976) with the same sign as the differences in mean streamflow. Moreover, using modern tree-ring analysis techniques, Fritts (1969) shows that precipitation over the upper Colorado Basin from 1900 to 1962 was from 20 to 40 percent above the 1631-1962 normal, and Stockton (1975), also using tree-ring techniques, showed that the 1900-1930 decade was the highest 30-year period of runoff at Lees Ferry in the past four centuries (Fig. 1).

Where good sets of tree-ring chronologies can be established, as in the Upper Colorado Basin, they have no peer for reconstruction of climate and streamflow. However, only a few species in select, stressed

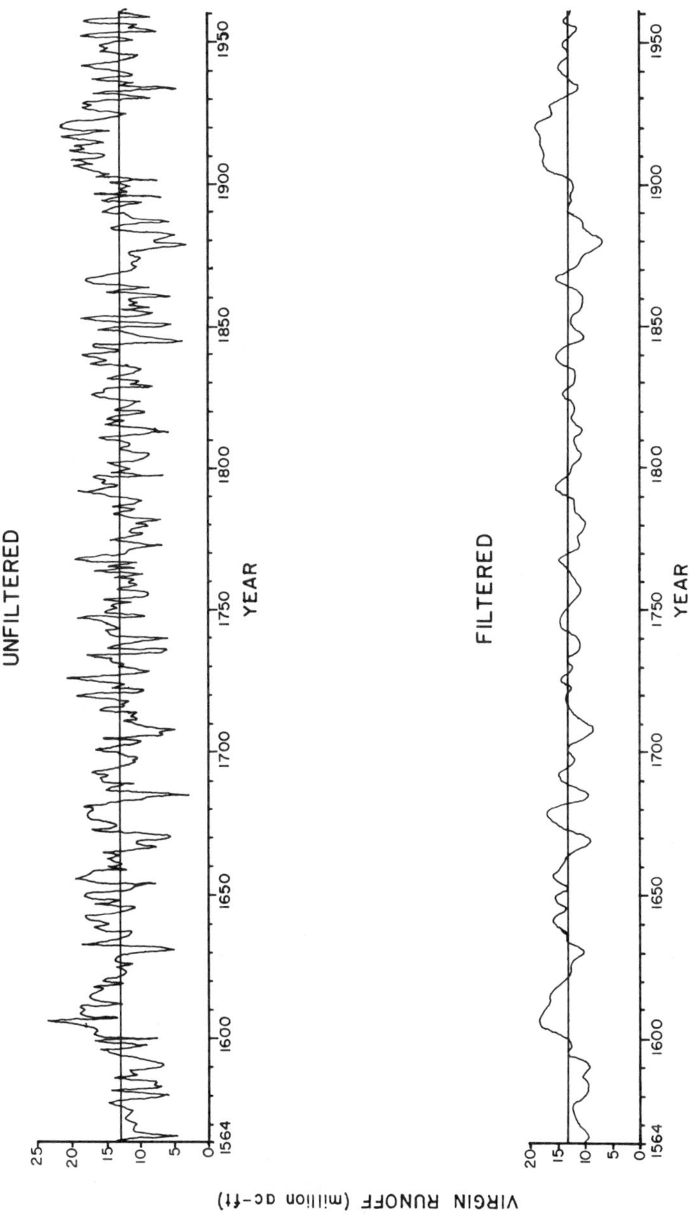

Fig. 1 Long-Term Reconstruction of the Annual Runoff at Lees Ferry, Ariz., Using Tree-Ring Data (after Stockton, 1975)

environments have been found to be useful for climatic reconstructions. In the absence of good tree-ring chronologies the available precipitation record is generally the next best alternative for back extension of the streamflow record. But in many undeveloped areas where the need is greatest, it is frequently found that the rainfall data are only available for the most recent 15-25 year period. Occasionally, longer records are found, but many of these have been decimated by wars and revolutions, and filling-in their gaps becomes a major challenge. It has been recognized as generally desirable to reconstruct the longest series of historic streamflow possible, hopefully at least 50 years. The reason for extension is to get better reliability in statistical parameters estimated from the data, such as the mean, standard deviation, lag-one serial correlation, and skew. Increasingly, practitioners are adding to this basic collection of statistics the unique drought-characteristic parameter known as the Hurst coefficent, and also the modern approach to detection of periodicity known as variance spectrum analysis.

However, long-term trends in the original or extended historic data series have seldom been investigated in water resources project analysis. If a trend exists and is detected, practice is to remove it prior to correlation and variance spectrum analysis, but otherwise to ignore its historic existence or the troublesome question of its future extrapolation. Even less attention - usually none - is given to the question of whether the historic sample (as extended) is representative of what may be expected in the 50-year planning period ahead.

Actually, considering the strong aberration in global climate that we now know occurred during the comparatively short streamflow sampling period, any probability distributions obtained from the data should be regarded in Bayesian fashion as <u>conditional</u> distributions, contingent on the strong climate aberration <u>that prevailed</u> during sampling. The question is, whether the same mean climate is expected to prevail over the next 50 years and if not, how should we adjust stochastic precipitation and streamflow models to incorporate any non-stationarity? This first focuses attention on the question of how might the streamflow and precipitation records be extended back further in time, so as to relate them in a risk-analysis approach to the <u>less abnormal</u> climatic conditions that prevailed over most of the last mellinnium. With long-term reconstructed records, it might then be possible to identify regional links between climate and water resources and more intelligently estimate model parameters, including trends, for the 50-100 years ahead. Had such an analysis been available when the Colorado River Compact was agreed upon, the economic history of the West and Southwest might have been very different. The 1976-1977 drought in that area indeed underscored the fragile water-resources foundation which that economy is based on. Some severe economic relocations and adjustments occurred, and these can be expected to intensify in the years ahead, as planners and decision-makers in both the public and private sectors realize that the droughts in the streamflow record are not (nor were they ever) sufficient models in themselves for estimating future drought-period water supply needs.

The technical competence and economic resources exist to overcome the problems that will unfold, as we adpt to the more probable state of less abundant future water supplies in the U.S. However, perhaps even more critical problems are being created in underdeveloped areas of the world where economic and agricultural needs are critical to political stability and improved social well being. Here large investments are being undertaken, based on climatologic and hydrologic records which are generally meager at best.

In the absence of adequate tree-ring chronologies and the availibility of only short-term or broken precipitation records, and only a few years of streamflow data, what else can be done in the way of record extension? An answer to this, in some parts of the world at least, may be the global pressure data base. Other climatological databases such as temperature and humidity may also be useful in this respect. The pressure database extends back to 1850 for much of the inhabited world and to 1750 for part of Europe (Lamb, 1969). Large areas of the globe are not covered, particularly in early years, but for many areas where development planning is underway, these databases may afford an excellent means for at least some back-extension of precipitation and streamflow records. In data-sparce countries, it is worth nothing that pressure data at stations thousands of miles from the location of interest may be indeed useful in extension and fill-in of local records, particularly for mimicking low frequency effects that are so important in water supply investigations. The various meteorological data banks are useful however for much more than single-station record extension.

Over the past few decades atmospheric scientists and specialists from a variety of other disciplines have completed a number of regional, hemispherical, and global reconstructions of past climates and circulations, leading to a better understanding of what took place historically and why. Activity in this area has been particularly intense in the 1970's, aided by vastly improved computer capabilities. Reconstructions have been developed over time spans of from 10^2 to 10^9 years, for areas that are regional to global in size. (National Academny of Sciences, 1975); of particular interest for water resources and development planning are the reconstructions of climates and circulations within the last millenium, since for the 50-year planning period ahead the prevailing climate will likely be one that has occurred during this most recent 1000-year period, possibly as modified by anthropogenic interference.

Climatic reconstructions have been approached mostly from the viewpoint of statistical analysis, supplemented by qualitative and sometimes a semi-quantitative relation to the global circulation. Some of these efforts will be noted below, since they embody analytical techniques that are directly applicable to the problem of streamflow record back-extension. Statistical approaches ignore fundamental physical relationships of mass continuity and atmospheric and oceanic dynamics; we are in fact fortunate that the statistical approaches exist, because the complexity of the physical systems involved and their interrelationships are exceedingly difficult and costly to model. Nevertheless, a considerable effort has been bent in the direction of deterministic modeling of the atmosphere with models of various levels of complexity, ranging from a simple global radiation balance to a multi-layered global atmosphere interacting with the oceans and continents (Schneider and Dickenson, 1974). These models have been instrumental in advancing knowledge of the general circulation, and also, giving approximate answers to what-if type questions related to recent and future anthropogenic effects. They potentially appear to also have an important role to play in long-term back-extrapolation of global climatologic and therefore hydrologic records, possibly in conjunction with developing a global tree-ring base and enhancing the global pressure data base.

Preview - In what follows, we first discuss the central concern of hydroclimatology, which can be described as modeling of long-term climatic fluctuations in water resources systems analysis. Some of the literature from the fields of hydrology, meteorology, climatology, oceanography, and statistics are reviewed, emphasizing accomplishments in attacking this problem, gaps between the various disciplines, and the lag in application to real problems of development planning.

One effect of hydroclimatology would be to close some gaps that exist between various closely related disciplines. Generally speaking, meteorology deals with short-term events, a day up to perhaps a week ahead. Climatology dwells on periods from years to centuries. Hydrometeorology was an appropriate name for the cross-discipline specializing in relatively short-duration major floods or 'Noah events'; by the same token Hydroclimatology would seem an appropriate name for a cross-discipline specializing in 'Joseph events', which are of multi-year duration. Hydroclimatology would be an area where atmospheric scientists, as it were, could come down to earth, and hydrologists could quantitatively recognize the atmospheric and oceanographic aspects of the hydrologic cycle in long-range development planning. There is room under this roof for a host of disciplines, including oceanography, fluid dynamics, geography, glaciology, statistics, and others. Some of these multidisciplinary aspects will be stressed in what follows, hopefully enough to encourage additional development and cross-fertilization and through that, improved operational ways for handling the 'Joseph event' in development planning.

3. JOSEPH EVENTS

Prelude - This section discusses Joseph events at a non-mathematical level and how they are measured and modeled. The discussion is skimmed from the hydrologic literature on the subject, which is extensive. To emphasize the peculiar nature of these models we will broadly use the term 'infrared noise' models in referring to them as a group.

The term 'infrared noise' is borrowed from Mandelbrot (1965), who studied very low-frequency phenomena in the communications field as well as in water resources. The usage here is based on the definition of the spectrum of a stochastic process as white, red or infrared noise. White noise amounts to equal power over the whole spectrum of frequencies: red-noise indicates more power in low frequencies, whereas infrared-noise models approach infinite power at near-zero frequencies. It is a useful trichotomy with respect to reproducing Joseph events, because the white and red noise models cannot explicitly model the very low-frequency effects. Apparently the infrared models have not yet received a great deal of attention in the atmospheric sciences, but as noted by Barry and Perry (1973) in a brief notation on one of these models, they would appear to have intrinsic value in studies of climatic 'change', which, as we shall note below, is only one step removed from the climatic fluctuations these models are meant to deal with. Excellent reviews and commentary on some of these models are given in Delleur, Tao and Kavvas (1976), Klemes (1974), and O'Connell (1974).

The Hurst Statistic - h. - Joseph events were first investigated in detail by Hurst (1951). He found that a useful statistic for measuring long-term fluctuations in geophysical time series is the range between the maximum and minimum of the cumulative departures from the mean. As related to the reservoir storage design problem, the range is the reservoir size required to deliver the mean annual flow from a given sequence on a firm annual basis. Note that this is an *a posteriori* estimate of required storage with respect to the historic sequence - i.e., it assumes that the entire sequence is deterministic with probability one. Any other random sequence with the same mean and standard deviation would give a different range (even the same set of values rearranged in some other sequence would give a different range, or storage requirement).

The storage required to deliver the mean flow (or some fraction of it) for a finite length of time thus uniquely depends on the sequential arrangement of the inflows. This corresponds to the notion conveyed in a colorful citation of Genesis by Mandelbrot and Wallis (1968), wherein the seven fat years preceded the seven lean years, and Joseph advised Pharoah (*a priori*) to lay up grain during the years of plenty, to tide Egypt over the years of famine. (We omit here a discussion of Joseph's unique forecast of 14 years.)

The theoretical expected value of the range for independent processes is $E(R) = 1.25SN^h$, where S is the standard deviation of the series of length N, and h, the exponent of N, has a value of 0.5. The exponent h has been named the Hurst statistic (Mandelbrot and Wallis, 1968).

Hurst's singular contribution from study of many hydrologic and geophysical time series was that h has an average emperical value of about 0.7 and frequently ranges up to nearly one. For storage analysis, this was a very important finding, for it indicated that from a statistical point-of-view such processes had properties not easily mimicked with conventional stochastic processes. Conventional Markov stochastic (red noise) models have an asymptotic value of h of 0.5. The mathematical problem has been variously considered as due to skew effects or long-term memory effects, or some combination. A great deal of effort has been directed toward solving the mathematical dilemma and while all the useful work done in this field cannot be covered, some of the more well-known developments will be cited. We will first discuss the relationship between Joseph events and trends in the historic sample and over the projected design period ahead.

Super-Josephs, Mini-Josephs and Trends - Joseph events can come in all sizes. For a finite project life, inflow sequence, and water demand, there is one *critical* period, if that is defined as the period over which the reservoir emptied completely, refilled and during which it met its full demand schedule. (In the context considered here this is a multiyear period, i.e., consideration is directed towards reservoirs designed to provide carryover storage for at least one year.) Over the full sequence there will be other ups and downs of course - Mini Josephs - but only *one* Joseph event is critical for reservoir design.

It is convenient to also consider a set of closely related statistics of the inflow sequence, and that consists of linear trends within the period. If the sequence is stationary, the short trends that occur within it will appear like superimposed sawtooths on longer trends, but all will average out over the period. The shorter trends are associated with none-critical mini-Josephs. Some combination of the longer of these trends will represent the critical Joseph event that governs reservoir sizing. If the process is non-stationery in the mean there will be another trend pervading the entire series, resulting in a change in mean level from beginning of the sequence to its end. Such trends do not always occur in streamflow or rainfall sequences, but they do occur frequently. They are usually evaluated as to magnitude with a simple least-squares regression versus time, although more elaborate techniques can also be used. Evaluation as to cause is another matter. Trends may be caused by changes in the micro-environment of the instrument, or possibly by fluctuations in the regional climatology over the period of the sample (we exclude from consideration here changes attributable to the works of man; it is presumed that by some means the records can be adjusted for these effects, however this is never an easy task).

Trends of duration less than project life are conveniently abbreviated as LPL trends; if there are any trends of length equal to or greater than the project life we designate them as GPL trends. LPL trends in a historic sample are presumably associated with Joseph events whose characteristics we wish to preserve in simulation.

GPL historic trends have a length equal to or exceeding either the sample length or the length of the project life. GPL trends must be evaluated and removed from the historic sample before using it for statistical investigations, or at least such investigations must be interpreted in light of GPL trends that the data contain.

The longer LPL trends must also be investigated, particularly considering the nature of the local climatic aberration - if any - during the historic sampling period. A decision must be made whether these trends should be considered as part of the 'Josephivity" to be retained in simulations, or should instead be considered with GPL trends as an explicit trend pervading the project life. This step has not been taken in project analysis to date (other than the assumption of zero trends). However it is related to the matter of model selection and is again obviously a place for multidisciplin-ary participation in modeling technology.

It is relevant at this point to zero in on an operational definition of climatic 'change' and climatic 'fluctuations'; herein, 'fluctuations' are associated with the LPL trends. The largest climatic fluctuation considered in water resources planning is equivalent to the design critical period or the design 'Joseph event'. Climatic change on the other hand is associated with a trend that is expected to pervade the entire project life - a GPL trend. This makes the operational definition dependent on project economic life - an acknowledged deficiency, since this could lead to a 'fluctuation' for one project being a 'climatic change' for another; but for the purposes of this paper we can ignore this inconsistency.

The infrared class of models have been historically motivated to capture an acceptable stochastic mimick of climatic <u>fluctuations</u>, i.e., the LPL trends or the 'mini-Josephs', and the 'design Joseph', or critical period. Since we have no <u>a priori</u> knowledge of the magnitude of the actual future critical Joseph event, Monte-Carlo techniques can be invoked with the infrared models to develop emperical probability distributions of critical drought events which are likely to occur over the project life. Embedding of a GPL trend of some shape - a climatic change - would of course significantly revise the emperical distributions of critical period droughts. At present it seems a moot question whether a 'climatic change' could be specified <u>a priori</u> if we accept the premise that fluctuations cannot. Perhaps there is some compromise relevant here. Court's studies (1972) showed that the best predictor of next year's rainfall or temperature is the median of the preceding 11, 13, or 15 years. Linsley (1972), in commenting on Court's observations, noted that this might be a significant factor to be considered in water resources planning.

Court elucidated (1976) that for any one station the optimum period for prediction the year after next was one year less than for predicting next year. Up to 10 years ahead, the optimum period decreased by one year for each year ahead added to the prediction period. Court speculated that his data displayed an average period of homogeneity of 15 years, and the more of this period that is used in anticipation the less that is available for prediction. Whether this would apply in a given case could be ascertained from a Court-type analysis of the available record; if so, stochastic generation might well be based on a model with an initial 5 to 10 year trend, with 'full uncertainty' taking over thereofter. (Some additional evidence of a 15-year period of climatic homogeniety comes from a recent study by Van Loon and Williams (1976), discussed in a later section.) There is an independent question here of the length of time between the completion of the analysis and the time the proposed project goes 'on line." Due to the political and economic factors involved, this may in fact be far more unpredictable than any climatic transient of the order of 10-15 years.

Let us now consider the infrared noise models in light of this discussion; most of these models are 'quasi-stationary', in that (except for one), they are founded on stationarity of the mean. Most are also basically annual models, since it is the annual time frame that is of importance in reservoir critical period analysis, rather than the monthly or daily time frame. Disaggregation techniques exist for operational consideration of the finer time increments, but these will not be discussed here (Valencia and Shaake, 1973).

The Infrared-Noise Class of Stochastic Models

Fractional Noise - Mandelbrot and Wallis (1968; 1969a, b) pursued Hurst's work at length and developed a modeling scheme that would impart the desired low frequency characteristic implied by $0.5<h<1.0$. This fractional gaussian noise and has the characteristic of generating so-called self-similar processes. These processes maintain a constant value of h throughout a sequence over all sub-partitions of the series of lengh L less than N. Fractional noise was the forerunner of a new infrared-noise class of processes characterized by high power at low frequency and capable of reproducing the Hurst effect - equivalently the Joseph event.

The original fractional noise models (Mandelbrot and Wallis, 1969a, b) presented many practical problems and were costly to use on the computer. Mandelbrot (1971) subsequently developed a fast-fractional noise model which basically consisted of a weighted sum of low frequency and high frequency Markov-Gauss processes which were parameterized by an appropriate value of h. Cost performance was greatly improved. McLeod and Hipel (1978) and Hipel and McLeod (1978a, b) present new algorithms with computer programs that are reportedly improved in efficiency, together with a detailed review of fractional noise.

ARMA (1,1) - The autoregressive-moving average of ARMA Model of order (1,1) (Box and Jenkins, 1970) was proposed and extensively examined by O'Connell (1971-1974) as a simpler substitute for the fractional noise model (as reported in Burges and Littenmaier, 1975).

The ARMA (1,1) model has been extensively tested and compared to the fast fractional gaussian noise model by Burges and Littenmaier (1975), and Littenmaier and Burges (1976). They note that mimicking a fractional gussian noise process is essentially one of trying to fit the parameters of the substitute model to the correlation function of fractional gaussian noise, which is given approximately as: $h(2h-1)\ |k|^{2h-2}$ where k is the lag.

In their studies they used both the log-normal 3-parameter and the Gamma as skew distributions for independent increments. There are advantages and disadvantages to each model and these authors conclude that the final selection should be based on the most important characteristics to be preserved in the generated traces. In some cases they found that ARMA model computer costs were about 1/3 that of fast fractional noise.

In a similar investigation Hipel and Mcleod (1978a, b) and Mcleod and Hipel (1978) found the ARMA Model statistically preferable to fractional noise and concluded that the latter is inherently less flexible than the Box-Jenkins models. They also present computer-coded algorithms for following Box-Jenkins modeling procedures.

Lag-One 3-Parameter Log-Normal - This model was first proposed by Matalas (1967) for synthetic generation of streamflow. It can be used as a monthly or annual model and is satisfactory for many purposes. Its advantage is that there is a considerable amount of experience available on its actual use. Strictly speaking, this is a red-noise model theoretically unable to reproduce the Hurst effect as N, the number of years, approaches infinity. However in finite simulations corresponding to project lives of practical interest it has been found that the Markov model can generate a rich spectrum of alternative futures with h greater than 0.5 (Wallis and O'Connell, 1973; Kilmartin and Burges, 1976).

This model may thus be adequate for simulations over periods within the practical range of interest for project planning, say 30 to 100 years. However, since h cannot be explicitly incorporated as a parameter, the adequacy of the model in a particular case must be judged after the simulation runs have been made and long-term properties of the generated sequences evaluated.

Time Varying Spectrum - Rodriquez-Iturbe, Dawdy and Garcia (1971) proposed that a non-stationary term be added to a stationary process model. This term was of the form $R_t \cos(Wt+A_t)$ where the amplitude R_t and phase A_t vary slowly with time and W is frequency. These parameters can be estimated using demodulation techniques for non-stationary series - basically a smoothing operation - as described by Granger and Hatanaka (1964). The Hurst coefficient is again not explicity modeled and more experience with selecting filters is still needed with this technique. However, the slowly-varying series that is fitted to the data would appear to have useful applications, particularly in areas where it is well established that rainfall (or streamflow) has an aperiodic tendency. Recent investigations using this technique are given in Anderson (1975) and in Andel and Balek (1977), and demodulation techniques are extensively discussed in Bloomfield (1976).

Broken-Line - This process was devised by Ditlevsen (1971) and explored in detail by Rodriquez-Iturbe, Mejia and Dawdy (1972), Mejia, Rodriquez-Iturbe and Dawdy (1972) and Mejia, Dawdy, and Nordin (1974). The model is of particular interest because of its relationship with the linear trend lines discussed earlier. A simple broken line process results from spanning a trend-line between time t_1 and t_2, with the ordinates at each time t_1 and t_2 determined randomly. The broken line process for a hydrologic series is simply a sum of such trend lines with randomly determined slopes but with different lengths (or 'effective' time periods). The short lengths are related to higher frequency effects and the longer lengths to low-frequency effects, i.e., long-term trends.

The Hurst effect can be explicitly modeled using parameter estimating techniques for mimicking fractional gaussian noise, much as described for the ARMA model above.

Shifting Mean - These models have been explored by Klemes (1974) and Potter (1976a, 1976b). Basically in this model the mean of a white noise process is shifted by Monte-Carlo determined increments for intervals that are specified by a selected probability distribution. This concept bears some relation to the concept of Almost-Intransitivity which Lorenz (1969) has advanced as an explanation of non-linear behavior in atmospheric processes. An almost-intransitive system will exist for some time at one level until by chance it acquires a state that permits a jump to another level. Operational details are given in Potter (1976b). Klemes and Potter both feel that the shifting mean is a more appropriate explanation of the Hurst effect; they would not invoke a requirement of self-similarity for the mini-Josephs on the various mean levels over the project length.

Discussion - The number of infrared models now available is partly a response to the cost of the original fractional noise models and partly to non-acceptance cf one or another tenet of that model. Much of this argument centers around whether to be acceptable a mathematical process must exhibit Hurst characteristics at all time scales or only over time scales the order of the length of the project. Some of the simulation experiments that have been done have examined behavior of various model over very long sequences, ranging from 1000 up to 10,000-20,000 years. In these simulations stationarity of the mean was assumed and the question as to the reality of this assumption has been raised (Klemes (1974), Anis and Lloyd (1975), and O'Connell (1976).

Since the planning period we are primarily interested in is only of 50 or 100 years duration, this order is considered the maximum simulation length of practical interest; invocation of longer periods for simulation would have no operational value, although it may be of some value in Monte-Carlo examination of the fundamental mathematical characteristics of the model. This is not an argument for or against a particular process but only a footnote that climatology obviously could make a vital contribution to the further development of the infrared class of models.

Closure - In Section 5 the discussion of hydrologic models will be continued, with emphasis on transfer functions for relating runoff to precipitation and precipitation to other climatic elements. First it is seems appropriate to take a closer look at 'Joseph events' and trends as manifested in the atmosphere, and some techniques that might be used to investigate them.

4. FAT YEARS, LEAN YEARS, AND THE GENERAL CIRCULATION

Prelude - Whereas the previous section was drawn mainly from the hydrological literature, this section is skimmed mainly from the climatologic literature. Lamb's encylopedic books (1972, 1978) are primary references on historical aspects, and excellent books by Barry and Perry (1973) and Stringer (1972) deal with atmospheric processes from the climatological viewpoint.

Despite the recent vintage of these primary references, a gread deal of additional work has been done is the past 6 years, particularly in reconstruction and analysis of historic climatological series for various regions.

This activity has been reported in numerous journals of various desciplines. We shall sample some important results here, but this should by no means be construed as a detailed survey of the work of this interlude. Barry (1978) gives an excellent summary of advances over the period 1960-1975.

We know that 'Joseph events' in streamflow must basically stem from 'Joseph events' in precipitation (leaving aside the works of man or major ecological catastrophies). These must in turn be triggered and maintained by severe, persistent anomalous behavior in regional, if not global, atmospheric circulations. Because of this, fields of pressure, temperature, moisture, wind, cloud cover, net radiation and pan evaporation or (derived) potnetial evapotranspiration are all likely to be affected in magnitude, and wind may also be afftected in direction. The data available for these fields are generally not as extensive or intensive as precipitation data, however, since many of these elements are conservative over great distances, fewer stations is not necessarily a major handicap. The major problems with these various data sets are: (1) that most of them do not extend back very far, and (2) that in many instances methods of recording or instrumentation have changed during the record period causing non-homogeneous jumps, which could be confused as climate fluctuations or changes.

Pressure and temperature data have the longest recorded series available. Of these, pressure is currently thought to have the greatest potential for use in back-extension of precipitation records. Speculations are offered below on what would appear to be desirable approaches for using these data series, which as noted earlier cover much of the world back to 1850.

This is nothing like an extension back over the last millennium, however, it would be a very useful extension indeed. Apart from the normal advantages accruing from a longer sample, the period back to 1850 includes the rising temperature trend of the Northern Hemisphere to the optimum of the 1940's and the decline since. This trend apparently began in the early 1800's from lows that were only slightly higher than the lowest of the past millennium - i.e., the little ice age of 1600-1750 ±.

It may thus be possible to determine whether trend effects on both sides of the 1940's optimum have also influenced a particular data set of interest.

In the Southern Hemisphere there is much less data available. However, based on studies in New Zealand (Salinger and Gunn, 1975) there has been an apparent opposite temperature trend over much of this same period, with a temperature minimum broadly between 1900 and 1940. Damon and Kunen (1976) have extended this study to include African, South American, and Antarctic stations for the period 1943-1974. They find no overall consistent trends between 0° and 45°S; instead they find opposing trends at different longitudes, e.g., a New Zealand warming, but a parallel cooling in subequatorial Africa. Between 45°S and 90°S they note a significant warming trend in most locations over the late 60's and early 70's. These scientists speculate whether the Southern Hemisphere warming may be due to increased global levels of CO_2 that are not reflected in the Northern Hemisphere data because of an overwhelming effect of particulate pollutants and greater volcanism.

Perhaps of equal or even greater significance than the global pressure data sets are dendroclimatic data that may potentially be available around the world but which have not yet been sampled, analyzed and cataloged.

Over the past few decades scientists at the Tree-Ring Research Laboratory of the University of Arizona and elsewhere (e.g., Fritts, et. al, 1971; Blasing and Fritts, 1976; Stockton, 1975; Fritts, 1976) have painstakingly nutured dendroclimatology from an interesting curiosity to an exacting science. Indeed the tree-ring chronologies developed for the Western U.S. can no longer be ignored in water resources planning in that region. Dendroclimatology, discussed again in Section 7 below, would appear to have a key role to plan in unlocking the past in the 30°N-30°S belt.

Atmospheric Joseph Episodes - Mandelbrot and Wallis (1968) defined their 'Joseph effect' as an amorphism for the fact that a long period of high or low precipitation can be very long indeed. Another characteristic is that from the climatological point of view, Joseph events tend to cover broad areas, since they are fundamentally caused by a weakening or strengthening of the major continental and oceanic centers of atmospheric action. The intensity of anticyclones which regulate the long Joseph wet and dry spells can be fundamentally measured by atmospheric pressure. The areas affected can be enormous, corresponding to the size of the dominant anticlyconic cells. Depending on the relative amount of zonal and meridional circulation, there may be alternating areas of drought and excess precipitation along any given latidunal belt. The subtropical highs interact with the circumpolar vortex, a wave-like arrangement of the isobars at higher latitudes.

In an analysis of the recent Sahel drought, Winstanley (1973) maintained that the strength of the vortex influences rainfall from Africa through the Indian subcontinent. Under conditions of strong zonal flow, monsoon rains extend well to the north in Africa and in India, whereas under a weak zonal flow regime (strong meridional flow regime) the northward penetration of the monsoon is suppressed (it is noted that Barry (1977, 1978) urges caution in accepting the vortex expansion concept entirely, and he cites several conflicting studies with respect to the cause of this drought).

Genesis does not tell us much concerning the circulation that accompanied the first (documented) 'Joseph event'. Reconstructions by Lamb (1972) for 4000 B.C. indicate that that was another time of a climatic optimum with temperatures and rainfall in England the highest in the past 8500 years. Even so, it would appear that the giant anticyclones regulating weather in the Nile watershed behaved then much as they do now. Based on our knowledge of the general circulation, we can speculate that weakening of the Azores high, perhaps accompanied by a strengthening of the South Indian Ocean high, resulted in unusually heavy precipitation in the Nile headwaters during the seven fat years, due to abnormal and presistent northward penetration of the intertropical convergence zone. Conversely, the seven lean years may have been caused by a reversal in the relative intensities of these two anticyclones, causing the monsoons to remain farther south than normal.

Indeed, in another recent study of the West African Sahel, Bryson (1974) also discussed the importance of the latitude of the Azores subtropical high pressure cell in determining Sahelian rainfall.

Citing Ilesanmi (1971), he notes that a southward change in latitude of the cell by 0.2 degrees causes a movement of the intertropical discontinuity by 0.6 degrees and shows that such a change could reduce annual rainfall by over 100 mm. Similar mechanisms apparently prevailed in the Nile Valley and headwaters at the time of the first recorded 'Joseph event'. The Azores high in fact regulates persistent wet and dry periods from Northern South America through the Mediterranean and eastward to the Tigris-Euphrates Valley and beyond.

Some Regional Pressure Relationships - Here we examine several important regional and even global studies on atmospheric pressure relationships, including interrelationships with temperature and precipitation. These results were selected to demonstrate the possible application of principles and techniques to 'Joseph event' analysis, from a statistical spatial-analysis point of view. They also touch upon or deal directly with historic regional secular trends and periodic phenomena.

Northern Hemisphere - Wagner (1971) performed a seasonal spectral analysis of sea level pressure for 226 grid points in the Northern Hemisphere spanning the period 1899-1964. Winter, Spring, Summer and Fall analyses were carried out and variation in spectral coefficients plotted. High spectral power was found in large but widely separated arid areas throughout the world in the spectral bank centered at 21 years.

In the Pacific, high spectral power was found in the bands centered at 5-6 years. In addition, Wagner observed that most of the data exhibited strong trends towards lower pressure over this period. Wagner also computed and mapped global variations in lag-one and lag-two autocorrelation coefficients for each season (e.g., autocorrelations between winters of subsequent years).

From 10° to 30° north these autocorrelations are highly significant throughout most of the world. For all seasons these characteristics are attributed by Wagner to oceanic temperature anomalies wich may act in turn as forcing functions, causing the atmospheric circulation to depart from normal in the same general sense for a number of years. The low frequency spectral bands that Wagner found which had power are related to the 'Joseph events' for water resources design. The long-term trends found by Wagner are related to climatic trends that should be considered in evaluating the historic samples and the planning period ahead.

Van Loon and Williams (1976) presented an analysis of Northern Hemisphere mean winter temperatures, pressures, geostrophic wind and meridional transport of sensible heat. Their data consisted of winter temperature (December, January, February) data for the years 1900-1972, at stations given in World Weather Records, and sea level pressure data resolved at 5° grid intersections for 20°N to 85°N (they note in passing a number of inhomogeneities in these data due to several shifts in data processing techniques during the period, but they note where results are affected). Their results are maps of 15-year linear trend coefficients for the periods 1899-14, --------, 1955-1969, and the two periods of 1900-41 and 1942-72. Trends are shown to be of the same secular duration and magnitude over vast areas. They conclude that the sign of a hemispheric trend depends largely on the sign of the trend in the most northerly latitudes. For 'Joseph event' quantification these results again indicate that space fields may be of significant value in estimating historic trends at specific sites of interest.

They found that when the trend in temperature was cooling the trend in pressure was rising. They also conducted a zonal harmonic wave analysis of the pressure trends at 55°N and found that waves No. 1 to 4 accounted for 80 percent of the variance in pressure at that latitude. Within each 15 year period however the percentage of the total variance accounted for by any one wave varied considerably. They found a considerable stability in wave phase ralative to Greenwich, varying for wave No. 3 only 2° of longitude over 24 years while the amplitude varied from 7 mb to 1.5 mb. A zonal harmonic wave analysis of this type would also have potential for assisting in the evaluation of historic regional climatic trends.

The Southern Oscillation - In the Southern Hemisphere a pressure phenomena occurs called the Southern Oscillatiuon (SO). This has been represented as an aperiodic surge of air mass between the Indian Ocean space and the South Pacific space. The phenomena is indicated by long-term multi-year surges in pressure at Darwin, Australia, and Jakarta, Indonesia, and opposite surges at Santiago, Chile and Easter Island in the Southeast Pacific. Rainfall, drought and pressure in these two widely separated areas are closely related, as are conditions in the intervening oceans and islands. Dry 'Joseph events' in Indonesia for example are associated with wet 'Joseph events' at the opposite side of the Pacific. The dry Josephs in Indonesia are accompanied by high pressures at Darwin-Jakarta and low pressures at Easter Island. There are similar relations with the Indian Ocean area and the lands surrounding it.

In the Southern Hemisphere, excepting the Southeast Pacific high, subtropical anticyclones tend to migrate, since there is very little land mass to constrain them south of 30°S. The SO seems to be related to the land-water surface ratio, which apparently is also why it is not as marked in the Northern Hemisphere. However, Berlage (1961) presented world wide maps inter-relating global correlation with 12-month moving averages at Jakarta (Figure 2). This map for the 10-year period 1931-39 shows a correlation of -0.9 with the Southeast Pacific and +0.7 with the Greenland ice cap area, West African Sahel region, and nearly all the lands bordering the Indian Ocean north of the equator. An interesting side-aspect of this map is the areal mapping of the line of zero correlation which passes between zones of positive and negative correlation. A zero correlation in a space field as well as small correlations on both sides have significance not considered in standard statistical tests (Cliff, et. al, 1975; this raises an important question as to the validity of some of the more common testing procedures when applied to space-fields). Such patterns in the pressure fields lend hope to the possibility of using space fields to assist in regional filling and extension when standard statistical inference procedures might cause us to reject relationships. The SO has been recently studied by Kidson (1975a, 1975b) using eigenvector techniques and by Trenberth (1975, 1976) who presents power spectra and cross spectra for Darwin and the Pacific Island of Apia, Tahiti and Easter Island, and also cross-spectra between these stations and the time series of coefficients P1 and P2, of the first and second eigenvectors.

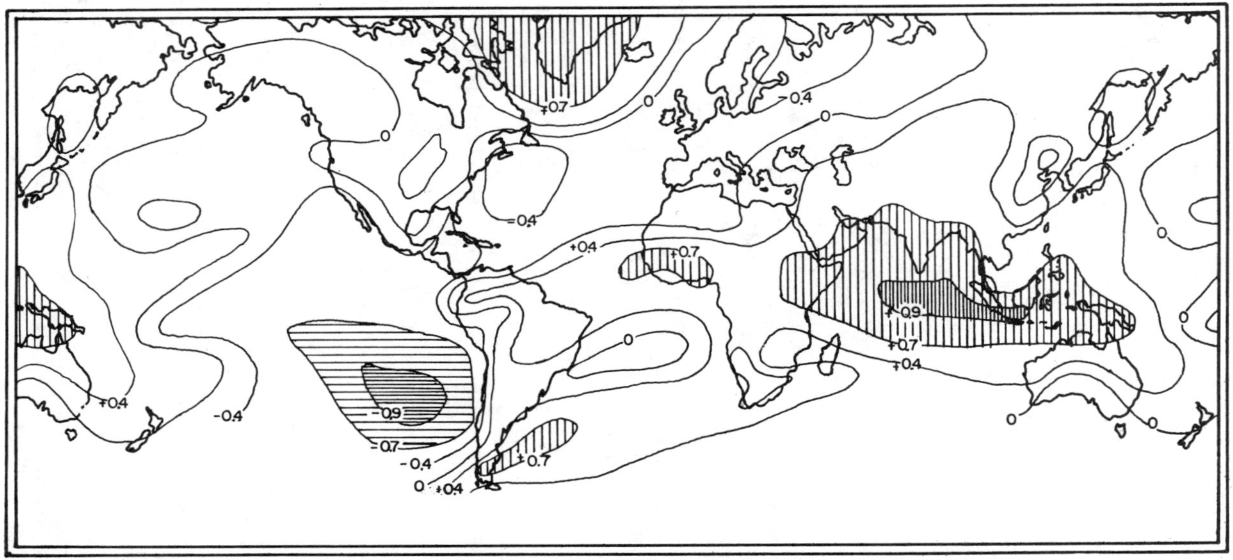

Fig. 2 Iso-correlates of twelve-monthly air-pressure anomalies with respect to Djakarta, January 1, 1931 to January 1, 1939, (After Berlage, 1961).

Although oriented towards the short-term forecasting problem, Trenberth's findings are significant for 'Joseph event' quantification in that region. He found that P1 and P2 accounted for 39 and 24 percent of the total variance in the monthly sea level pressure anomaly fields. He also found that much of this variance was contained in periods of more than one year.

There are important oceanic-atmospheric feedback mechanisms (Berlage, 1966; Bjerknes, 1969) of the Southern Oscillation; while not yet well-understood (Ramage, 1975), the enormity and continuity of the Indo-Pacific Oceanic Basin and its heat storage capacity insures that it plays a key role in regulating long-term climatic fluctuations in large parts of surrounding continents, if not globally (Berlage, 1966).

Wyrtki (1973) shows a strong relation between 12-month running means of sea surface temperatures (SST) off the Central American coast and sea level difference across the Equatorial countercurrent, as measured at several Western Central Pacific Islands, obtaining a correlation of 0.79. The Mid-Pacific Sea level differences are an indicator of the magnitude of the countercurrent, which averages from 40 megatons/second in the Western Pacific to 10 megatons/second in the Eastern Pacific. Wyrtki shows that periods of exceptionally high transport in the Western Pacific are accompanied by warm El Nino currents several thousand miles and about 3 months lag downstreams. Namias (1973) in turn obtained a correlation of .48 between Wyrtki's index of the countercurrent and the strength of the 700-MB westerlies between 20 and 35°N, with somewhat higher values obtained using lags up to 5 months. He concludes that when the upper subtropical westerlies are strong the trade winds are weak, inducing a strong countercurrent and warm water anomalies 8 months later off Central America; conversely, weaker subtropical upper westerlies are associated with a weak countercurrent and colder waters off Central America. The warm waters are in turn associated with wet periods and the cool waters with periods of drought.

The El Nino phenomena has recently been extensively summarized by Wyrtki (1975). The SO has similarly been scrutinized in detail by Julian and Chervin (1978), who used a global climatic model (discussed further below); their results suggested that the general circulation is more sensitive to changes in tropical SST than to changes in mid-latitude SST.

It seems plausible that the El Nino-Southern Oscillation phenomena may be at the root of many if not most Joseph events, around the world.

Pacific Space - Predictability of the atmosphere is intrinsically limited to about two weeks (Lorenz, 1975). Thus multi-year high-pressure anomalies causing extended severe drought must in some way be the result of an interaction with forcing systems external to the atmosphere, such as for example, variations in solar irradiance, vulcanism, snow-ice fields, ocean temperature and a number of other possible factors. Global 12-month running mean pressure correlations as found by Berlage (1961) could not otherwise occur. None of these external mechanisms or their interactions are well understood. However, in recent years oceanic aspects have received a great deal of attention and some important inter-relationships have been partly unveiled, as noted above for the SO.

Persistence in trough patterns are often not traceable on daily weather maps, but exhibit long-term persistence on mean monthly maps, where their global progress can be mapped over periods up to 12-18 months. Namias and his associates at Scripps have studied in detail post-war Pacific sea surface temperature data (from over 10 million ship reports) and inter-relationships with lower trophosphere pressure and temperature anomalies (Namias, 1971). While we cannot review this vast work in depth here, we note some of their findings that seem relevant to Pacific-space droughts.

Monthly SST data display a non-Markovian memory, with correlation die-away much less rapid than with atmospheric data. Cold-season to cold-season correlations can be quite high, with low correlation during intervening warm seasons, due to warm season covering of cooler waters with a shallow layer of unrepresentative water, followed then by a stirring of the deeper waters again in the ensuing stormy season.

Depending on the sign of the SST departure from normal, evaporation and subsequent precipitation downstream will generally be greater or less than normal; however, the relation is highly non-linear. In studying relations between patterns of SST anomalies and overlying 1000-700 mb thicknesses, little or no relation was found at low SST anomalies. But at abnormally high SST departures, high correlations were obtained with the thickness of the overlying air mass. Thus multi-year persistence of very cold temperatures in the North Pacific could induce abnormal cold-season air mass stability for several years in a row, which in turn could induce a multi-year drought in areas affected by the stable air mass. (In the precipitation or the runoff phases of the cycle this type of behavior could be elicited from several of the infrared class of statistical models). Since SST anomalies are carried along by the currents around the Pacific gyres, their continued inducement of drought over several continuous years would seem to require not only abnormally high SST anomalies but anomalies covering an unusually large area; this is borne out by a number of recent studies.

Weare, Navato and Newell (1976) studied seasonal and non-seasonal emperical orthogonal functions (EOF's) of Pacific sea surface temperature based on marine observations averaged over areas of size 10° longitude by 5° latitude. They found that the second seasonal component -S2- (which explained 4.6 percent of the variance) and the first non-seasonal component -NS1- (which explained 23.1 percent of the variance), were excellent indexes of both El Nino and Darwin sea level pressure. They also noted agreement between NS1 and EOF's of temperature, pressure and precipitaion by Kidson (1975), for the period 1951-60. The hemispherical coherence displayed by NS1 moreover indicates an association between El Nino off Peru and SST throughout the Pacific; for example, they show a positive coherence betwwn El Nino and warm water off of California.

This investigation pulled together many threads linking aperiodic phenomena in SST throughout the Pacific and clearly demonstrated the atmospheric linkage. While much remains to be learned, at this point we can hazard with low probability of error the statement that long, intense, large-scale droughts and wet spells in the Pacific space and bordering regions are coherent with sea surface temperature, salinity of the upper ocean, and with each other; moreover, since the Pacific space accounts for half the globe, coherence in low frequency phenomena can be expected from physical considerations alone. Berlage (1966) identified an even more apt division of the world into a land hemisphere and a water hemisphere (Figure 3) with the center of the water hemisphere near New Zealand. In Berlage's analysis of the Southern Oscillation, he showed that "...natural fluctuations of the general atmospheric and

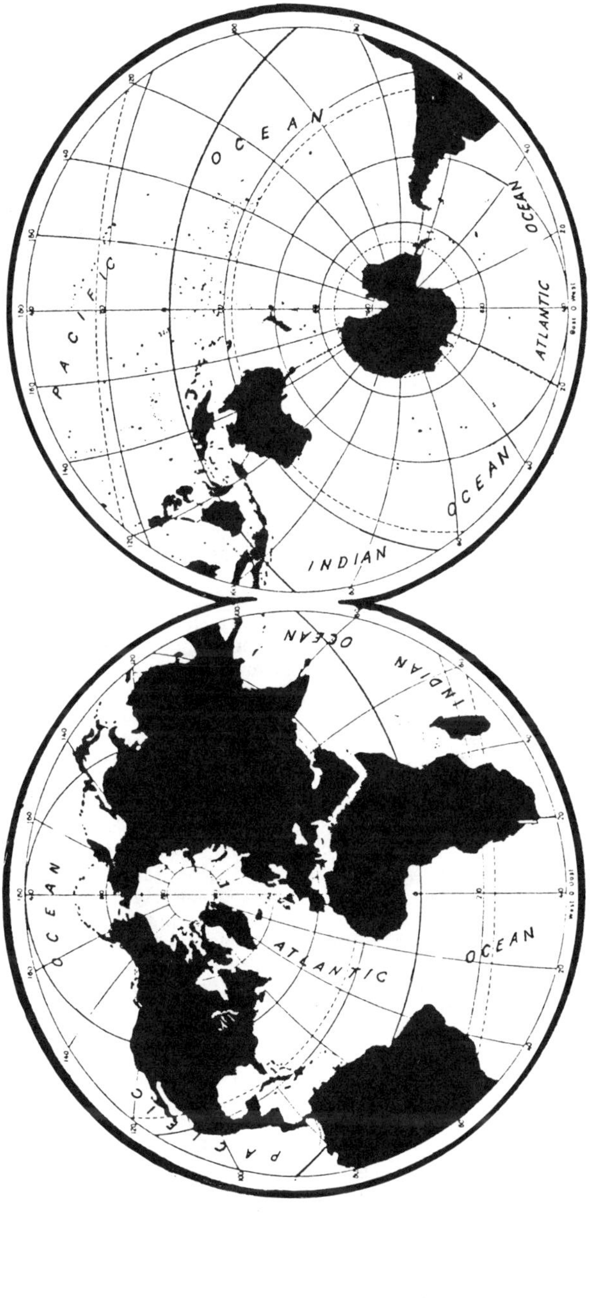

Fig. 3 Land and water hemispheres. (after Berlage, 1966)

hydrospheric circulation over and through the water hemisphere... act as a dominant feature of world weather".

The study of principal components of the Pacific trade winds by Barnett (1977) lucidly demostrates the hemispherical-interralations between the two trade wind fields in a time and space domain. Moreover, his analysis of the principal components of the anomalies (i.e., the residuals after subtracting the matrix of long-term means at each grid intersection) and his interpretation of them in a time and space domain has special Joseph-event significance.

Barnett interpreted the variation in the first principal component of the zonal wind anomalies as representing changes in areas covered by high wind values rather than a change in the values themselves. He concluded that large scale fluctuations occur in both the strength and area of the entire trade-wind field.

With this in mind a routine correlation analysis among a collection of fixed stations in the trade winds may indicate that a good deal of statistial noise exists when in fact the correlation may be very high. The problem here is that we are actually dealing with the largest space and time scales of turbulence. At this level we have very high Lagrangian correlations (moving with the system) but low Eulerian (stationary) correlations. The ease of the latter type of analysis makes it convenient and indeed statistically 'justifiable' to mechanically conclude that no long-term semi-global or global inter-relationships exist, when in fact they do. Thus Berlage's analysis of global correlation between 12-month moving averages of pressure cannot be dismissed in regions where the indicated correlation is near zero. The trace of this isopleth is simply a mean Eularian location of the string of points where zero correlation occurs; when a large drought-causing anticyclone dominates a region the mean Eulerian correlations will be meaningless.

This noise effect has led to a considerable underestimation until now of inter-relationships in global Joseph event probabilities. When it is understood that these events are globally inter-related we can begin to appreciate that there may indeed be a natural if aperiodic 'rythm' in global economic extremes (good and bad) induced by climatological extremes hitting many places simultaneously and perhaps is different ways. 1972 was a well-documented year of this type. The Australian-Indonesian drought was the longest and most intense on record. The Sahelian drought was at its peak. The El Nino severely cut back the Peruvian fishery, giving rise to a shortage of fish meal that later led to increased soybean plantings in the U.S. The El Nino anomaly also gave rise to unstable air masses which moved northward into the Caribbean where they became partly responsible for Hurricane Agnes, which then ravaged the Eastern U.S.; at the same time severe drought affected Venezuela, Western Africa and India. Kanamitsu and Krishnamurti (1978) contrasted the 1972 200-mb northern-hemisphere air flow with 1967, a more or less normal year, and showed the global interconnections at the 200-mb level.

1976-77 was another period of intense near-global anomalies, some of which seemed to have persisted into the winters of 1977-78 and 1978-79. Kirwan, et al (1978), report on 1976/1977 North Pacific SST as measured on 16 drogues set out over a grid 35°N to 45°N and 154°W to 160°W and note the existence of an extremely severe cold temperature anomaly which persisted after the drogues left the area, as confirmed by ship observations. They state this anomaly was the strongest ever reported in this area, an area whose SST anomalies according to Namias (1969) can be related to downwind temperatures and precipitation patterns in North America. Namias (1978a, 1978b), analyzed the near-global northern hemisphere anomalies that occurred during this same period and showed that very strong air-sea interactions occurred both in the Atlantic and Pacific; he also noted that the cold waters which persisted in the Central North Pacific were associated with a strong Aleutian low and a persistent El Nino along the equator.

Central Atlantic-Carribean - A final study we shall discuss here is by Hastenrath (1976), which appears to have special significance for 'Joseph event' hydroclimatological analysis. Hastenrath studied pressure, wind and sea surface temperature for 10 extreme dry years and 10 extreme wet years in the Carribean area by mapping departures from normals over the area 25°N-25°S 10°E to 100°W.

He found very strong regional patterns that might enable a forecast of a wet or dry rainy season the preceding winter. The area included extends from the Eastern Pacific to the African coast and reflects the action of the Azores high, the South Atlantic high and the Southeast Pacific high. While Hastenrath's study was limited to averages of single year extreme events, the methodology is intriguing for possible analyses of individual multi-year Joseph events. Hastenrath also made regional correlation studies with precipitation in various parts of South America, The Central U.S. and Africa, and found significant spatial relationships, similar to the global relations observed for pressure by Berlage (1961). For example, the correlation with precipitation in the Central U.S. is -0.32; with Equador/Peru sea surface temperatures -0/43; and with Chad precipitation +0.43.

Summation - The investigations noted are a small but important sample from the climatological literature. Most do not explicitly study multi-year persistence, except as implied in some of the spectral and trend results. However, they have direct value and if redirected towards the exploration of 'Josephivity' they would appear to have considerable potential for enhancement of our knowledge of atmospheric phenomena triggering of these events. As they stand these studies and many like them in various meteorological and geographical journals are extremely useful for examining historic trends of hydroclimatic elements in particular regions of interest.

Climatic Modelling - Schneider and Dickenson (1974) (herinafter S-D), discuss various categories of climatic models and the state-of-the-art at length. Computerized climatic models have been developed at all levels of complexity, ranging from simplified one-dimensional to three-dimensional models with from 1 to 8 horizontal atmospheric layers, and incorporating various levels of oceanic, terrestial, and cyrospheric detail. The latter are referred to as Global Climatic Models (GCM's) and have resolutions of typically a few degrees of latitude and longitude horizontally and several kilometers vertically. They are time-dependent models based on momentum and thermodynamic relations. They describe the detailed evaluation of the state of the atmosphere at selected time scales. S-D discuss a number of problems that have been encountered with time and space resolution of various meteorological fields.

The more simple models are based on energy balance relationships for one-dimensional equilibrium studies. These have been extended to include time dependency and simplified 2-dimension representations of the earth's surface. S-D describe increased understanding of atmospheric phenomena that has already been gained with these relatively simple models.

Zonally-symmetric dynamic models are of far greater complexity than the energy models. S-D summarize the equations for momentum, continuity, hydrostatic and thermodynamic balance used in this class of model, and discuss them at length. This class of model would appear to be of particular potential value for future 'Joseph event' investigations, since they explicity deal with the giant eddy systems. We have seen that Joseph events are related to persistent anomalous behavior of these systems. We cannot take the space to go into detail on these models here; the reader is referred to S-D for a very complete summary and bibliography. However, it seems worth noting some of the remarks S-D made in contemplation of the future of climatic modelling, at least as it might pertain to hydroclimatology. The GCM's are and will remain very costly to operate even on future generations of computers. S-D see their principal value as a surrogate laboratory for the atmosphere itself and as a means for testing the assumptions of and improving the simpler models. For the simpler classes of models S-D see a need for an almost continuous hierarchy of models of increasing physical and mathematical complexity, and a continual interplay between: (1) the observed statistics of the real atmosphere on many time scales, (2) the derived statistics of GCM's, and (3) the results of statistical dynamic models at climatic and others of lesser complexity. Apparently some of the simpler statistical-dynamical models could be of utility for evaluation of Joseph events and trends in the not-too-distant future.

5. BACK-EXTENSION OF THE RUNOFF RECORD

Prelude - This section discusses the problems of putting together relationships between basin runoff and precipitation, with stress on investigation and modelling of LPL and GPL trends and low-frequency effects associated with 'Joseph events'.

The Historic Data Samples - The historic streamflow and precipitation data samples figure into the problems of hydroclimatology in several ways. They are the sources for statistical parameters; they are the sources of information on historic trends, and the best relationships found between them will be used to extend the streamflow data for the full length of the precipitation series.

Both raw data series must be laundered with various techniques such as outlier detection and modulation, detrending, filling-in and extending. The data sets finally used for stochastic modelling may thus appear somewhat different from the unlaundered raw data. Since the estimated economic viability of an entire project plan will rest on this foundation, the importance of a reliable set of data at this point cannot be over-emphasized. It is possible to be partially objective on deciding when this point has been reached but the final judgement will likely still be subjective. The various homogenizing techniques are generally standard. Trend analysis, spectral analysis, and Hurst range analysis are particularly important techniques for detecting the low frequency characteristics that are of prime interest. We shall skip the details of these techniques here (Kilmartin and Burges, 1976) and center attention on basin processes, transfer functions, Type II errors, and recent techniques gleaned from the literature which appear promising.

Basin Processes - Basins as hydrologic units vary greatly in runoff response to precipitation, depending on vegetation, topography, nature and depth of soil mantle, geology, whether the channel recharges a groundwater aquifer or is fed by one, and other factors (Linsley, Kohler and Paulhus, 1958). Response also depends on antecedent precipitation, or how wet the basin is prior to a storm, whether temperatures are above or below freezing, and ice and snow conditions in the basin. Responses can be catagorized in the time domain as short-term for periods of a few hours to a few days, and as responses for longer periods of time of months up to a year, over which the basin is considered as an integrator of the shorter term processes. Basins with significant groundwater components will themselves exhibit reservoir storage characteristics with carryover affects of one or more years in the streamflow.

The precipitation events - the storms themselves - can be frontal systems, thunderstorms, or various tropical systems depending on the basin's location. The size of the various storm systems may completely envelope a particular basin or only affect a small part of it. The local intensity at any point in the basin during a particular event will vary by an order-of-magnitude with simultaneous intensities elsewhere in the storm.

It is remarkable that despite these complexities it is frequently possible after several trials to establish meaningful statistical relations (generally, transfer functions) between the precipitation records at a few 8" cans located somewhere in or near the basin and the runoff at the gage. The bigger the basin of course the more 'cans' are required to maintain the same quality of relationship; otherwise blips in the streamflow record may not be accompanied by blips in the precipitation record. Also, there is an optimum interval of data summation for determining the best relationships with the time series that are available.

The time scale problem for hydrologic models is somewhat similar to the time scale problem for climatic models. For modelling a set of short multi-year streamflow trends (i.e., LPL trends) which might characterize a 'Joseph event', monthly creven annual resolution will usually be sufficient, and for a GPL trend in streamflow annual resolution is usually sufficient. These remarks apply to statistical models of the historic data, as well as to stochastic models to be parameterized from the historic data.

For discharge record extension and fill-in using precipitation data, it may be possible to model on a monthly time frame, depending on the length of the concurrent series, the model used, and the statistical reliability of the derived model parameters. But if the concurrent record is short - too short for reliable statistical estimation of parameters, a shorter time frame may be necessary, perhaps down to days, in order to properly model the effect of higher-frequency processes on the lower frequency basin processes.

Transfer Functions - Most of the work done with the infrared-noise models has been with streamflow data. Multivariate versions for some of the infrared noise processes exist and are discussed in the reference previously cited, however these versions were generally formulated with a view towards maintaining appropriate covariance relationships among regional streams. Use of the infrared noise models for precipitation and other atmospheric processes would appear to be feasible, but there is not much in the literature on such applications, at least at the scale of interest here, and this author has found nothing at all in the literature on transfer functions specifically built around any of the infrared noise processes.

In what follows we note for reference a few of the better known transfer function techniques in hydrology. Most of these models could also be used for stochastic modelling of streamflow as an output process, if desired, using an appropriate model to generate stochastic rainfall inputs.

The term transfer function is used broadly here to cover a wide assortment of algorithms and mathematical processes that have been developed for transforming precipitation to streamflow. On the one hand are deterministic watershed models such as the Stanford model (Crawford and Linsley, 1966) which models all significant basin processes on a semi-continuous basis (5 to 60 minute intervals depending on the process); on the other hand are statistical-mathematical system models incorporating varying degrees of partial synthesis of some of the physical systems. In between are models of all degrees of complexity. Where the simultaneous precipitation and runoff records are very short but include at least 3-5 years of streamflow, the deterministic watershed model has found favor. Where there is a sequence of 15-20 years of streamflow and precipitation data it is likely that one of the statistical-mathematical models will be sufficient. Even for the short record situations where watershed models seem most appropriate, non-linear mathematical system models with small time increments have been advocated as a less costly substitute. The status of these models is detailed in Amorocho (1973) and an application is given in Rao and Rao (1974). A few of these are noted, particularly those that have or probably would be useful in modelling the Joseph phenomena.

Matalas (1963) and Yevdjevich (1963) independently developed statistical transfer functions in which annual runoff was modeled as a moving average of effective annual precipitation. Effective precipitation is precipitation less all losses except the outflow; i.e., the volume of outflow equals the volume of effective precipitation in the long run. Both these models were explicitly developed to handle multiyear carryover. Matalas compares second, third and fourth moments of the distributions of runoff and precipitation; the skewness of runoff is dependent on the moving average coefficients, the carry-over period and the skewness of the effective precipitation. Runoff second and third moments are less than those of the precipitation, due to damping, but fourth moments may be higher. The serial correlation coefficients for the runoff are a function only of the effective precipitation moving-average coefficients and the carry-over period, and are independent of the skewness of the precipitation. Yevdjevich shows how the moving average coefficients can be empirically derived from flow recession curves taken from the stream discharge records. Both of these models appear to be useful for extension of annual series of runoff from effective precipitation.

The conventional least-squares regression model has been frequently used as a runoff-rainfall model with various easily implemented transforms (Chow, et al; 1975; Kilmartin and Peterson, 1972). But the regression coefficients are linear and this may mask important non-linear processes that are then attributed to noise. The Box-Jenkins (1970) series of transfer-function models afford a rich variety of capabilities incorporating all the many possible combinations of the ARIMA model. They could be applied on any time scale desired and with a variety of transforms. Kavvas and Delleur (1975) report on detailed investigations of the univariate versions of the Box-Jenkins models, as applied to monthly rainfall. While the Box-Jenkins transfer functions have received considerable use in relating water quality to streamflow, examples relating streamflow to precipitation on the space and time scales of interest here are sparce in the literature. The Box-Jenkins transfer function could apparently include the Matalas-type model as a subset. Clarke (1974) presents an analysis of a 19-KM^2 well-instrumented experimental basin, in which he makes use of Box-Jenkins transfer function techniques for forecasting, and provides an excellent example of the mathematical and numerical details of model-fitting as recommended by Box-Jenkins.

The State-Space Modeling Approach also appears to have potential for relating runoff to effective precipitation (Chow, et al, 1975, Szollosi-Nagy 1976; Todini and Bouillot, 1976). This might be a particularly useful approach with a fairly large multi-site region characterized by a vector of input stations (pressure for example) and another vector of output stations.

Another model that might be used for streamflow extension is the spectral regression model presented by Hamon and Hannon (1963). In the multivariate context the regression coefficients would be determined as the ratio of the cospectrum to the cross-power spectrum and each would essentially impart to the runoff the appropriate contribution from its frequency band. Hamon and Hannon consider the methods robust, depending only on the data, and easily capable of handling non-stationarities. The procedure has appeal for possibly capturing the low frequency relations that are 'Joseph event' related.

Periodicities, Joseph Events, Scoffers, and Type II Error - As we have seen the 'Joseph events' we deal with in water resources planning can be characterized as being elements of processes having high spectral power at low frequencies, or finite correlations at long lags. Due to the shortness of records and the development of rigorous testing procedures for the significance of periodograms and correlation coefficients, many early claims of periodicities and long-term cycles were discounted. Most of the mathematically derived testing procedures are oriented towards the null hypothesis, of not making a "Type I" error, i.e., erroneously asserting a trend or periodicity when there is none. Due to the mathematical difficulty of constructing Type II error testing procedures, proponents of low-frequency effects, such as low-frequency cyclic phenomena, have been more or less on the defensive for many years; as characterized by Dressler (1973), they have been subject to a great deal of scoffing down through the years, beginning with no less a scoffer than Lord Kelvin himself.

Today it is possible to overcome the Type II error problem with Monte Carlo techniques, as demonstrated by Wallis and Matalas (1971). They found the Type II errors of the well known Anderson Test to be indeed significant. Under this test for example with a sample size of 25 and a Type I error parameter of 0.05 one would reject a true correlation coefficient of 0.3 about 78 percent of the time (the Type II error). Another factor has indeed been the size of the small samples that have been available. Since most of the procedures for estimating correlation coefficients are based on asymptotic assumptions, the estimates are biased for small, that is, for nearly all samples. Wallis and O'Connell (1972) present Monte Carlo results indicating that standard estimating procedures consistently underestimate true correlations. For example with a sample of size 30 and a true correlation of 0.8 the average estimate of the correlation was 0.7. Obviously the sample size factor has also played into the hands of the scoffers.

Perhaps even more basic to the Joseph event Type II error matter is that the linear correlation coefficient derives from the expected value of cross products and is estimated from their averaged sum over the applicable collection of sample cross-products. This step averages out correlations (auto and cross) in Joseph years with correlations in non-Joseph years. But as previously noted, investigation of Joseph years alone indicates much stronger than average correlations during these years (e.g., Hastenrath, 1976); one could expect that conversely, in non-Joseph years, correlation should be weaker than average with more of a tendency towards white noise.

Thus the scoffer who lumps all years of a hydroclimatological time series and mechanically declares the process to be white noise with a Type I error probability of 5 percent should examine his single-population assumption more closely. There may be justifiable statistical evidence of at least two populations in such a series and possible three; the calculated average correlation coefficient may thus be merely a computational artifact of the sample instead of a meaningful parameter of the true process or processes. There is now strong evidence moreover that a sound (if not yet well understood) physical basis exists for hypothesizing multiple populations: Joseph events frequently have their roots in anomalies of sea surface temperature and associated interactions with the atmosphere. It is not fundamentally reasonable to expect a region's precipitation volume to maintain the same statistical distributions when moisture-source sea-surface temperatures vary sufficiently to change the size and the intensity of the large-scale centers of action.

The above discussion is not to be construed as a plea to kick out the scoffers and the Type I error criterion, but as a caveat against the ready acceptance it has attained for denying the existence of low-frequency effects. Historically, much of the scoffing has been aimed at the short-term prediction problem, where it may indeed still be well-founded. But for long range planning which must consider the likelihood of Joseph events, it is <u>not</u> conservative for project analyses to go overboard in protecting against a Type I error. After all, for water-supply, a white-noise world would be the rosiest of all!

<u>Summing Up</u> - There are a variety of tools available for back extension of streamflow records using precipitation as a process input. Many of these have received a good deal of developmental effort, although there is little in the literature on the effectiveness of transfer of low frequency effects from the input to the output process. After the parameters are determined for the gross data sets it is usually assumed that any low-frequency effects are fully imparted by the estimated constant coefficients. The Hamon-Hannon model appears to be the only one that could attack this problem explicitly. There is a need for systematic comparisons of these models and examples of applications, stressing low-frequency effects. Additional statistical procedures that

appear to have potential are cited in Section 6 where transfer functions relating precipitation to atmospheric processes are discussed.

6. BACK EXTENSION OF THE PRECIPITATION RECORD

General - Backward extension of the precipitation record using pressure data has apparently not been attempted too often. The author used monthly pressure from three widely separated stations (Darwin, Honolulu, Tainan) in the Indo-Pacific space for a backward and forward extension of rainfall at several Indonesian stations. In that case a regression against 3 and 5 month moving averages was used to obtain the long-term trends, and a noise term was added to restore the local monthly variance.

There is a rich meteorological literature on use of pressure and other fields for operational prediction of rainfall (i.e., for 1 to 10 days ± in advance). Seldom do such studies deal with the time scales of interest here, because of low signal-to-noise ratios encountered. However it would appear that predictor variables found locally successful should at least warrant an analysis for low frequency spectra. This points to a basic dichotomy between the interests of the meteorologist and those of the "hydroclimatologist". The meteorologist is concerned with predictions of events a few hours or a few days ahead. He is attempting to make a point prediction. The hydroclimatologist on the other hand is generally not concerned with the forecasting of a low-frequency event within a project-length time period ahead, but only defining its probability distribution over that period. However if he decides on incorporation of specific trends in his "future" model (including a zero trend) that would in effect be a forecast, unless the trend direction and magnitude are considered as completely governed by chance and explicitly included in the generation scheme, in a decision-analysis context.

It would appear that many of the statistical techniques discussed for runoff-precipitation transfer functions would also be applicable for extending rainfall, in addition to the various techniques discussed in Section 4. Basically however this would be a new endeavor with much to be learned through the empirical approach of data analysis, problem exposition and "cut and try".

Some Recent Developments - In addition to the techniques cited in Sections 4 and 5, some relatively recent developments in statistical analysis also seem worth noting.

In Geophysics a new high-resolution spectral technique has recently been formulated called maximum entropy spectral analysis. This technique was developed by J. P. Burg and has been extensively reviewed by Ulrych and Bishop (1975). They show some very interesting relations between optimum autoregressive model parameters and maximum entropy spectra. While confidence limits have not been developed for this technique its extremely high resolution could probably be accepted in some cases if correlated with other results. Currie (1974) shows a meteorological application indicating temperature peaks at many U.S. stations and also global averages corresponding to the 10.5-year sunspot cycle.

Principal component methods (also called emperical orthogonal functions) were noted earlier. They have recently been applied by Kidson (1975a, b) to analyses of tropical pressure, temperature and rainfall data for the period 1951-1960. Brillinger (1975) discusses techniques for spectral analysis of principal components. Trenberth (1975, 1976) demonstrates these techniques in exploration of the temporal and spatial variations of the Southern Oscillation. Principal component analysis would appear to be of considerable value in studying Joseph events in atmospheric variables, with high spectral power showing up at low frequencies in the various principal component time series. A relatively high "h-regime" (i.e., nearer 1 than 0.5) would apparently be indicated by high spectral power in the low frequencies of the first principal component coefficients. On the other hand if power were mostly in the high frequency range we should expect a low "h-regime".

Brier and Meltesen (1976) pointed out an interesting aspect of principal component analysis that arises from a joint consideration of the eigenvectors of the normal spatial correlation matrix and its transpose. The transpose amounts to a regional mean autocovariance matrix that measures autocorrelation between years over the region. This involves a time-varying regional mean from which deviations are measured, rather than constant station means. The time-varying mean might be viewed as yielding a physically-stationary "Lagrangian" auto-correlation function for the region being analyzed. Also associated with regional mean autocovariance is a regional mean spectrum. These concepts are surficially explored in Kilmartin (1977).

Kavvas and Delleur (1975) report on the analysis of daily rainfall occurrences using statistical techniques developed for analysis of series of events (Cox and Lewis, 1966). They discuss long-term trends and cycles in modelling rainfall occurrences and possible atmospheric mechanisms that could be responsible. Dobson and Stewart (1974) apply these techniques to the study of trends in the occurrence of tropical cyclones. It would appear that these techniques would be indeed useful for detecting trends and aperiodicities in the occurrence of events and deducing explanations of non-stationarities in the total amounts of precipitation.

Finally, the recent study by Tyson, Dyer, and Mametse (1975) on 1880-1972 rainfall in South Africa is cited as an example of an excellent general regional investigation of trends, periodicities, and geographical inter-relationships. This was a univariate study of rainfall but it would be an excellent starting point for a hydroclimatologist to begin in making judgements on local streamflow trends, particularly if supplemented by a Kavvas-Delleur-type study of trends and periodicities in precipitation occurrences.

Summing Up - The atmospheric studies cited in Section 4 and elsewhere above show conclusively that trends and aperiodicities do exist in the historical meteorological data and that many of these can be shown to be interrelated. In particular, such phenomena exist on the time scale of 'Joseph events' and we conclude they must be the basic causal factors of such events in streamflow. In fact, the evidence on this may be sufficient to cause one to question any raw data that do not indicate LPL and GPL trends in conformance with like-duration regional trends in other meteorological fields.

It would appear that development of: (1) Joseph event detection techniques for application in historic atmospheric data analyses, and (2) appropriate transfer functions relating streamflow and precipitation, should both proceed on a number of modelling 'fronts'. Note that Schneider and Dickenson (1974) reach a similar conclusion with respect to climatic modelling. We are not yet at the point where a cookbook approach can be taken to these problems, and we may never get to such a point, but we should strive to improve our understanding of these events and overall planning for them. Under-estimating such effects in the planning of major economic ventures could create socio-economic problems on a broad scale.

7. BACK EXTENSION WITH DENDROCLIMATOLOGY

The remarkable climatic reconstructions achieved by dendroclimatologists were noted earlier. What was not noted is that these scientists have pushed some of the statistical tools required in hydroclimatology to new levels of elegance, particularly multivariate techniques such as multiple regression and principal components analysis (Fritts, et.al, 1971). In this field, transfer functions are used to first model tree-rings as functions of climatic elements and then to model climate based on tree-ring data. Stockton (1975) has specifically developed techniques for back-extrapolation of streamflow (Colorado at Lees Ferry to 1600), and Stockton and Fritts (1973) used tree-ring techniques for extending the record of Lake Athabasca levels back to 1810.

Because of the potential length of the dendrochronologies as compared to the pressure data, there would appear to be an opportunity to extend the global pressure data base back in time using the dendrochronologies much as Fritts, et. al, (1971) have already done for the Western U.S., possibly in conjunction with an adaptation of a statistical-dynamical global spectral model. This seems like a particularly promising approach for data-sparce tropical regions where suitable tree-ring data may be locally difficult to find because of the general lack of dormancy. Invocation of atmospheric dynamics for long-distance extrapolation of tree-ring reconstruction from one continent or region to another would seem worth exploring, particularly in view of global studies of pressure oscillations such as Berlage's (1961, 1966). Moreover in some regions it may be just as good or better to argue directly from tree-rings to precipitation or some other index of drought; Stockton and Meko (1975) for example reconstructed the Palmer Drought Severity Index (Palmer, 1965) for the period 1700-1963 for the central and western U.S. using tree-ring data.

8. A NICHE IN PROJECT PLANNING

Ackermann, et. al. (1978) summarized a recent report of the Office of Science and Technology Policy in which the climate and water supply problem linkage was recognized. However, this report deals more with long-term research recommendations rather than application of the existing data base and tools to present problems. Long-range research of this type is of course needed, but condensing our present collection of hydroclimatological tools and data sets into a form that is useful for project analysis should also rate a high priority. To argue we cannot predict climatic change begs the question for one actively engaged in project design. Even if he ignores the matter he implicitly assumes a zero climatic trend. This assumption would appear naive at least and probably even dangerous. It is one thesis of this paper than we can move ahead now in a significant way with the tools and data base that we now have.

To date hydroclimatology has not been awarded a specific technical or budgetary niche in project planning. Hydrologists have long struggled with the problem of record extension, of course, but with little attempt to relate basin Joseph events to global atmospheric telleconnections. Where such relations exist the potential for record extension and detection of infrared signals can be greatly expanded to include the global pressure data base and potentially a global-scale dendrochronological data base. The former exists in a form at the present time but the latter does not. A joint international effort is needed in order to develop and make available appropriate data sets for project planning.

Although pressure data back to 1850 exist for a network that includes most of the inhabited parts of the globe, there are significant inhomogeneities in the historic data sets, as noted by Van Loon and Williams (1976) (these data are available on tape from the National Weather Records Center). This is a particularly serious situation that needs attention; chasing down the inhomogeneities in some of the older records in the tropical countries is time consuming but particularly important to be sure that we are drawing the right conclusions based on the data. The older and the more isolated of these sites contain the most valuable information since they figure so importantly in the early grids, however they are also most likely to be inhomogeneous.

For example, pressure data at Bombay and Madras, India exhibit sharply declining trends for over 100 years. Such trends are undoubtedly nonsense and are probably due to changes in instrumentation, sites, or observation procedure; however, this information is nowhere cataloged for use by a busy project analyst.

It would be highly desirable for a catalog of homogenized monthly station pressure histories to be developed (in the 30°N - 30°S belt at least, and perhaps everywhere), complete with remarks on instruments, times of daily recordings, changes, conversion factors, graphical plots, and statistics. Attention should be given to reconstructing sea-level conversion constants and comparative studies undertaken to assure that the sea level reductions are not biased.

After the original data have been reviewed and homogenized it would be desirable to have reconstructed regional sequential mappings of various meteorological parameters extended back as far as possible. To accomplish this, dynamic models may be useful in extending limited databases back further in remote areas than currently considered feasible with statistical techniques alone.

A realistic short-term goal would be a global set of historic extended homogeneous monthly time series of pressure, or pressure functions and possibly other climatic variables back to 1850, resolved at 1° latitude-longitude intersections. These data could then be used for local extension. Thorough documentation of the homogenization effort would be required, with review of results by a competent group of scientists. This would lend the authority to the data strings that is a necessity for a sound acceptable basis in project analysis. Publication of the documentation and data sets (in hard copy and magnetic form) and wide-scale distribution would permit their use by water resources and agricultural planners around the world.

Dendroclimatology is still under-utilized in project analysis but it is potentially our most valuable source of information on infrared noise in the climatic record. A similar international cataloging as for pressure data must first await more intensified efforts at national levels to develop sets of dendrochronologies. The techniques for accomplishing this work are well documented (Fritts, 1976). WMO should encourage meteorological services of member countries to move ahead with development of their tree-ring data base in each climatic zone, wherever possible.

CONCLUSIONS

This paper set out to demonstrate a need for hydroclimatology as a new cross-discipline. The main tasks of project-oriented hydroclimatology would be: (1) to extend project streamflow records and other project-related climatic data as far back as possible, using original and extended precipitation data gotten through transfer function relationships with tree-ring data or meteorological space time fields, with emphasis on preserving low-frequency events important in water resources design; (2) to investigate trends in all available local historic data and reconcile them with regional trends, correlations, spectra, and R/S (as index of local 'Josephivity'; (3) to process local historic streamflow traces that are suitable for parameter estimation for use in stochastic modelling; and (4) to determine what non-random trend parameters (including zero) should be considered in the stochastic flow sequences for some or all of the planning period ahead, based on the latest LPL/GPL trends found in the historic records.

Statistical tools exist for accomplishing these analyses, however, there is evidently a need for improvement in methodologies for analysis of climatological space-time fields. The infrared signals are necessarily weak in short records, and Type I error statistical inference procedures frequently label them as white-noise manifestations when global telleconnections show they are not.

Regional meteorological data series exist for accomplishing extensions back 80-100 years at most locations, however, they are difficult to access, and prior to use they must be extensively 'laundered'. Documented tree-ring chronologies outside the North American continent are rare. An international effort is needed to develop a global network of dendrochronologies, which could in turn possibly provide a 400 year or more space-field time-series that would yield an even richer series for local data extension and infrared signal or 'Joseph event' detection.

ACKNOWLEDGEMENT

Appreciation is extended to Dr's. Arnold Court, Roger Barry, Ray K. Linsley, and Charles W. Stockton, who provided valued comments on an early version of this paper that was presented at the 1976 Winter Meeting, American Geophysical Union, San Francisco, California.

BIBLIOGRAPHY

Ackermann, William C., David J. Allee, Jaime Amorocho, Yacov Y. Haimes, Warren A. Hall, Richard A. Meserve, Ruth Patrick and Philip M. Smith, "Scientific and Technological Considerations in Water Resources Policy", Transactions, Amer. Geophys. Union, 59(6), 1978.

Amorocho, J., "Non-linear Hydrologic Analyses", Advances in Hydroscience, (ed. V. T. Chow), Vol. 9, Academic Press, N.Y., 1973.

Andel, J., and J. Balek, "Model of the Disturbances in Hydrological Sequences Based on the Method of Demodulation", Hydrological Sciences Bulletin 22 (1), 163-174, March 1977.

Anderson, Malcom G., "Demodulation of Stream-Flow Series", Journal of Hydrology, 26: 115-121, 1975.

Anis, A. A. and E. H. Lloyd, "Skew Inputs and the Hurst Effect", Journal of Hydrology, 26; 39-53, 1975.

Barry, R. G., Personal Correspondence, 1977.

____________, "Climatic Fluctuations During the Periods of Historical and Instrumental Record", in Climatic Change and Variability - A Southern Perspective, Ed. A. B. Pittock, L. A. Frakes, D. Jenssen, J. A. Petersen, J. W. Zillman, Cambridge Univ. Press, New York, 1978.

____________, and A. H. Perry, Synoptic Climatology Methods and Applications, Methuen & Co., London, 1973.

Berlage, H. P., "Variations in the General Atmospheric and Hydrospheric Circulation of Periods of a Few Years Duration Affected by Variations of Solar Activity", Ann. of The New York Acad. of Science, 95, 345-367, 1961.

____________, "The Southern Oscillation and World Weather", Koninklijk Nederlands Meteorologish Instituut, Mededeligen en Verhandelingen No. 88, Staatsdrukkerij/'s-Gravenhage, Netherlands, 1966.

Blasing, T. J., and H. C. Fritts, "Reconstructing Past Climatic Anomalies in the North Pacific and Western North American From Tree Ring Data", Quaternary Research, 6, 563-579 (1976).

Bloomfield, Peter, Fourier Analysis of Time Series, An Introduction, John Wiley, New York, 1976.

Box, George E.P. and Gwilym M. Jenkins, Time Series Analysis, Forecasting and Control, Holden-Day, San Francisco, 1970.

Bradley, Raymond S., "Secular Changes in Precipitation in the Rocky Mountain States", Monthly Weather Rev. 104, 513-523, 1976.

Brier, Glenn W., "Long Range Prediction of the Zonal Westerlies and Some Problems in Data Analysis", Rev. Geophys., 6: 525-551, 1968.

____________, and Gayle T. Meltesen, "Eigenvector Analysis for Prediction of Time Series", Jnl. of Applied Meterology, 15 (12), 1307-1312, 1976.

Brillinger, David R., Time Series Analysis and Theory, Holt, Rinehart and Winston, Inc., New York, 1975.

Bryson, R. A., "A Perspective on Climate Change" Science, 184(4183), 753-760, 1974.

Burges, Stephen J. and Dennis P. Littenmaier, "Operational Comparison of Stochastic Streamflow and Generation Procedures", Tech. Rpt. No. 45, Charles W. Harris Hydraulic Laboratory, Dept. of Civil Engineering, University of Washington, Seattle, Wa. 1975.

Chow, Ven Te, Dong Hee Kim, David R. Maidment and Taylan A. Ula., "A Scheme for Stochastic State Variable Water Resources Systems Optimization", Res. Rpt. No. 105, Water Resources Center, University of Illinois, Urbana, Illinois, 1975.

Clarke, R. T., "The Representation of Experimental Catchment Data", Proceedings of the Warsaw Symposium, July 1971, Intl. Assn. Scientific Hydrology, Published 100, 1974.

Cliff, A. D., P. Haggett, J. K. Ord, K. Bassett & R. Davies, Elements of Spatial Structure, Cambridge University Press, New York, 1975.

Court, Arnold, Discussion in "International Symposium on Uncertainties in Hydrologic & Water Resources Systems", Vol. 3, University of Arizona, Tucson, Arizona, 1972.

____________, Personal Correspondence, 1976.

Cox, D. R. and P. A. W. Lewis, The Statistical Analysis of Series of Events, Methuen & Co., Ltd., London, 1966.

Crawford, N. H. and R. K. Linsley, "Digital Simulation in Hydrology: Stanford Model IV", Technical Report No. 39, Department of Civil Engineering, Stanford University, Stanford, California 1966.

Currie, Robert G., "Solar Signal Cycle in Surface Air Temperature", Jnl. Geophys. Res. 79 (36), 5657-5660, 1974.

Damon, Paul E. and Steven M. Kunen, "Global Cooling? No, Southern Hemisphere Warming Trends May Indicate Onset of the CO_2 Greenhouse Effect", Science, 193 (4252), 447-453, 1976.

Delleur, J. W., P. C. Tao, and M. L. Kavvas, "An Evaluation of the Practicality and Complexity of Some Rainfall and Runoff Time Series Models", Water Resources Res., 12 (5), 953-1046, 1976.

Dressler, A. J., "Some Problems in Coupling Solar Activity To Meterological Phenomena", in "Possible Relationships Between Solar Activity and Meterorological Phenomena", Greenbelt, Ma., SP-366 Symposium Goddard Space Flight Center, p. 187-197, NASA, Washington D.C., 1975.

Ditlevsen, Ove., Extremes and First Passage Times with Applications in Civil Engineering, Tech. University of Denmark, Copenhagen, 1971.

Dobson, Annette J. and J. Stewart, "Frequencies of Tropical Cyclones in the Northeastern Australian Area", Australian Meteor. Mag. 22(2), 27-36, 1974.

Fritts, Harold C., "Tree-Ring Analysis: A Tool for Water Resources Research", Transactions of the American Geophysical Union, 50:22-29, 1969.

____________, Terence J. Blasing, Bruce P. Hayden, and John E. Kutzbach, "Multivariate Techniques for Specifying Tree-Growth and Climate Relationships and for Reconstructing Anomalies in Paleoclimate", Journalof Applied Meterology, 10 (5): 845-864, 1971.

____________, Tree Rings and Climate, Academic Press, New York, 1976.

Granger, C. W. J., and Hatanaka, M., Spectral Analysis of Economic Time Series", Jnl. Geophys. Res. 68(21), 6033-6041, 1963.

Hastenrath, Stefan, "Variation in Low-Latitude Circulation and Extreme Climatic Events in the Tropical Americas", Jnl. Atmos. Sciences, 33, 202-215, 1976.

Hurst, H. E., "Long Term Storage Capacity of Reservoirs", Trans. Am Soc. Civil Engrs. 116, 770, 1951.

Ilesanmi, O. O., Jnl. Appl. Meteor. 10,882. 1971.

Julian, Paul R., and Robert M. Chervin, "A Study of the Southern Oscillation and Walker Circulation Phenomenon", Monthly Weather Review, 106 (10), 1433-1451, 1978.

Kavvas, M. Levant and Jacques W. Delleur, "The Stochastic and Chronologic Structure of Rainfall Sequences-Application to Indiana", Tech. Report No. 57, Water Research Center, Purdue University, West Lafayette, Indiana, 1975.

Kanamitsu, Masao and T. N. Krishnamurti, "Northern Summer Tropical Circulation During Drought and Normal Rainfall Months", Monthly Weather Review, 106(3), 331-347, 1978.

Kidson, J. R. "Eigenvector Analyses of Monthly Mean Surface Data", Monthly Weather Review, 103 (3), 177-186, 1975a.

__________, "Tropical Eigenvector Analysis and the Southern Oscillation", Monthly Weather Review, 103 (3), 187-196, 1975b.

Kilmartin, Ronald F. and Stephen J. Burges, "Hydrology For Analysis of Multiple Reservoir Systems", ASCE Conference - Fort Collins, CO 1976.

__________________ and R. J. Peterson, "Rainfall-Runoff Regression with Logarithmic Transforms and Zeros in the Data", Water Resource Res., 8(4) 1096-1099, 1972.

_______________, "Regional Mean Autocavariance" AGU Western Conference, San Francisco, CA 1977.

Kirwan, A. D., Jr., N. L. Guinasso, Jr. and G. McNally "Sea Surface Temperatures in the North Pacific Through the Winter 1976-1977, From Nimbus 6, Jnl. Gephys. Res. 83(C11), Nov. 20, 1978.

Klemes, V., "The Hurst Phenomena: A Puzzle?", Water Resources Res. 10, 175-188, 1974.

Lamb, H. H., "Climate, Past Present and Future", Vol. I 1972, and Vol. II, 1978. Methuen, London.

_____________, "Climatic Fluctuations", in World Survey of Climatology, Vol. 2, General Climatology, 173-249, H. E. Landsberg, Ed., Elsevier, New York, 1969.

Linsley, Ray K., discussion in "International Simposium on Uncertainties in Hydrologic and Water Resources Systems", Vol. 3, University of Arizona, Tucson, Arizona, 1972.

_____________, Max A. Kohler, and J. L. H. Paulhus, Hydrology for Engineers, McGraw Hill, 1968.

Littenmaier, Dennis P. and Stephen J. Burges, "An Operational Approach to Preserving Skew in Hydrologic Models of Long Term Persistence", in press, 1976.

Lorenz, E. M., "Climatic Change as a Mathematical Problem", Jnl. Appl. Meterol., 9, 325-329, 1970.

__________, "Climatic Predictability", in The Physical Basis of Climate and Climate Modelling, GARP Pub. No. 16, World Meteorological Org., 1975.

Mandelbrot, B., "Time Varying Channels, 1/F Noises and the Infrared Catastrophy, or" Why Does the Low Frequency Energy Sometimes Seem Infinite", Cong. Rec. 1st IEEE Ann. Commun. Conv. Boulder, CO., 1965.

____________, "A Fast Fractional Noise Generator", Water Resource Res., 7(3), pp. 543-553; 1971.

____________, and J. R. Wallis, "Noah, Joseph and Operation Hydrology", Water Resources Res., 4(5), pp. 909-918, 1968.

_________________, "Computer Experiments with Fractional Gaussian Noises Parts 1, 2, and 3", Water Resources Res., 5(1), pp. 228-267, 1969a.

______________, "Some Long Run Properties of Geophysical Records", Water Resources Res., 5(2), pp. 321-334.

Matalas, N. C., Statistics of a Runoff-Precipitation Relation, U.S. Geological Survey, Prof. Paper 434-D, Government Printing Office, Washington, D.C., 1963.

______________, "Mathematical Assessment of Synthetic Hydrology", Water Resources Res., 3(4), pp. 937-945, 1967.

Mejia, Jose M., David R. Dawdy and Carl F. Nordin, "Streamflow Simulation 3. The Broken Line Process and Operational Hydrology", Water Resources Res. 10(2), 1974.

______________, Ignacio Rodriguez-Iturbe, and David R. Dawdy, "Streamflow Simulation 2. The Broken Line Process as Potential Model for Hydrologic Simulation", Water Resources Res. 8(4), pp. 931-941, 1972.

Namias, J., "Seasonal Interactions Between the North Pacific Ocean and Atmosphere during the 1960's", Monthly Weather Review, 97(3), 173, 1969.

__________, "Response of the Equatorial Countercurrent to the Subtropical Atmosphere", Science 181, 1245-1247, 1973.

__________, "Multiple Causes of the North American Manual Winter 1976-77, Monthly Weather Review, 106(3), 279-295, 1978a.

__________, "Recent Drought in California and Western Europe", Rev. Geophys. & Sp. Phys. 16(3), 435, 1978b.

National Academy of Sciences, "Water and Choice in the Colorado Basin-An Example of Alternatives in Water Management", Washington, D.C., 1968.

______________________________, "Understanding Climate Change, A Program for Action", Washington D.C., 1975.

O'Connell, P. E., "A Simple Stochastic Model of Hursts' Law", International Symposium on Mathematical Models in Hydrology, Warsaw, pp. 327-358, 1971.

________________, "Skew Inputs and the Hurst Effect-A Comment", Journal of Hydrology 31 (1976, pp. 185-191)

Potter, Kenneth W., "Evidence For Nonstationarity as a Physical Explanation of the Hurst Phenomenon", Water Resources Res 12(5), pp. 1047-1052, 1976a.

______________________, "A Stochastic Model of the Hurst Phenomena" Non-Stationarity in Hydrologic Processes", Ph.D. Thesis, John Hopkins University, Baltimore, MD., 1976.

Palmer, Wayne C., "Meteorological Drought", Research Paper No. 45, Weather Bureau, U.S. Department of Commerce, Washington D.C,, 1965.

Ramage, C. S., "Preliminary Discussion of the Meteorology of the 1972-73 El Nino", Bull. Am. Met. Soc., 56(2), 234-242, 1975.

Rao, A. R., and R. G. S. Rao "Comparative Analysis of Estimation Methods in Nonlinear Functional Models of the Rainfall Runoff Process", Tech. Report No. 56, Purdue University Water Resources Center, West Lafayette, Ind., 1974.

Rodriguez-Iturbe, Ignacio, David R. Dawdy and Luis F. Garcia, "Adequacy of Markovian Models with Cyclical Components for Stochastic Streamflow Simulation", Water Resources Res. 7(5), pp. 1127-1143, 1971.

________________________, Jose M. Mejia and David R. Dawdy, "Streamflow Simulation 1. A new look at Markovian Models, Fractional Gaussian Noise and Crossing Theory", Water Resources Res. 8(4), pp. 921-974, 1972.

Salinger, M. J., and J. M. Gunn, "Recent Climatic Warming Around New Zealand", Nature, 256 7/31/1975, pp. 936-938, 1975.

Schneider, Stephen H. and Robert E. Dickenson, "Climatic Modeling", Rev. of Geophys. and Sp. Phys. 12(3), pp. 447-493, 1974.

Stringer, E. T., "Techniques of Climatology", and "Foundations of Climatology", 2 vols. W. H. Freeman, San Francisco, 1972.

Stockton, Charles W. "Long-Term Streamflow Records Reconstructed from Tree Rings", Papers of the Laboratory of Tree Ring Research, 5, the University of Arizona Press, Tucson, Arizona, 1975.

___________________, and Harold C. Fritts, "Long-Term Reconstruction of Water Level Changes for Lake Athabasca by Analysis of Tree Rings" Water Res. Bull., 9(5), 1006-1027, 1973.

___________________, and David M. Meko, "A Long-Term History of Drought Occurrence in Western United States as Inferred from Tree Rings, Weatherwise, 245-249, December, 1975.

Szollosi-Nagy, "An Adaptive Identification and Prediction Algorithm for the Real Time Forecasting of Hydrological Time Series", Hydrologic Sciences Bulletin , 21(1), 163-166, 1976.

Todini, E., and D. Bauillot, "A Rainfall-Runoff Kalman Filter Model", in System Simulation in Water Resources, G. C. Vansteenkiste, Ed., North Holland, Amsterdam, 1976.

Trenberth, Kevin E. "A Quaisi-biennial standing wave in the Southern Hemisphere and Interrelations with Sea Surface Temperature", Quart. Jnl. Roy. Met Soc., 101, pp. 55-74.

____________, "Fluctuations and Trend in Indices of the Southern Hemisphere Circulation", Quart, Jnl. Roy. Met. Soc., 102, pp. 26-75.

Tyson, P. D., T. G. J. Dyer and M. N. Mametse, "Secular Changes in South African Rainfall: 1880 to 1972", Quart Jnl. Roy. Met Soc., 101, pp. 817-833, 1975.

Ulrych, Tad J. and Thomas M. Bishop, "Maximum Entropy Spectral Analysis and Autogressive Decomposition", Rev. Geophys. & Sp. Phys. 13 (1), pp. 183-200, 1975.

Valencia, R. D., and J. C. Shaake, "Disaggregation Processes in Stochastic Hydrology", Water Resources Res., 9(3), pp. 580-585, 1973.

Van Loon, Harry and Jill Williams, "The Connection Between Trends of Mean Temperature and Circulation at the Surface: Part I, Winter", Monthly Weather Review, 104, pp. 365-380, 1976.

Wagner, A. James, "Long Period Variations in Seasonal Sea-Level Pressure Over the Northern Hemisphere", Monthly Weather Review, 99, pp. 49-66, 1971.

Wallis, J. R. and N. C. Matalas, "Correlation Analysis Revisited", Water Resources Res., 7(6), pp. 1448-1459, 1971.

Wallis, J. R. and P. E. O'Connell, "Firm Reservoir Yield-How Reliable are Historical Hydrological Records", Hydrologic Sciences Bulletin, 18(3), pp. 347-365, 1973.

________________, and P. E. O'Connell, "Small Sample Estimation of p_1", Water Res. Res., 8(3), pp. 707-712, 1972.

Weare, Bryan C., Alfredo R. Navato, and Reginald E. Newell, "Emperical Orthogonal Analysis of Pacific Sea Surface Temperatures", Jnl. Phys. Oceanography, 6(5), 1976.

White, R. M., D. S. Cooley, R. C. Derby and F. A. Seaver, "The Development of Efficient Linear Statistical Operators for the Prediction of Sea-Level Pressure", Jnl. of Meteorology, 15, pp. 426-434, 1958.

Winstanley, Derek, "Rainfall Patterns and General Circulation", Science 180, pp. 66-65, 1973.

Wyrtki, K., "Telleconnection in the Equatorial Pacific Ocean", Science 180, 66-68, 1973.

__________, El Nino- The Dynamic Forcing of the Equatorial Pacific Ocean to Atmospheric Forcing", Jnl. of Physical Oceanography, 5(5), 572, 1975.

Yevdjevich, V. M., "Fluctuations of Wet and Dry Years, Part I Research Data Assembly and Mathematical Models", Hydrology Papers No. 1, Colorado State University, Fort Collins, CO. 1963.

PRECIPITATION FORECASTING AT A NATIONAL CENTER

DAVID A. OLSON*

INTRODUCTION

The National Meteorological Center, located in Camp Springs, Maryland, about 10 miles southeast of Washington, routinely supports all meteorological activities in the U. S. and globally too, as a World Meteorological Center component of the WMO. Many of the activities of the NMC support both functions. Some of these activities are very routine, but if not performed satisfactorily, can seriously disrupt non-routine operation. Activities include data collection, editing, processing and communication. Perhaps the more important functions are analyses, prognoses and specialized forecasts which are best accomplished by a national center--these include world-wide aviation forecasts, generalized medium-range forecasts in the 3-5 and 6-10 day range and quantitative precipitation and heavy snow forecasting. Many of the basic or primary and routine functions either have been or are being automated, and include data processing, plotting, analysis and of course the running of the mathematical models which describe atmospheric motion and may run as far as 10 days in the future. Much processing or computations are accomplished on the 360/195 system while communications are handled by 360/40 computers.

The Quantitative Precipitation Branch (QPB) of NMC's Forecast Division has prepared and issued forecasts of precipitation amounts (QPF) and snowfall amounts since the early '60's. The prediction of precipitation amounts is one of the most difficult and challenging aspect of meteorology, yet is an area with promise of great economic benefit. The successes which have occurred have in large part been due to the assigning of this task to a small highly specialized group of forecasters.

PRODUCTS

The basic products are 24-hour forecasts of areas of 0.25, 0.50, 1.00, 2.00 etc inches of precipitation. Early each morning these forecasts are issued covering a 48-hour period--Day 1 and Day 2 (Fig. 1). The Day 2 forecast is updated each afternoon. It is possible to further define the Day 1 forecasts into either 6- or 12-hour increments. Forecasts for areas likely to get 4 inches or more of snow during the 6-18 hour period are prepared on 6-hourly cycles through the cold season. In support of the national flash flood effort, forecasts for areas

*Note: Author is currently Chief, Quantitative Precipitation Branch, National Meteorological Center, Camp Springs, Maryland.

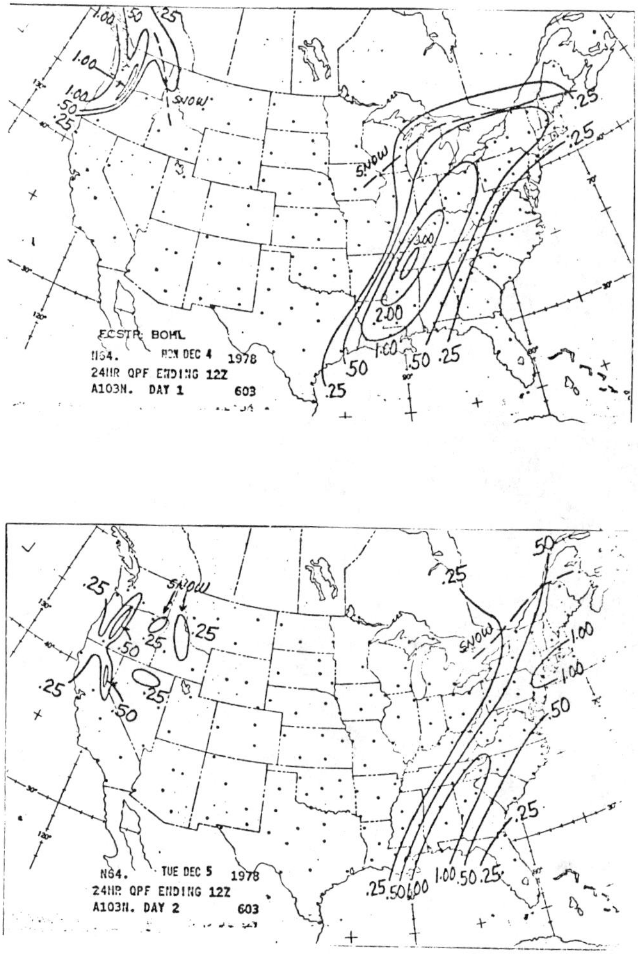

Fig. 1. Sample of subjective Day 1 and Day 2 QPF's for 24-hour periods.

where rainfall could equal or exceed the flash flood guidance values are issued and updated each day. Routine forecasts are available on either facsimile in graphic format or in a coded version over teletype. During or preceding critical river flooding, special forecasts covering four 6-hour periods are prepared and issued for use in river forecasts.

PHILOSOPHY

The forecast process resolves observational and objective numerical guidance into consistent predictions of precipitation amounts. The goal is to forecast patterns as realistically as possible, taking into account effects of the larger scale orographic features. In the case of snow, which tends to occur in similar patterns in association with certain types of meteorological features, this is very feasible. Although rainfall tends to associate in similar patterns with certain features, the amounts show considerable variance--both within a storm and between different storms. Because of these differences, the forecaster applies considerable professional meteorological judgment and ingenuity in formulating his forecasts into the greatest feasible detail. Because of the difficulty in specifying in advance the detailed structure of a storm, rainfall 24-hour forecasts tend to be area-averaged. A forecaster may feel great confidence in occurrence of a 3" area, yet lack total confidence in location--his best judgment could be to leave this out of the graphics and either mention it in his discussion--or include it in the flash flood potential forecast--if the criteria is below the 3". The intent of the flash flood or heavy rainfall outlook is to indicate areas where rainfall maximum are likely to occur and which cannot be specified spatially and temporally to a professional degree of accuracy.

FORECAST PROCEDURE

Procedures do not differ greatly for the different products. Twice each day our complex mathematical models begin their analysis of the initial state of the atmosphere and projection to 48 hours and beyond, providing output to meteorologists at either 6- or 12-hour increments. At the same time the forecaster is evaluating the initial state, relating observed rainfall to associated features and projecting the near future on the basis of observational data. As the objective output becomes available, it is necessary to evaluate it in terms of consistency--with the initial state and within the model itself, in terms of potential modification--or correction--to the circulation fields. Unlike earlier models, current NMC models do not display consistent forecast errors, therefore modifications must be based on sound physical reasoning, consistent with initial data, or must be made to one or more forecast parameters to produce a consistent product. In basic terms the work of the quantitative precipitation forecaster is aimed at specification of initial and predicted distributions of the upward component of the wind field and moisture. Convection, which is handled poorly by the models, must be incorporated and is basically added where smaller scale features (i.e. meso-scale system) are anticipated. Incidentally, the smaller-scale features are not predicted at all by the models. However, since these systems are driven by the larger scale, there is some limited success in looking for smaller-scale features. Large-scale systems are predicted with fair success routinely by numerical methods.

VERIFICATION

For purposes of verification, a 'best' analysis is constructed using all readily available real-time rainfall data incorporating radar and/or satellite data wherever there are gaps in observed rainfall data. Predicted, observed and correct areas are determined areally. These derived figures are used to calculate various quantities. Although much of the verification work is done manually at present, we are actively working to convert the different data sets into a uniform consistent set that can be manipulated easily within the computer. This will lead to an objective but comparable areal verification to the present system, and permit comparison between the numerical and subjective forecasts. Figure 2 (a,b) present the annual Threat Scores (=Area Correct divided by sum of Area Forecast + Area Observed-Area Correct) for 0.50" and 1.00". These curves show a slow improving trend since 1970, especially for the UPDATE and DAY 2 forecasts. Improvements are generally attributed to improved circulation forecasts by the numerical models in the 24-60 hour range, improvements which have been largely due to reducing grid distance. The Threat Score is a unique quantity whose use has been largely confined to NMC, where its general use is accepted as a single measure of forecast accuracy or quality.

Other measures of forecast performance, which can be computed readily, are the Postagreement--or correctness percentage of the forecast and the Prefigurance, which is the ratio of the observed events which were predicted correctly. Fig. 3 readily shows the effects of scale on predictability. This is a graph of 10-year means of Postagreement, with the minima and maxima forming an envelope. The purpose of this graph is to show the drop in skill or predictability that occurs when the warm season precipitation becomes largely due to small-scale convective systems. The values on the graph are a little misleading, as the Postagreement for the past two winter seasons has been between .50 and .55 for 0.50" and the summer has also been higher at about .35. Curves for 1.00" and for Prefigurance show similar characteristics. In the past 2 winter seasons, Prefigurance for 0.50" has been about .65 and for 1.00" a little above .50. Interpreted, this means that about 65% of the 0.50" or greater rainfall and 50% of the 1.00" or greater rainfall were predicted correctly.

FUTURE WORK

In these days of advancing technology in virtually all areas, it is a difficult proposition to project what the future holds. The development and introduction of satellite sounding data are likely to result in eventual improved numerical circulation forecasts--which will be increasingly important in projecting forecasts further into the future. At present our regional fine mesh model runs too early to establish the best available initial state over the data sparse oceanic areas and the models can be extremely sensitive to small errors. An experiment will be undertaken to determine the feasibility and accuracy of an isohyetal Day 3 forecast. On a parallel basis, the possibility of extending snow forecasts another 12 hours will be explored.

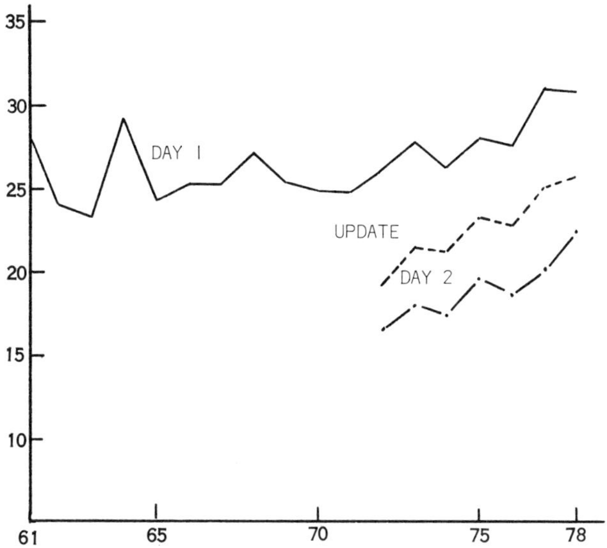

Fig. 2a. QPB annual Threat Scores for 0.50" rainfall predictions.

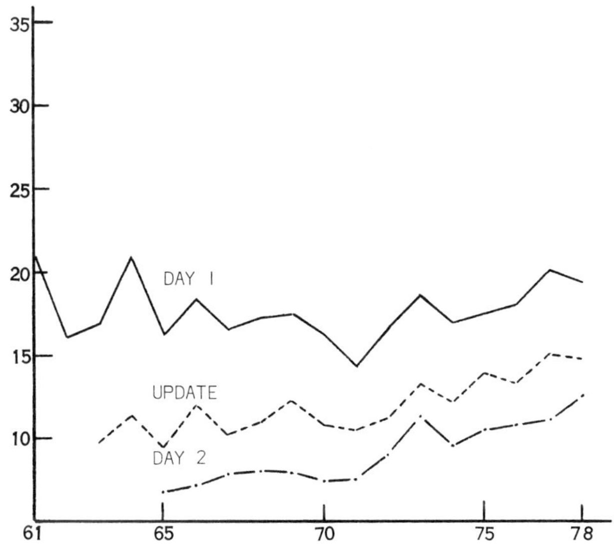

Fig. 2b. QPB annual Threat Scores for 1.00" rainfall predictions.

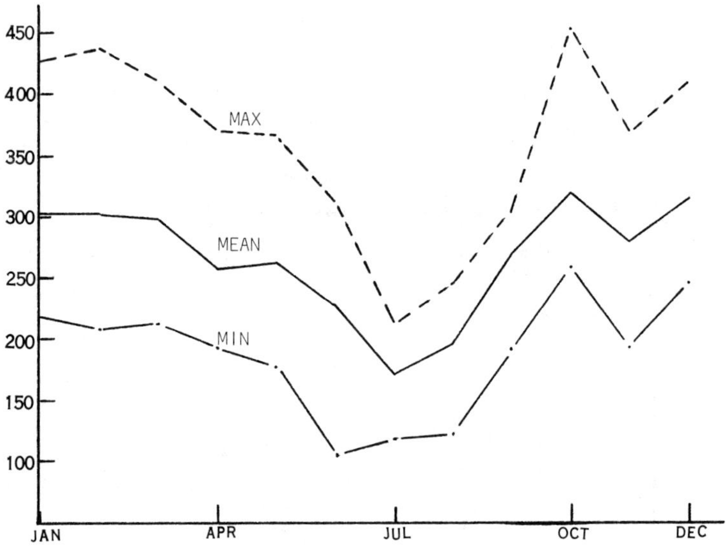

Fig. 3. Monthly variability of mean, maximum and minimum Post-agreement (Percent correct) for the period 1966-75.

It is becoming an operational necessity to establish an observed rainfall data base on a nationwide scale. Other than the obvious users such as hydrology and agriculture, it is essential to have a ground truth available to rainfall forecasters that they can use to correlate with various atmospheric phenomena. Much of the data is already available--in differing formats and places. An effort is being made to bring it together in a uniform consistent package and then analyze it to various subsets of our numerical model grid system.

Research needs to continue in the area of intense small-scale convective rain systems--particularly with emphasis on the prediction of these flash flood producers with several hours lead time. Forecasts need to and can be improved with greater understanding through research, particularly in regard to the magnitude and duration of the rainfall.

Although an automated version of precipitation amount probability is currently available, it would perhaps be a worthwhile undertaking to investigate the utility of subjective probability estimates.

Finally, the future holds promise for improved rainfall forecasting for both large and small scale due to increased emphasis both numerically and subjectively.

ENVIRONMENTAL INFORMATION FROM SEDIMENTS

by

David P. Adam[1]

INTRODUCTION

This meeting recognizes the tremendous importance to mankind of hydrologic systems and their behavior. To take full advantage of our water resources without exposing ourselves to unnecessary risks requires that we predict how both natural and man-modified systems are likely to behave in the future; such forecasts should be based on a knowledge of present and past behavior. Only rarely (perhaps never) are instrumental data adequate for long term predictions; such records must be supplemented by longer-term data from historical records, tree rings, and other proxy data that provide only indirect records of past hydrologic behavior.

Proxy data may often be obtained from sediments. Under favorable conditions, sediment may accumulate more or less continuously in a given spot for thousands of years, initially including whatever happens to be washed in, blown in, or chemically precipitated, including both mineral grains and biological remains. Post-depositional processes may alter the character of the sediment, but traces of the environment of deposition are likely to persist. If the sediment accumulates continuously (and I am assuming that this happens, otherwise I would look somewhere else), the result is a sedimentary sequence that may be interpreted as a series of environmental samples taken at a point through time.

SITE SELECTION

This paper is primarily concerned with reviewing data of interest to hydrologic forecasters; I have therefore limited it to continental environments, and also to the last several thousand years (the geologic Holocene). Readers interested in pursuing the subject beyond those limits may wish to consult the recent treatise by H. H. Lamb (1972, 1977).

Among continental environments, lakes and marshes are particularly suitable for study because they act as sediment traps. Sediment, once deposited, is likely to remain in place; the environment is also likely to be chemically reducing, so that organic

[1]Geologist, U.S. Geological Survey, 345 Middlefield Road, Menlo Park, Calif.

materials are preserved. Many of the techniques described below can be applied to other sedimentary environments as well, although difficulties may arise because of discontinuities in the record, because of either erosional events or post-depositional changes in the sediment.

Suitable sites should respond sensitively to hydrologic fluctuations. Site sensitivity may vary according to the study techniques being used. Over the past several thousand years, hydrologic changes are likely to have produced primarily local changes, rather than the broad regional changes that characterize glacial/interglacial climatic shifts. Sites that respond primarily to local changes are thus likely to provide the best hydrologic records. It is not usually possible, however, to tell in advance just how a site has responded to past environmental fluctuations, or which variables in the sedimentary record will provide the best information. One must take advantage of whatever is found.

SITE SAMPLING

If a site is to be regarded as providing a series of samples through time, it must be sampled carefully. Where possible, samples should be taken from cleared exposures, either in outcrops or from excavation walls, where stratigraphic relations can be observed. Where this is not possible, as in lake and marsh deposits, continuous and undisturbed cores should be taken, with as large a diameter as possible. In particular, one should avoid augering through samples that are to be analyzed for microfossils, as samples may easily become contaminated with material dragged by the auger blades from other levels; drilling mud can also introduce contaminants unless precautions are taken. Whenever possible, the person who is going to analyze samples should collect them; data are much easier to interpret when one has seen the field evidence.

PHYSICAL EVIDENCE

Physical evidence for past hydrologic behavior may be found in grain size distributions and in mineralogy. Grain size distributions reflect the energy available for transport of sediment; if bedrock lithology stays constant, as is likely for the past several thousand years in most localities, then changes in grain size distributions must represent changes in the effectiveness of erosion, transport, and deposition of sediment. This concept assumes that erosion, transport, and deposition are by running water; other modes of erosion may include wind deflation, avalanches, landslides, ice rafting, and glacial abrasion. These modes may usually be distinguished from each other by careful field and laboratory observations.

When a deposit is formed of minerals precipitated from solution, it is possible to place limits on the conditions under which precipitation took place by observing which minerals were formed (Smith, in press).

Other information can sometimes be derived from the stratification of the sediments. In some lakes with active glaciers in their headwaters, as well as in meromictic lakes (those with no dissolved oxygen in their bottom waters to sustain burrowing organisms), annual layers (varves) may be preserved. By counting these layers, one can get a much more accurate estimate of the age of individual layers than is possible using other methods; also, changes in the thickness of the layers from year to year are a result of environmental variations. If the causes of the thickness variations can be determined, one may then establish a record of the past behavior of the causative factors. For glacial varves, for example, thick layers are produced by warm summers.

Another interesting aspect of stratification is that loose sediment layers not yet deeply buried may be offset or distorted by seismic activity. Disturbed layering in sediments may thus record past seismic activity in areas where modern seismic activity is low. Where large hydrologic systems are being constructed, it is essential to know whether an area is seismically active or not, as is shown by the history of the proposed Auburn Dam.

BIOLOGICAL REMAINS

Studies of organic remains in sediments are diverse, and most workers specialize in one or a few types of remains. The number of workers varies among specialties but is generally low. Many workers recognize an informal distinction between microfossils and macrofossils. Microfossils require high magnification for proper study, are usually not visible to the naked eye, and range in size from roughly half a millimeter down to a few micrometers. Macrofossils are not, as one might expect, dinosaurs and elephants, but rather, small remains that can be seen with the naked eye and studied under a dissecting microscope; sizes range roughly from a quarter of a millimeter to a centimeter or so. In addition to macro- and microfossils, there are also biogeochemical indicators of past environments, including such things as amino acids, chlorophyll degradation products, and plant pigments.

Microfossils

Microfossils include pollen and spores and various kinds of algae, including diatoms, chrysomonads, *Pediastrum*, *Botryococcus*, and *Coelastrum*. Their study generally involves concentrating the remains of the organisms, followed by identification and counting of several hundred organisms per sample using a microscope. Two major concentration methods are used. For siliceous remains (diatoms, chrysomonads, phytoliths and sponge spicules), the sample must be thoroughly oxidized to remove organic material; for pollen samples, mineral matter is removed using mineral acids, followed by further treatments to remove cellulose and humic acids. The two treatments are incompatible, each one destroying what the other concentrates, so duplicate samples are required when both methods are to be used. Samples need not be large; in suitable sediments, one or two cubic

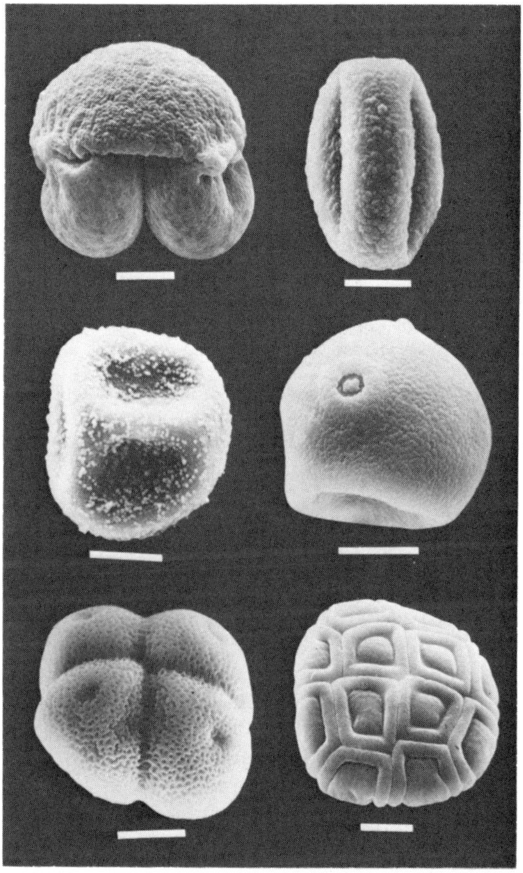

Figure 1.--Scanning electron microscope photographs of some typical pollen grains. Clockwise from top right: oak, bermuda grass, acacia, cattail, juniper, and pine. Scale bars are 10 micrometers long.

centimeters can provide ample material.

Pollen

Palynology is the study of pollen and spores. Pollen grains resist decay and can be recovered in great numbers from suitable sediments, which are usually fine grained and reduced. Pollen grains are morphologically distinctive, although usually not at the species level. Examples of several types of pollen grains are shown in figure 1. A rough model of the system that concerns us is shown in figure 2. Climate, vegetation, and the hydrologic system are interdependent, although external factors also influence them. What we are really interested in here is a record of the past behavior of the hydrologic system; what is available to us is a pollen diagram, which is a distorted record of the past behavior of the vegetation.

An example of a pollen diagram is shown in figure 3. Each vertical curve represents a variable, and each bend in the curve represents data point derived from a sample. The frequencies of the variables, expressed as percentages of a total sum defined by the observer, vary with depth, and thus with time.

Because of the complexity of the relation between the pollen diagram and the vegetation history that produced it, interpretation of a pollen diagram such as figure 3 is beset by many hazards. If we

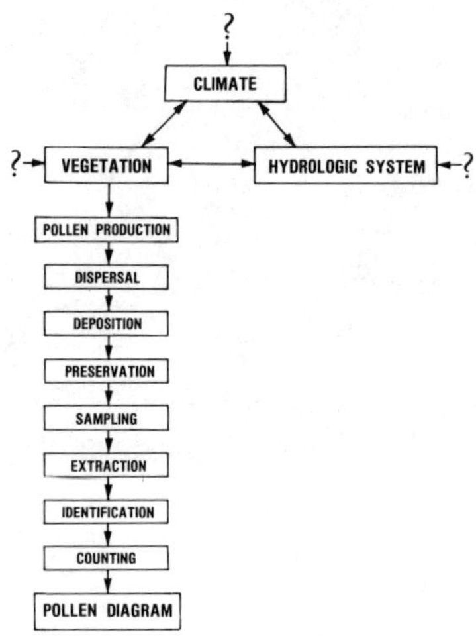

Figure 2.--A simple model showing the relations that must be taken into account when interpreting a pollen diagram in terms of the history of the environment that it reflects.

look again at figure 2, each of the steps shown between the vegetation and the pollen diagram involves some transfer function, and the exact nature of these transfer functions is generally not known. For example, different species produce different amounts of pollen per plant, and dispersal mechanisms and their effectiveness vary between species. At each stage shown in figure 2, biases may arise, and the investigator must be aware of these potential sources of error. In spite of these difficulties, however, consistent records have been obtained over large areas, such as northern Europe, where many sites have been studied.

Certain types of algae, such as *Pediastrum* and *Botryococcus* (figure 4), survive the pollen extraction procedure and may be analyzed along with the pollen, as is shown on figure 3. In that pollen record, the fluctuations in numbers of algae provided the most sensitive hydrologic monitors. During two wet intervals that correspond to glacial advances of the Neoglacial interval, or "Little Ice Age", the pond from which the core was taken stayed wet long enough in the summers for blooms of *Pediastrum* and *Botryococcus* to appear. At other times, the pond dried up each summer before water temperatures rose high enough to support blooms (Adam, 1975).

Diatoms

Diatoms are algae that normally live sandwiched between two siliceous cell walls known as frustules; some typical examples are shown in figure 5. Over 20,000 known species of freshwater diatoms are known; the environmental tolerances of many forms are relatively well known, and this makes them potentially quite useful as indicators of past water conditions. However, much more work has been done on the distributions of modern forms than on their fossil record of the last several thousand years, and there is a shortage of workers studying young fossil materials.

Chrysomonads

Chrysomonads are flagellate algae that spend a resting stage of their life cycle inside a siliceous cyst. These cysts are very small, ranging between 2.5 and 30 micrometers in size. Because they are so small, they can be studied effectively only by using a scanning electron microscope. Some typical examples are shown in figure 6. Unlike the diatoms, which spend the active part of their lives within

---

Figure 3 (facing page).--A partial pollen diagram showing plots of the relative frequencies of selected pollen types and algae versus depth in a 210-cm core from Pearsons Pond, San Mateo County, California. Depth scales at the sides are in centimeters; frequencies are expressed as percent of total pollen. Two curves are shown for each variable. The percentage scale applies to the left curve (darker shading); the right curve is a 4x exaggeration of the left curve. Data recalculated and redrawn from Adam (1975).

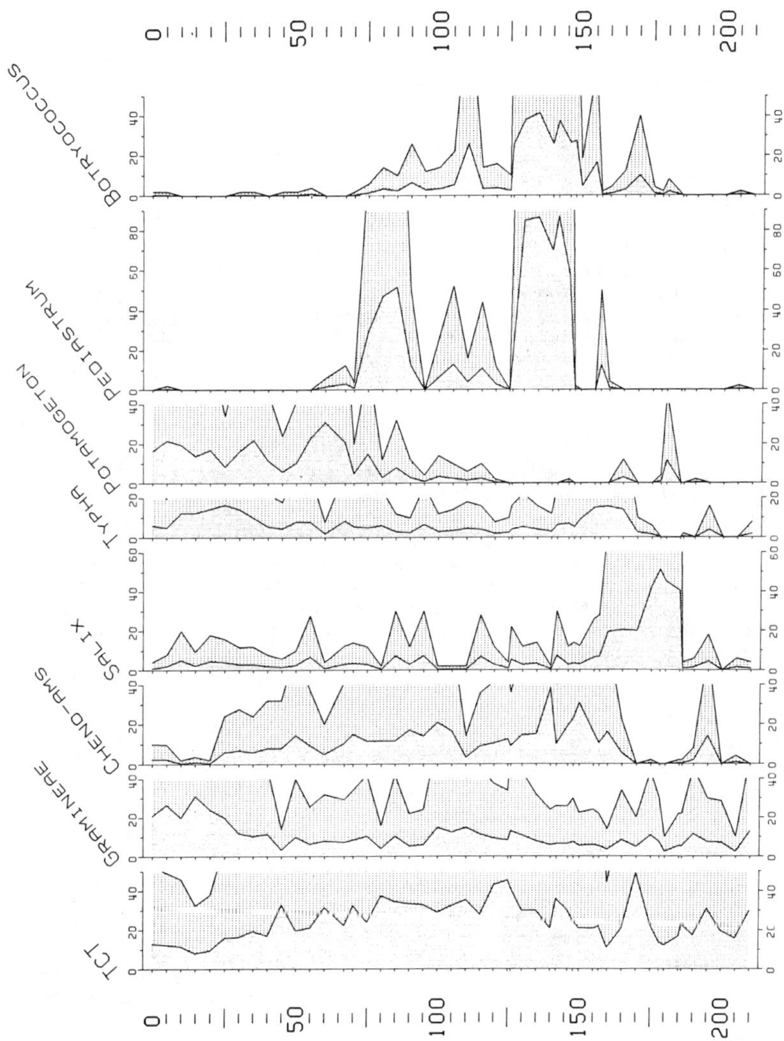
BOTRYOCOCCUS
PEDIASTRUM
POTAMOGETON
TYPHA
SALIX
CHENO-AMS
GRAMINEAE
TCT
0
50
100
150
200

their frustules, the chrysomonads create cysts only when they enter a resting stage, presumably in response to adverse conditions. They should thus reflect somewhat different aspects of the environment from the diatoms and are potentially useful environmental indicators. However, they have received very little study, and it is in general not known which algal species produce which cysts.

Phytoliths

Many plants, especially the grasses, produce internal siliceous particles known as phytoliths. These have been recovered from sediments on occasion, but taxonomic problems have kept them from being generally accepted, and there are very few workers investigating phytolith taxonomy and distribution.

Sponges

Sponges, inlcuding freshwater species, produce siliceous elements called spicules. Each species produces more than one kind of spicule, and some kinds cannot be distinguished at the species level.

Cladocera

Cladocera, or water fleas, are small crustaceans that are common in the freshwater zooplankton. They are recovered by washing sediment through fine screens and are useful environmental indicators, but interested investigators are scarce. A useful discussion of the use of animal remains from sediments, including the cladocera, may be found in Frey (1964).

MACROFOSSILS

Remains of both plants and animals are found as macrofossils. Because they are relatively large, it is usually difficult to obtain large numbers of specimens for quantitative studies. On the other hand, it is often possible to identify macrofossils to the species level, and this makes interpretation of individual specimens easier than with many microfossils, notably pollen.

Animal remains

Fish bones and scales are sometimes recovered from sediments. Fish scales are potentially very useful as environmental tools because of the growth rings they display. Fish are poikilotherms whose growth rate depends upon temperature, and fish scales have been used to infer fluctuations in water temperature for the last several thousand years in Clear Lake, California (Casteel and others, 1977). However, it is not possible to tell where a given scale grew on the fish, so large numbers of scales are needed to apply this method, and this requirement restricts its applicability.

The teeth of small mammals are distinctive and are sometimes found in sediments; other bones are more difficult to identify but are still often useful.

The shells of gastropods are sometimes found in great numbers in sediments, but under acid conditions they are destroyed, along with other calcareous remains. The taxonomy of freshwater gastropods is difficult, so interpretation of results is somewhat uncertain. Workers are few, but the study of gastropods is potentially quite useful.

The ostracods are another group with calcareous remains found in lacustrine sediments; they have small shells that resemble beans. Like the freshwater gastropods, their taxonomy is difficult, and workers in this field are scarce.

Insects are sometimes found in organic, silty sediments, and they can, under favorable conditions, supply very precise information about the environment. It is not yet known, however, whether this group of organisms will be useful over the relatively short time interval of interest to us here.

Plant remains

Plant macrofossils are usually seeds, leaves, or wood. They are recovered from sediments by washing on sieves, followed by sorting under a dissecting microscope. Where sediment is silty, this process is quite easy, but in peaty sediments it is very time-consuming to disaggregate a sample, as well as to separate identifiable remains from unidentifiable plant fibers. An important aspect of plant macrofossils is their carbon content, which may permit radiocarbon dating.

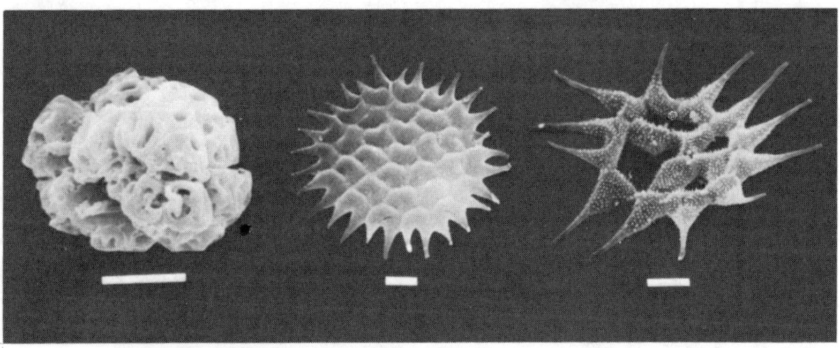

Figure 4.--Scanning electron microscope photographs of Botryococcus (left), Pediastcum boryanum (center), and P.simplex (right) from Clear Lake, California. Scale bars are 10 micrometers long.

Figure 5.--Scanning electron microscope photographs of selected freshwater diatoms. Scale bars are 3 micrometers long.

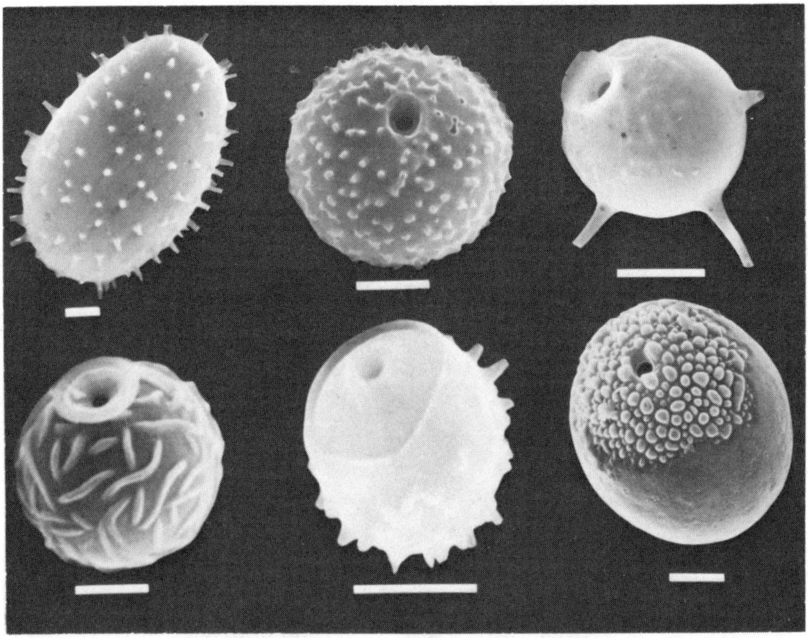

Figure 6.--Scanning electron microscope photographs of selected chrysomonad cysts. Scale bars are 3 micrometers long.

One may also simply observe changes in the relative amounts of plant remains versus other materials through time. For example, in the pollen diagram shown in figure 3, the bottom and top thirds of the core contained abundant coarse plant debris, but the middle third did not. While the middle part of the core was being deposited, the pond was well enough sealed to hold water and deep enough to drown the vegetation around its margins when it was full; under these conditions, the algal record responded sensitively to variations in rainfall. The bottom of the core reflects conditions just after a landslide formed the pond; at that time its bottom was poorly sealed so that water could percolate out of the basin, and marshy vegetation grew in the depression. By the time of deposition of the top of the core, the depression had filled in enough that emergent vegetation covered the pond, eliminating the open-water habitat required by the algae (Adam, 1975).

ESTABLISHING TIME CONTROL

In order for a climatic record to be of maximum use in developing forecasts, there must be adequate time control throughout the record. With the unusual exception of annually layered sediments, such control is generally not available, and the dating methods that can be applied are frustratingly imprecise. Records are usually dated either by methods that give a direct age estimate from the sediment itself, or by correlation of properties of the sedimentary record with other similar records for which time control is already established.

The primary direct dating method for the last several thousand years is radiocarbon (C-14) dating. This technique is based on the observation that living organisms contain a known fraction of their carbon as C-14, and that once an organism dies, it ceases to exchange carbon with the atmosphere. The radiocarbon that it contains then decays exponentially, with a half-life of about 5700 years; by measuring the radiocarbon activity of a sample, one may thus estimate its age. The overall time control that has been established through thousands of radiocarbon determinations is satisfactory, but there are many sources of potential errors, and many analyses have given results that could not be reconciled with other knowledge of the samples that produced them. Radiocarbon "dates" are the most useful time control available under most circumstances, but they cannot be accepted blindly.

When radiocarbon dates are available and appear to be valid, it is a common practice to assume uniform sedimentation rates between dated levels, even when the environmental record suggests that changes occurred. The ages of samples between dated levels are then estimated by interpolation (for example, Adam, 1975). This technique may make the best of the available data, but it does introduce some doubtful assumptions into the time scale for events dated by interpolation. If such events are of critical importance, they should be dated independently if at all possible.

Methods of correlation vary greatly in their precision. The best correlations of individual time horizons are those provided by airfalls of volcanic ash; for geologic purposes, the occurrence of the same ash in two sedimentary sections establishes their exact temporal equivalence. Most other techniques rely on curve matching, with the matching done either against known data or against a theoretical model. Particularly in the latter case, there are significant opportunities for indulging in circular reasoning. The uncertainties involved in curve matching may range from decades to millennia.

Further discussion of the difficulties of establishing reliable time control in the fossil record is beyond the scope of this paper. However, the evaluation of probabilities for future events depends critically on such time control, and developers of forecasts should be wary of possible sources of error and uncertainty in the time control of paleoenvironmental data.

CONCLUSION

In this brief overview, I have tried to draw attention to the wealth of environmental information that is available in sedimentary deposits. Properly studied, they can provide proxy environmental records extending thousands of years into the past. These records often show that substantial environmental changes have occurred, and there is every reason to expect that similar changes will continue to occur, superimposed on any changes we bring about through our own activities. As man strives to utilize his environment fully, he becomes increasingly vulnerable to decreases in resource availability and environmental productivity that are a natural consequence of environmental variability over time. Long-term planning for hydrologic systems requires a thorough understanding of the patterns of climatic change, and the study of sediments has a significant role to play in developing that understanding.

I hope I have shown above that paleoenvironmental studies should be of significant interest to hydrologic planners and engineers. In conclusion, I would like to point out how the engineering profession may help to provide such studies. Paleoenvironmental workers are usually based at universities, and field budgets are generally low. Mechanized equipment in particular is almost unheard of. Many engineers, on the other hand, routinely collect continuous sediment cores and make sizable excavations in the course of site investigations. These cores and trenches represent valuable opportunities to collect paleoenvironmental samples, yet such opportunities are usually wasted. What I now ask of you is that engineers who work in sediments and paleoenvironmental workers establish a rapport, so that when critical cores or excavations become available, the opportunity to study them will not be lost. With such an effort, we can more quickly provide the results that are important to all of us.

REFERENCES CITED

Adam, D. P., 1975, A late Holocene pollen record from Pearson's Pond, Weeks Creek Landslide, San Francisco Peninsula, California. Journal of Research of the U.S. Geological Survey, vol. 3, no. 6, pp. 721-731.

Casteel, R. W., Adam, D. P., and Sims, J. D., 1977, Late Pleistocene and Holocene remains of *Hysterocarpus traski* (tule perch) from Clear Lake, California, and inferred Holocene temperature fluctuations: Quaternary Research, v. 7, p. 133-143.

Frey, D. G., 1964, Remains of animals in Quaternary lake and bog sediments and their interpretation: Ergebnisse der Limnologie, Heft 2, 114 p. + 2 plates.

Lamb, H. H., 1972, Climate: present, past, future. Vol. 1: Fundamentals and climate now: London, Methuen, xxxi + 613 p.

Lamb, H. H., 1977, Climate: present, past, future. Vol. 2: Climatic history and the future: London, Methuen, xxx + 833 p.

Sims, J. D., 1974, Searching for prehistoric earthquakes in lake sediments: Earthquake Information Bulletin, v. 6, No. 1, p. 3-9.

Smith, G. I., in press, Subsurface stratigraphy and geochemistry of late Quaternary evaporates, Searles Lake, California: U.S. Geological Survey Professional Paper 1043, in press.

GEOMORPHIC EVIDENCE OF FREQUENT AND EXTREME FLOODS

James C. Knox[1]

Introduction

Hydrologic planning often is confronted with the problem that inadequate hydrologic data exist to provide a reasonable estimate of the magnitudes of potential floods and their recurrence intervals. The purpose of this paper is to illustrate the use of geomorphic properties of alluvial channels and floodplains as supplementary sources of information on magnitude and frequency characteristics of floods.

The physical properties of channel morphologies and of sediments in channel margins and floodplains provide a measure of both high frequency and low frequency flood flows. River channels tend to be adjusted to prevailing hydrologic conditions. The magnitude of a flood with a recurrence interval between about one and three years can usually be quite accurately estimated from the "bankfull stage" morphologic dimensions of a river channel. In contrast, the high energy conditions that are associated with large floods often produce unusual erosional or deposition features that may be preserved in the geomorphic and stratigraphic record. These geomorphic and stratigraphic features can be used to estimate magnitude and frequency characteristics of the large floods. Because the design of many engineering structures and floodplain zoning require knowledge of the probable maximum floods, as well as frequent low magnitude floods, geomorphic evidence of former floods can serve as a needed extension of hydrologic records that typically are too short for reliable inferences.

Estimating Magnitudes of Floods from Channel Morphology

In most regions the size of a channel cross section is strongly correlated with the volume of water transported during flood flows that recur frequently (Gilbert 1877:113; Rubey 1952:130; Wolman and Miller 1960:66; and Dury 1974:109). The shape of a channel cross section is highly correlated with the textural characteristics of sediment loads transported during high stage flows (Lane 1937:135-138; Leopold and Maddock 1953:19-21; Schumm 1960, 1968:37-41). The

[1]Professor, Department of Geography, Science Hall, University of Wisconsin, Madison, Wisconsin 53706.

channel and floodplain system therefore develop morphologic properties that reflect magnitude and frequency characteristics of floods and sediment yields. Because channel morphologic features are strongly correlated with flood flows, they can be used as a supplementary source of information to estimate flood magnitude and frequency characteristics when recorded streamflow data are unavailable. The strongest statistical correlations between flood magnitudes and channel features tend to be associated with floods that recur more often than once in about five years. The magnitudes of low magnitude and high frequency floods usually are related to the magnitudes of large but low frequency floods. For this reason strong and statistically significant correlations also occur between channel features and low frequency floods, even though the low frequency floods greatly exceed the dimensions of the typical channel cross section. Dunne and Leopold (1978:644), for example, reported that channel width at bankfull stage displays strong statistical correlations with discharges of other recurrence intervals. They found correlation coefficients were typically between 0.75 and 0.93 when discharges of recurrence intervals of 2, 10, 25, 50, and 100 years were correlated with the bankfull channel width associated with a discharge of 1.5 years recurrence interval.

Bankfull Stage and High Frequency Floods

Most geomorphologists identify the bankfull stage as the dominant channel forming discharge. Numerous analyses representing many diverse sites reveal a modal value of 1.5 years recurrence interval for the bankfull stage. Williams (1978) has shown that bankfull stage at certain sites may deviate from the modal value of 1.5 years recurrence interval because of specific physiographic and hydraulic conditions. A certain level of disagreement has been introduced because different investigators have employed different definitions for bankfull stage. Williams (pp. 1141-1146) has reviewed the many definitions of bankfull stage and identified at least eleven different methodologies.

The tendency of the bankfull stage to associate with the 1.5 year flood in the annual flood series or approximately the 1.0 year flood in the partial duration series is the most convenient hydrologic association available for approximating a flood magnitude that has a high probability of occurrence. In this case, care must be taken to ensure that the "true" bankfull channel is identified.

The bankfull channel that most often approximates the 1.5 year flood stage is the active channel as defined by Riggs (1974:53) and by Dunne and Leopold (1978:612). The upper limits of the active channel can sometimes be identified by morphologic evidence such as breaks in slope on channel banks. More often, it is necessary to use other evidence such as changes in bed material characteristics at the channel margins or changes in vegetation at the channel margins. The channel cross section associated with a flood of recurrence interval that is approximately 1.5 years usually is represented by the area below the tops of active point bars or berms along the channel margins. For example, in Figure 1, the bankfull channel filled by a flood with a recurrence interval of 1.5 years is the cross section below line "b." The line "a" connects the bank tops of a former floodplain surface that has been converted to a terrace in response to an

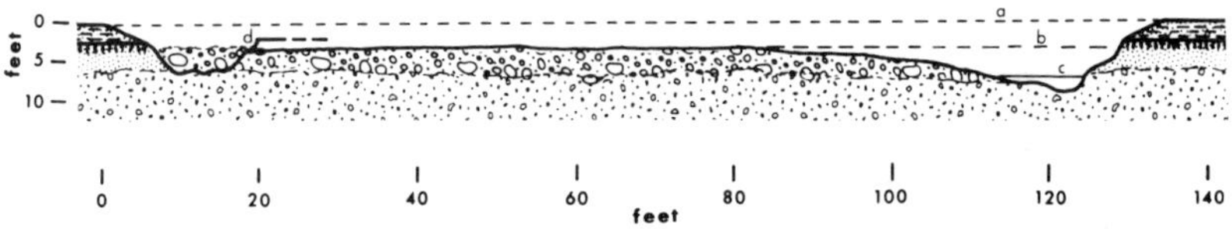

Figure 1. Typical tributary channel cross section in the headwaters of the Platte River, Grant County, Wisconsin. The cross section below line "a" contains large floods of recurrence intervals 50-60 years. The cross section below line "b" is the contemporary "bankfull" channel that contains a flood discharge with a recurrence interval of about 1.5 years. The channel that contained the approximately 1.5 year flood in presettlement time is denoted by line "d." The increase in size of the bankfull channel from presettlement times to the present is a result of increased flooding associated with the conversion of the landscape from a mosaic of native prairie and forest to fields and pastures. (For a detailed discussion, see Knox 1977.)

\- - -

increased incidence of large floods related to agricultural land use.

Care must also be exercised to ensure that channel cross sections which are selected to estimate flood discharges are not undergoing active metamorphosis. For example, a large flood that occurs shortly before the time of selection of channel sites can enlarge channel cross sections and result in over-estimation of the magnitude of the 1.5 year flood. In humid climate regions, after a large flood, a channel system within the period of a few years recovers a stable channel form that is adjusted to low magnitude and high frequency floods, providing other large floods do not occur (Costa 1974). In arid regions the length of time for channel recovery is longer. In very arid regions, where moisture and vegetation are scarce, flood discharges which erode the bed and banks of channels can leave a virtually permanent imprint on the landscape (Wolman and Gerson 1978:198). Riggs (1974:53) suggested the following criteria should be met when selecting a channel cross section that is to be used for estimating potential flood discharges of a specified recurrence frequency.

(1) Channel shape should be uniform throughout.
(2) The bed and banks should be of a material that has permitted the channel to develop into a normal size and shape for the flow regimen.
(3) Channel banks should appear to have been permanent for some years.

Discharge Estimation

Several empirical formulas have been developed to estimate discharge in a specified channel cross section. One of the more widely tested formulas that is known to produce reasonable estimates is the Manning equation:

$$Q = \frac{1.486}{n} AR^{0.66}S^{0.50}$$

where Q is discharge in cubic feet per second, A is channel cross section area in square feet, R is the hydraulic radius in feet, S is the dimensionless slope of the energy grade line, and n is a roughness coefficient. A detailed discussion of methodology for application of the Manning equation, including selection of appropriate roughness coefficients, has been presented by Barnes (1967). Example studies that illustrate the use of the Manning equation to estimate bankfull discharge include Brown (1971), Pickup and Warner (1976), and Williams (1978).

The Manning equation requires detailed survey data on a channel cross section and involves estimating the roughness coefficient. If there are sufficient stream gaging sites in a region to define relationships between channel geometry and discharge, then it may be possible to estimate discharge directly from only one morphologic parameter of the channel. For example, Leopold and Maddock (1953) introduced the term "hydraulic geometry" to describe the power function relationships between selected channel parameters and river discharge. They showed that channel properties such as width, depth, flow velocity, slope, hydraulic roughness, and sediment transport vary systematically with changes of river discharge. Hydraulic geometry relationships can be exploited in regions of homogeneous physiographic and climatic environments to directly estimate discharges from single channel parameters. A number of studies have revealed that channel width is one of the more reliable channel parameters for estimating discharges. Channel top width, for example, commonly increases with the 0.5 power of increasing discharge in the downstream direction. Therefore, discharge magnitudes can be estimated quickly at ungaged sites by substituting measured channel widths into an equation or nomogram that relates channel widths with discharge magnitudes.

The reliability of discharge estimates from single parameters of channels has been shown to be nearly as good as that achieved with equations that involve conventional watershed variables. For example, Hedman (1970) reported that the standard error of estimate for mean annual runoff predicted from channel dimensions was 38 and 29 percent respectively for perennial and ephemeral streams. These standard errors compare closely with a standard error of estimate of 33 percent for mean annual runoff predicted for the same region with conventional watershed variables such as drainage area, surface storage index, and mean annual precipitation.

Because the magnitudes of relatively frequent floods are an especially important influence on channel cross section area, the measured cross section capacity at the bankfull stage also is a very useful parameter for estimating magnitudes of high frequency floods. Brown (1971:307) used discharges estimated by the Manning equation as an index of channel capacity in streams of Minnesota, Kentucky, and Tennessee. He found that the bankfull discharge estimates explained 94 percent of the variance in the one-year flood, 88 percent of the variance in the ten-year flood, and about 70 percent of the variance in the maximum flood of record. These results are in agreement with Dunne and Leopold's (1978:644) findings that show stream width alone can explain between 85 and 55 percent of the variance of floods with recurrence intervals between about 2 and 100 years.

Most researchers have found that the strength of the correlation between channel parameters and floods of a given recurrence interval tends to decline as the recurrence interval becomes less frequent. In an attempt to improve the reliability of prediction of magnitudes of low frequency large floods, Riggs (1974:53) found it useful to differentiate between the active-channel section and the whole-channel section. Riggs defined the whole-channel section by breaks in bank slope at the edges of the valley flat and by the lower limits of permanent vegetation. In Figure 1, the whole-channel section is the cross section area below the dashed line "a." Riggs presented graphical illustrations which showed that whole-channel widths, when calibrated, produced very accurate approximations of the fifty-year floods in basins of northwestern United States, Colorado and Nevada, and Alaska. The U.S. Geological Survey has experimented with the use of channel parameters to estimate stream discharges of various return frequencies in several regions of the western United States. Dunne and Leopold (1978:646) have presented a listing of several recent U.S. Geological Survey publications that pertain to these western U.S. channel studies.

Transferability of Channel Data

In the processes of calibrating channel morphologic data to predict flood discharges, caution must be taken to ensure that environmental conditions are comparable between the region in question and the region that provides the basis for calibration. Climatic and physiographic conditions represent two principal classes of watershed variables that strongly influence the magnitudes and frequencies of floods. The hydrologic effects of climate are readily apparent through variations of temperature, precipitation, and runoff. The hydrologic effects of physiography often are more subtle, but are of equal importance in affecting magnitudes and frequencies of floods. Because of the more subtle influence on physiographic variables on channel properties and floods, the physiographic effect is illustrated below.

The hydrologic influence of physiographic factors can be illustrated by comparing magnitudes of the mean annual flood (recurrence interval 2.33 years) in small Midwest watersheds with different geomorphic histories but which now experience nearly identical climatic conditions (Figure 2). Note that in Figure 2A a ten-fold variation in the size of recorded mean annual floods may occur for a specified drainage basin size. The large variation from the best-fit line of Figure 2A is a result of mixing sites from physiographically diverse drainage basins. Different geomorphic histories have resulted in differences in drainage density, hillslope and channel gradients, watershed relief, and surficial materials, which in turn, produce different runoff characteristics. When the relationship between drainage area and mean annual flood is reexamined for sites with homogeneous physiographic conditions, the scatter of points from the best-fit line is greatly reduced (Figures 2B and 2C). Sample drainage networks and related statistics for the two regions represented in Figures 2B and 2C are presented in Figures 3 and 4. The drainage network of Figure 3 illustrates the low drainage density, low relief, gentle slope, and other characteristics which favor low intensity surface runoff that is typical of the glaciated region of north-central Iowa. Figure 4, in contrast, represents an unglaciated region of southwestern Wisconsin that has high drainage density, high relief, steep slopes, and other characteristics which favor high intensity surface runoff.

The important response of channel morphologic features to the

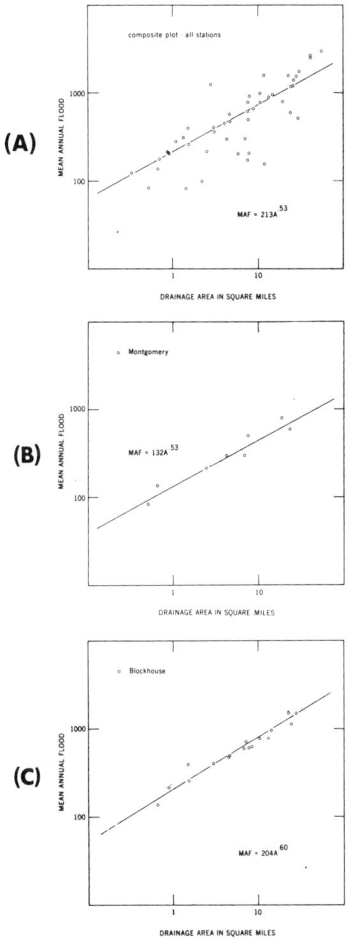

Figure 2. The relationship between mean annual flood and drainage area for Iowa and southwestern Wisconsin drainage basins. Mean annual flood (recurrence interval 2.33 years) was determined from recurrence interval curves for annual flood series in physiographic regions representative of study basins. Base data were from tabulations given in Schwob (1966), Ericson (1961), and in various U.S. Geological Survey Water Supply Papers.

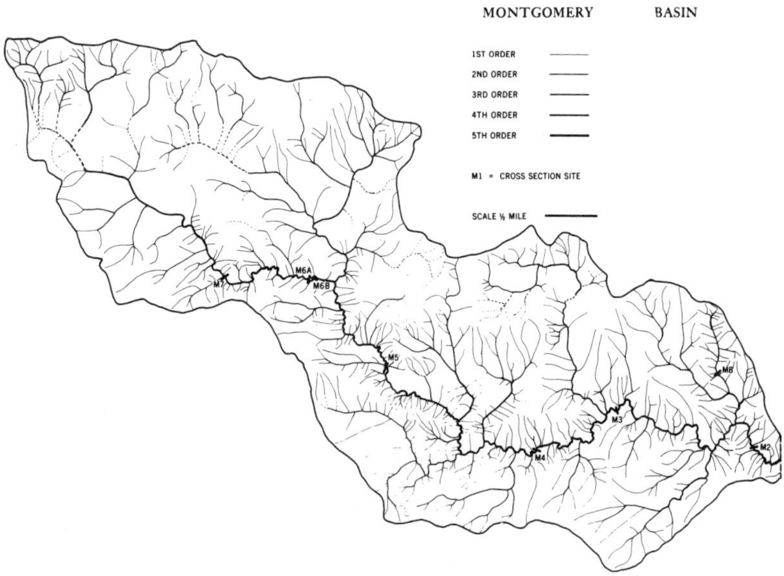

Figure 3. Montgomery basin is located on the Boone East and Mackey, Iowa 7 1/2 minute U.S. Geological Survey topographic maps. The drainage network was delineated by extending channels through contour crenulations (10 foot contour interval). Most of the upland is represented by gently undulating depositional topography associated with continental glaciation about 13000-14000 years ago. Local relief is usually less than 20 feet per square mile on the upland areas. Except for immediately adjacent to the entrenched main trunk channel, drainage density is relatively low for most of the watershed. The average frequency of unbranched first order channels is 43 per square mile. Their average length is about 1200 feet and their average gradient is 2%.

\- - -

physiographic conditions described above for Figures 2, 3, and 4 are illustrated by comparing flood magnitudes and channel dimensions for equivalent drainage basin areas in the two regions. Evaluations for a drainage basin size of 10 square miles reveal that the glaciated region represented by Figures 2B and 3 is associated with a mean annual flood of about 450 cubic feet per second, whereas the unglaciated region represented by Figures 2C and 4 is associated with a mean annual flood of about 800 cubic feet per second. From field survey data (Knox:1970), the associated channel cross sections at 10 square miles drainage area are approximately 33 feet wide and 3.8 feet deep at the glaciated basin site versus 57 feet wide and 3.9 feet deep at

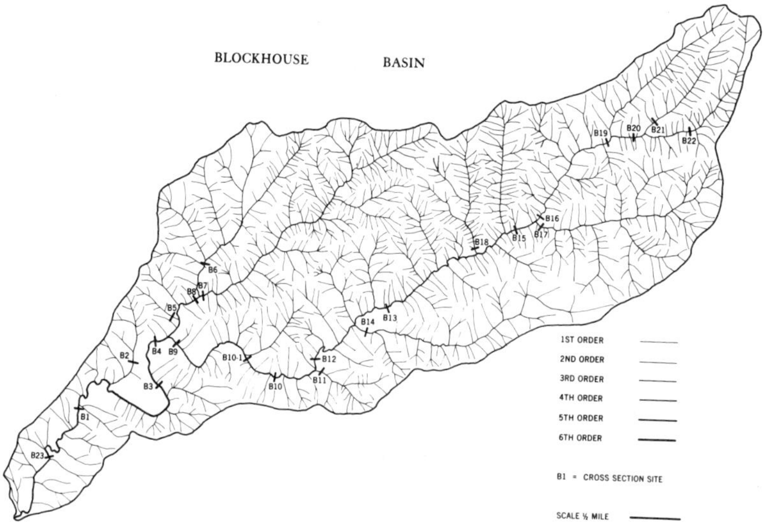

Figure 4. Blockhouse basin is located on the Platteville and Dickeyville, Wisconsin 7 1/2 minute U.S. Geological Survey topographic maps. The drainage network was delineated by extending channels through contour crenulations (10 and 20 foot contour intervals respectively on the Platteville and Dickeyville maps). The drainage basin is typical of the highly dissected unglaciated terrain of southwestern Wisconsin. Drainage density is relatively high and drainage patterns are affected by the fracture orientations in the underlying limestone and dolomite bedrock. The average frequency of unbranched first order channels is 71 per square mile. Their average length is about 850 feet and their average gradient is 7%. Local relief typically exceeds 100 feet per square mile and may exceed 200 feet per square mile along the main trunk channel.

\- - -

the unglaciated basin site.

High Frequency Floods and River Meanders

It has long been known that the wavelengths of river meanders in alluvium tend to increase downstream at about the square root of increasing discharge at the bankfull stage (Dury 1964). Therefore, where it is impossible to collect data on channel cross section dimensions and when no stream gaging data are available, a rough approximation of flood magnitudes which approximate the bankfull stage can be obtained by measurement of meander wavelengths from aerial photography or topographic maps, if available. Dury (1976:226) proposed the following equation to describe the stereotype relationship between meander wavelength (L) in feet and the 1.5 years bankfull discharge (Q) in cubic feet per second:

$$L = 15.179Q^{0.55}$$

Carlston (1965), working with a data set that was more geographically restricted than that developed by Dury suggested that meander wavelength is related to the mean annual flood (recurrence interval 2.33 years) according to the equation:

$$L = 106Q^{0.5}$$

where (L) is meander wavelength in feet and (Q) is discharge of the mean annual flood in cubic feet per second. A recent example of the application of these equations to estimate discharges in the upper Amazon tributaries of South America has been presented by Baker (1978: 222).

Although estimates of flood discharge from meander wavelengths offer a means for approximating magnitudes of high frequency floods when no other data are available, the method is often unreliable because of the independent influence of sediment load on meander geometry. Schumm (1968:44-45), for instance, has noted that tenfold variations in meander wavelength can occur for a given bankfull discharge. Schumm found that for United States rivers, high silt-clay channels tend to plot near or below Dury's equation line and that low silt-clay channels generally plot well above Dury's equation line.

Estimating the Magnitude and Frequency of Extreme Floods

Thus far, discussion has focused on the use of channel morphologic features to estimate the magnitudes of floods with recurrence intervals between about 1 and 50 years. The design of many engineering structures and decisions on floodplain zoning often requires estimates of the probable maximum flood (Benson 1973:30-31). The probable maximum flood usually is estimated from the probable maximum rainfall. Records of floods and rainfall frequently are too short to reliably estimate such extreme events. Geomorphological phenomena associated with floodplain stratigraphy and alluvial surfaces provide an alternative methodology for evaluating magnitude and frequency characteristics of extreme floods.

Large floods commonly produce unique erosional and depositional features that may be preserved in the alluvial stratigraphy of a floodplain. In addition, an extremely large flood may erode landforms or topple trees, whose ages provide an approximation of the minimum recurrence interval of the flood. Excellent reviews of the application of geomorphological phenomena to estimate recurrence intervals of extreme floods have been provided by Costa (1974; 1978). Costa (1978: 628) lists the following methods as useful in determining the frequency of extreme floods:

(1) tree ring analysis of floodplain trees destroyed or scarred by floodwaters;
(2) radiocarbon dating of organics in deposits eroded by recent floods;
(3) radiocarbon dating of organic debris in older flood deposits;
(4) pollen analysis and radiocarbon dating of organic materials

in lenses of fine-grained alluvium buried by coarse flood deposits;
(5) radiocarbon dating of soil horizons or organic material in slack-water deposits;
(6) radiocarbon dating or archaeological dating of soils developed in older floods deposits; and
(7) dating of landforms eroded or buried by floods based on changes in quantitative soil properties versus time.

Application of the above methodology requires that the investigator be able to recognize erosional and deposition effects of floods and also be able to approximate their dates of occurrence. Because of the long time spans that are often involved with estimation of recurrence intervals of extreme floods, tree rings, soil development or other degrees of weathering, radiocarbon dating and archaeological dating can all be applied, where available, to estimate ages. Description of the many characteristics of alluvial landforms and stratigraphy that are evidence of past extreme floods is beyond the scope of this short paper. Interested readers who desire detailed examples of geomorphic evidence of extreme floods are referred to papers by Baker (1977), Costa (1974; 1978), Gupta and Fox (1974), Harrison and Reid (1967), Helley and LaMarche (1973), Jahns (1947), Patton and Baker (1977), Ritter (1975), Sigafoos (1964), and Stewart and LaMarche (1967). Since many of the techniques involve extraction of information from alluvial sediments, a brief discussion follows to illustrate how sedimentary sequences record flood occurrences.

Stratigraphic Evidence of Floods

The transportation of alluvial sediments requires that a certain threshold of applied stress be exceeded. Transport thresholds for fine sediments are very low, but transport thresholds for very coarse sediments may only rarely be exceeded during periods of extreme flooding accompanied by high velocities and deep flows. Therefore, the presence of coarse textured sediments overlying fine textured sediments in certain stratigraphic sequences may represent deposition from a former large flood. The textural characteristics of these deposits can be related to flow conditions that, in turn, can be incorporated into hydraulic equations to estimate the magnitude of the related flood event.

Deposits of coarse sediments from large floods represent channel bed sediments that have been lifted to higher alluvial surfaces (e.g., tops of point bars, berms, and floodplains) during high stage flows. The deposition of coarse sediments on the higher surfaces interrupts fining-upward stratigraphic sequences that are typical of alluvial deposits produced by lateral accretion. For example, Figure 5 illustrates a fining-upward sequence across the surface of a point bar. The fining-upward sequence results from a differential in flow velocity between the outside and inside portions of the meander bend. As the meander bend in Figure 5 shifts laterally with further erosion of the concave bank and further deposition on the convex point bar, a vertical fining-upward sequence will also be produced in the point bar. However, the occurrence of a large flood can interrupt the fining-

Figure 5. View upstream in meander bend of McPherson Branch, Grant County, Wisconsin. Note the gradation from coarse gravels at the lowest elevations on the outside of the meander bend to fine silts at the top of the point bar on the insider of the meander bend. The fining-upward sequence is typical of lateral accretion deposits.

- - -

upward sequence as is shown in Figure 6. In succeeding years, as small and frequent floods deposit fine sediments on the lateral bar shown in Figure 6, the coarse sediments will be buried and surficial evidence of the large flood will no longer be apparent, except in excavations or in eroded stream banks. Figure 7, for example, represents at least five occurrences of coarse gravels from floods that were successively buried by fining-upward sedimentary sequences.

Extremely large floods often produce striking discontinuities in sedimentary sequences because they have the available energy to transport sediments that are not moved during floods of lesser magnitude. Fluvial responses to a flood of about 100 years recurrence interval are illustrated in Figures 8 and 9. Figure 8 is a downstream view showing the very coarse bedload sediment that was deposited on top of a fine-grained point bar. Also note in Figure 8 that several large blocks of peat-rich silt are interspersed with the coarse sediments. These blocks are derived from severe erosion of the basal portions of upstream channel banks. They are further evidence of the high energy conditions associated with the large flood. Figure 9 shows that the thick unit of coarse sediments from the large flood are deposited on a point bar that, in comparison, has a finer texture and is mantled with grassy vegetation.

Figure 6. Burial of fine-textured point bar sediments by gravel, cobbles, and small boulders, Blockhouse Creek, Grant County, Wisconsin. I have estimated the recurrence interval of the related flood to be between 10 and 20 years.

\- - -

Flood Stage Estimation

The particle sizes of the coarsest sediments in an alluvial deposit present an index of a river's competency to transport bedload sediment. Geomorphologists commonly estimate stream competency in terms of shear stress, as expressed by the DuBoys equation for boundary shear:

$$\tau = \gamma RS$$

where τ is the critical mean shear stress for initiating particle transport, γ is the specific weight of the water and sediment mixture, R is the hydraulic radius, and S is the energy gradient for the channel reach (Baker and Ritter 1975). The application of the DuBoys equation in geomorphic studies usually involves the assumption that

Figure 7. Cross section through a point bar, Platte River, Grant County, Wisconsin. The gravel horizons interrupt fining-upward sequences of sands, silts, and clays. I associate the gravels with occurrences of moderate magnitude floods with recurrence intervals between 5 and 20 years.

- - -

the specific weight of water (62.4 lb/ft^3) is a reasonable approximation of combined water and sediment mixture, and that the energy gradient is approximately parallel to the channel bottom slope and the water surface slope. Engineers tend to be somewhat critical of the DuBoys equation for predicing tractive force on the channel bed because of assumptions related to its use and because some feel that it inadequately represents the many friction and turbulent forces of resistance in the channel cross section (Bogárdi 1974:80-84; Graf 1971:124-129). Although DuBoys' original concept of sediment movement on the channel bed is now known to be incorrect, his equation has proven very useful in characterizing flow conditions and in evaluating sediment transportation potential (Bogárdi 1974:83).

As noted above, geomorphologists have found the DuBoys equation particularly useful for estimating stream competence to transport bedload sediment. Conventional usage of the equation involves relating bed sediment particle size, expressed by the mean length of the intermediate axis of the 10 largest particles from a deposit, to the calculated shear stress on the channel bed. Recently, Baker and Ritter (1975) compiled data from several sources to determine a general relationship between the intermediate axis particle size and mean

Figure 8. Point bar deposition related to July 1978 flood, Brush Creek, Monroe County, Wisconsin. The estimated recurrence interval of the flood was about 100 years at a U.S. Geological Survey gaging station located about 20 miles downstream.

\- - -

shear stress for coarse bedload material transported in rivers, canals, and flood channels. They computed shear stress by using maximum depth, rather than hydraulic radius, and they used 62.4 lb/ft^3 as an approximation of the specific weight of the water sediment mixture. Their regression line fit to the plotted data was:

$$D = 0.498\tau^{0.54}$$

where D is the particle size intermediate axis in feet and τ is the shear stress in pounds per square foot on the channel bed that is necessary to initiate particle movement. Baker and Ritter reported that the correlation coefficient for the above equation was 0.92.

Rearranging the terms in the stream competency equation provides a method for estimating magnitudes of shear stress associated with the former movement of coarse particles located in flood deposits. In turn, if an energy gradient (water or channel bed slope) can be approximated for the flood which deposited the coarse sediments, then the shear stress estimated in the above procedure can be incorporated into the DuBoys equation to solve for the depth of flow during the flood.

The above illustration was restricted to methodology for determination of former flood depths. It is worthwhile to note, however,

Figure 9. Point bar deposition related to July 1978 flood, Brush Creek, Monroe County, Wisconsin. The above site is the downstream end of the point bar shown in Figure 8. Note that the pre-flood point bar is vegetated and capped by sediments with finer textures than represented by the coarse point bar sediments of the 1978 flood.

- - -

that flood depths derived from the DuBoys equation can be input into other hydraulic equations to estimate flood discharges. An obvious example is the Manning equation, which if used, would require the additional estimation of a roughness coefficient for the former flood. Clearly, there are many assumptions and estimates involved with these indirect calculations of former flood characteristics. Nevertheless, these procedures provide an important alternative method for estimating potential magnitudes of former floods for which there is no written record. If deposits from former large floods can also be dated, as described earlier, then very important hydrologic information is provided about the magnitude and frequency characteristics of large and relatively rare flood events.

Summary and Conclusions

Geomorphic properties of alluvial channels and floodplains offer an important supplementary source of information on magnitude and frequency characteristics of floods. The morphologic dimensions of alluvial channels have strong statistical correlations with low magnitude, high frequency floods. The strong relationship can be exploited to estimate magnitudes of high frequency floods in regions where river gaging data do not exist. Because the morphology of alluvial channels in most regions is shaped by floods which tend to recur frequently, channel morphology has limited usefulness in estimating potential magnitudes of extremely large and infrequent floods. Magnitudes of potential extreme floods are best estimated from the physical properties of alluvial deposits that are related to former extreme floods. Especially useful in the reconstruction of former depths of large floods is the relationship between stream competency, as measured by the DuBoys equation, and the size of the largest bed-load particles transported during floods. Because extreme floods frequently result in the burial of datable materials (radiocarbon, archaeological, etc.), it also is often possible to date their approximate time of occurrence. Because the time span associated with alluvial sediments under investigation may spread over hundreds of years, or even a few thousand years, the dated flood events provide an important understanding of the recurrence intervals of large and extreme floods. These methods also greatly assist in the determination of the true recurrence interval of an unusually extreme flood that may be contained in a short record associated with a period of stream gaging. As with most techniques and methodologies for estimation of flood characteristics, there are many assumptions and conditions associated with the use of the geomorphic methods. Wherever possible, results from geomorphic methods should be cross-checked against results from other methods.

References

Baker, V.R. 1977. Stream Channel Response to Floods, With Examples from Central Texas. Geological Society of America Bulletin 88: 1057-1071.

__________. 1978. Adjustment of Fluvial Systems to Climate and Source Terrain in Tropical and Subtropical Environments. In A.D. Miall, ed., Fluvial Sedimentology, Calgary: Canadian Society of Petroleum Geologists, pp. 211-230.

Baker, V.R., and P.F. Ritter. 1975. Competence of Rivers to Transport Coarse Bedload Material. Geological Society of America Bulletin 86:975-978.

Barnes, H.H., Jr. 1967. Roughness Characteristics of Natural Channels. Washington: U.S. Geological Survey Water Supply Paper 1849, 213 pp.

Benson, M.A. 1973. Thoughts on the Design of Design Floods. In E.F. Scholz, V.A. Koelzer, and K. Mahmood, eds., Floods and Droughts, Fort Collins: Water Resources Publications, Proceedings of the Second International Symposium in Hydrology, pp. 27-33.

Bogárdi, J. 1974. Sediment Transport in Alluvial Channels. Budapest: Akadémiai Kiadó, 826 pp.

Brown, D.A. 1971. Stream Channels and Flow Relations. Water Resources Research 7:304-310.

Carlston, C.W. 1965. The Relation of Free Meander Geometry to Stream Discharge and Its Geomorphic Implications. American Journal of Science 263:864-885.

Costa, J.E. 1974. Response and Recovery of a Piedmont Watershed from Tropical Storm Agnes, June 1972. Water Resources Research 10: 106-112.

__________. 1978. Holocene Stratigraphy in Flood Frequency Analysis. Water Resources Research 14:626-632.

Dunne, T. and L.B. Leopold. 1978. Water in Environmental Planning. San Francisco: W.H. Freeman & Co., 818 pp.

Dury, G.H. 1964. Principles of Underfit Streams. Washington: U.S. Geological Survey Professional Paper 452-A, 67 pp.

__________. 1974. Magnitude-Frequency Analysis and Channel Morphometry. In M. Morisawa, ed., Fluvial Geomorphology, Binghamton: State University of New York, pp. 19-121.

__________. 1976. Discharge Prediction, Present and Former, From Channel Dimensions. Journal of Hydrology 30:219-245.

Ericson, D.W. 1961. Floods in Wisconsin: Magnitude and Frequency. Madison, Wisconsin: U.S. Geological Survey Open File Report, 109 pp.

Gilbert, G.K. 1877. Report on the Geology of the Henry Mountains: U.S. Geographical and Geological Survey of the Rocky Mountain Region. Washington, U.S. Geological Survey, 160 pp.

Graf, W.H. 1971. Hydraulics of Sediment Transport. New York: McGraw-Hill Book Company, 513 pp.

Gupta, A. and H. Fox. 1974. Effects of High-Magnitude Floods on Channel Form: A Case Study from the Maryland Piedmont. Water Resources Research 10:499-509.

Harrison, S.S. and J.R. Reid. 1967. A Flood-Frequency Graph Based on Tree-Scar Data. Proceedings of the North Dakota Academy Sciences 21:23-33.

Hedman, E.R. 1970. Mean Annual Runoff as Related to Channel Geometry of Selected Streams in California. Washington: U.S. Geological Survey Water Supply Paper 1999-E, 17 pp.

Helley, E.J., and V.C. LaMarche. 1973. Historic Flood Information for Northern California Streams from Geological and Botanical Evidence. Washington: U.S. Geological Survey Professional Paper 485-E, 16 pp.

Jahns, R.H. 1947. Geologic features of the Connecticut Valley, Massachusetts as Related to Recent Floods. Washington: U.S. Geological Survey Water Supply Paper 996, 158 pp.

Knox, J.C. 1970. Stream Channel Adjustment to Physiographic Factors in Small Drainage Basins: Iowa and Southwestern Wisconsin. Iowa City: University of Iowa, Department of Geography, Ph.D. Dissertation, 321 pp.

__________. 1977. Human Impacts on Wisconsin Stream Channels. Annals of the Association of American Geographers 67:323-342.

Lane, E.W. 1937. Stable Channels in Erodible Materials. American Society of Civil Engineers, Transactions No. 102:123-194.

Leopold, L.B., and T. Maddock. 1953. The Hydraulic Geometry of Stream Channels and Some Physiographic Implications. Washington: U.S. Geological Survey Professional Paper 252, 57 pp.

Patton, P.C. and V.R. Baker. 1977. Geomorphic Response of Central Texas Channels to Catastrophic Rainfall and Runoff. In D.O. Doehring, ed., <u>Geomorphology of Arid and Semiarid Regions</u>, Binghamton, New York: Publications in Geomorphology, pp. 189-217.

Pickup, G. and R.F. Warner. 1976. Effects of Hydrologic Regime on Magnitude and Frequency of Dominant Discharge. Journal of Hydrology 29:51-75.

Riggs, H.C. 1974. Flash Flood Potential from Channel Measurements. International Association of Scientific Hydrology, Symposium of Paris, IAHS-AISH Publication No. 112, pp. 52-56.

Ritter, D.F. 1975. Stratigraphic Implications of Coarse-Grained Gravel Deposited as Overbank Sediment, Southern Illinois. Journal of Geology 83:645-650.

Rubey, W.W. 1952. Geology and Mineral Resources of the Hardin and Brussels Quadrangles (in Illinois). Washington: U.S. Geological Survey Professional Paper 218, 175 pp.

Schumm, S.A. 1960. The Shape of Alluvial Channels in Relation to Sediment Type. Washington: U.S. Geological Survey Professional Paper 352-B, pp. 17-30.

__________. 1968. River Adjustment to Altered Hydrologic Regimen: Murrumbidgee River and Paleochannels, Australia. Washington: U.S. Geological Survey Professional Paper 598.

Schwob, H.H. 1966. Magnitude and Frequency of Iowa Floods. Ames, Iowa: Iowa Highway Research Board Bulletin 28.

Sigafoos, R.S. 1964. Botanic Evidence of Floods and Flood Plain Development. Washington: U.S. Geological Survey Professional Paper 485-A, 35 pp.

Stewart, J.H., and V.C. LaMarche. 1967. Erosion and Deposition in the Flood of December 1964 on Coffee Creek, Trinity County, California. Washington: U.S. Geological Suvey Professional Paper 422-K, 22 pp.

Williams, G.P. 1978. Bank-Full Discharge of Rivers. Water Resources Research 14:1141-1154.

Wolman, M.G. and R. Gerson. 1978. Relative Scales of Time and Effectiveness of Climate in Watershed Geomorphology. Earth Surface Processes 3:189-208.

Wolman, M.G., and J.P. Miller. 1960. Magnitude and Frequency of Forces in Geomorphic Processes. Journal of Geology 68:54-74.

AUGMENTATION OF HYDROLOGIC RECORDS USING TREE RINGS

Charles W. Stockton and William R. Boggess[1]

INTRODUCTION

Any statistical analysis of hydrologic data is handicapped when the available records are of relatively short duration as is generally true in the western hemisphere. This is because the available records are not necessarily a random sample of the infinite population of events and consequently any statistical descriptions are likely to be in error to some extent.

Recent work by Stockton (1975), Stockton and Jacoby (1976), Stockton (1977), Mitchell, Stockton and Meko (1979), has shown that tree-ring data can be used to extend available hydrologic records backward in time, thereby providing a longer record from which to more accurately estimate the three most common statistics used in hydrology: the mean, the variance, and the first order correlation. In addition, records reconstructed from tree-ring data series can provide information on (a) longest periods of sustained high or low flows; (b) the representativeness of the historical record in comparison to the long-term record reconstructed from tree-ring data; and (c) the long-term persistence as expressed by the Hurst Coefficient. Among the useful features of tree-ring series as repositories of hydrologic information are their great number; their longevity; the critical fact that the information they contain is annually cumulative; and the absolute dating control. Thus, tree-ring data can be an important source of added information to the hydrologist, provided that hydrologic inferences based on such data can be supported by acceptable statistical controls.

When hydrologic phenomena are analyzed statistically it is usually assumed that the finite record of events represents a random sample from an infinite population, and the occurrence of each event is governed by some probability distribution. Any change in the hydrologic regime with which a given record of events is associated results in a change in the probability distribution.

For practical purposes, a probability distribution is described by the mean ($\overline{X}$), a measure of central tendency; the variance (S^2), a measure of the average spread of the events about the mean; and the skewness (G), a measure of the asymmetry of the distribution of the events about the mean. In most cases these three parameters uniquely define a probability distribution and are useful for describing hydrologic phenomena. For many annual runoff and tree-ring series, the variables are normally distributed (skewness approaches zero) and the probability distribution is completely described by the mean and variance.

[1] Respectively, Assistant Professor and Research Associate, Laboratory of Tree-Ring Research, University of Arizona, Tucson, AZ 85721

In the first order autoregressive models of runoff time series, the first order autocorrelation (r_1), a measure of short term persistence in a series, is used along with the mean and variance. The population values of these statistics are usually unknown and must be estimated from the existing record of observations. Wallis, Matalas and Slack (1974) have performed Monte Carlo experiments involving hundreds of thousands of trials on each of several probability distribution functions for various values of n and γ (the population coefficient of skew). These results show that the distributions of $\overline{X}$, S, and G are functions of n, γ, and the probability distribution function. While $\overline{X}$ is an unbiased estimate, S and G are biased estimates of the population values. The biasness in S and G are functions of n, γ and the probability distribution. They present corrective algorithms for providing approximately unbiased estimates. Errors in the estimates of the population parameters owing to shortness of observed records are preserved in any synthetic series that is generated from the available data. Rodriguez-Iturbe (1969) showed that if the length of an annual runoff record is 40 years or less, there may be an error of 2% to 20% in estimation of the mean, from 15% to 60% in the estimation of the variance, and as much as 200% in the estimation of the first order autocorrelation. The high error in the autocorrelation is probably related to the inadequacy of short records in estimating the low-frequency persistence in climatologic data, which Mandelbrot and Wallis (1968) have dubbed the "Noah and Joseph effects" after the well-known Biblical calamities. The "Noah effect" means that extreme precipitation tends to be very extreme, the archetype being the 40-day rains that resulted in inundation of the entire earth (Gen. 7:11-21). In contrast the "Joseph effect" means that a period of unusually high (or low) precipitation is commonly an extended one, so named after the widespread famine of seven years' duration that Joseph had predicted from Pharoah's dream (Gen. 41:51-57).

Fiering (1962), Matalas and Jacobs (1964), and Julian and Fritts (1968) have demonstrated the use of correlation techniques for augmenting hydrologic records. In each of these cases, a single record was used to augment another. Fiering (1963) also approached the problem using multiple linear regression, where several independent variables were used to predict a dependent variable. He showed that a better estimate of the mean could be obtained in the multivariate case if

$R^2 \geq q_i / n_1 - q_i$ where

R = the combined correlation coefficient
q_i = the number of variables included in the prediction equation
n_1 = the length of the record to be extended

In the case of the variance, the variance of the reconstructed record is a better estimate if the relative information ratio I (the ratio of the variance to that estimated from the original record) exceeds 1. When I exceeds unity, it implies that the variance of the estimate made from the original record is larger than that of the estimate made from the combined record, and therefore a more precise estimate is computed from the combined data. As a general rule the estimate from the longer series is more reliable if R (the correlation coefficient) exceeds 0.80 (Table 3 of Fiering, 1963; p.2 of Matalas and Jacobs, 1964). However, Matalas and Jacobs (1964) point out that these requirements can be reduced and that the parameters estimated from the longer series are an unbiased estimate if a noise factor is added to the estimated values.

Hurst, Black and Simaika (1965) used tree rings as well as other data to obtain long-term estimates of the Hurst coefficient, k. Mandelbrot and Wallis (1969) also used tree-ring data to demonstrate that long-term climatic data possess an average k value of .73. In addition, tree-ring data were used by Hipel (1975) to secure long-term estimates of parameters used in Box-Jenkins modeling of hydrologic records and by Lettenmaier and Burges (1978) to test for climatically induced nonstationarity.

The basis for comparing hydrologic records with tree-ring series is the hypothesis that each responds to a common climatic signal or signals that permit prediction of annual runoff from the annual ring-width index. Indexing of the measured ring widths is necessary to convert the nonstationary ring width series to a stationary time series (Matalas, 1962). This is done by converting the raw ring widths to indices by first fitting a curve to the raw data and then dividing each actual ring width by the corresponding curve value. This treatment first removes the growth trend and then equalizes or brings all ring-width curves to a uniform mean value (a value of 1). When plotted against time, the indices represent a new time series known as a standardized ring-width chronology. These procedures are discussed in detail by Fritts (1976, Ch. 6).

As shown by Figure 1, precipitation (a), temperature (b), and evapotranspiration (c) influence the water balance and directly affect both runoff and tree growth. In the case of tree growth these variables, especially temperature, have physiological influences not directly related to the water balance. The seasonal distribution of the variables (d) influences both runoff and tree growth; on the latter the monthly distribution includes at least a 14-month period -- from the July prior to the growing season in which the ring is formed to the July concurrent with the growing season (Fritts *et al.*,1971). Spatial distribution of precipitation and temperature (e) within large watersheds may influence both the annual runoff regime and the variability in growth of trees from site to site. The noise component represents both the model's inability to adequately describe the two series and the differences in the way the two series respond to climatic inputs.

Of major concern in reconstructing hydrologic series from tree-ring records is the difference in persistence within each of the two series - e.g. how much do events of the previous year or years influence the current year? In many studies, differences in persistence have been resolved by using lagged dependent variables on the right-hand side of the reconstruction equation (Johnston, 1963). Unfortunately, this causes the residuals to be dependent upon residuals of prior reconstructed values. Also, the regression coefficients tend to be biased although they have the properties of consistency and efficiency if the residuals are normally distributed (Johnston, 1963). Another remedy would be to use a matrix of the tree-ring data, lagged up to three times, and extract principal components from this supplemental matrix. The covariation in this matrix can be decomposed by extracting the eigenvectors, the amplitudes of which yield a new set of uncorrelated variables. To account for the differences in persistence, these amplitudes may be lagged in certain ways with the runoff data, and multiple regression may be used to weight the respective series.

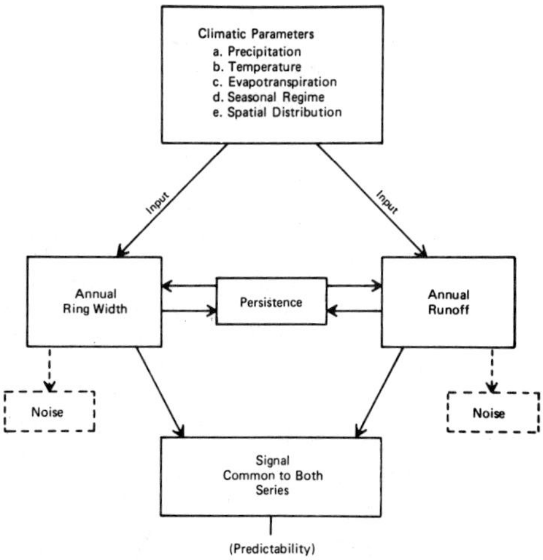

Figure 1. Schematic diagram of relationship between ring-width series and annual runoff series for medium and large watersheds

Climatic Response Function

Fritts (1976) has shown how the tree-ring response to climatic inputs can be recorded in ring widths through time. This is illustrated in a greatly simplified form by Figure 2. The climatic factors, precipitation and temperature, coupled with atmospheric elements, carbon dioxide and wind movement, are reflected in the ring width for the current year (t); for the year t + 1 (bud development and sugar and hormone carryover); and for year t + k through leaf, root and fruit processes. The effect of soil moisture and food storage carryover is superimposed upon the climatic carryover as reflected by the tendency for a rather significant autocorrelation in ring-width series. Fritts et al., 1971) described a method for modeling the response of trees to different climatic variables. Their method, which provides a means of determining the importance of monthly temperature and precipitation throughout the 14-month period prior to actual growth, uses multiple linear regression to predict ring-width indices from the amplitudes of eigenvectors of monthly precipitation and temperature along with variables representing the persistence within the ring width series. That is, the tree-ring indices are fit to the model:

$$y_t = \theta_1 \xi_{1t} + \theta_2 \xi_{2t} + \dots + \theta_p \xi_{pt} + \phi_1 y_{t-1} + \dots + \phi_3 y_{t-3} + e_t, \qquad (1)$$

where

y_t = standardized ring-width in year t

Θ_p = least squares coefficient for variable ξ_p

ξ_{pt} = amplitude for year t of p eigenvectors extracted from a correlation matrix of climatic variables

Θ_{t-n} = least squares coefficient for variable y_{t-n}

y_{t-n} = standardized ring-width index at time t-n

e_t = error component

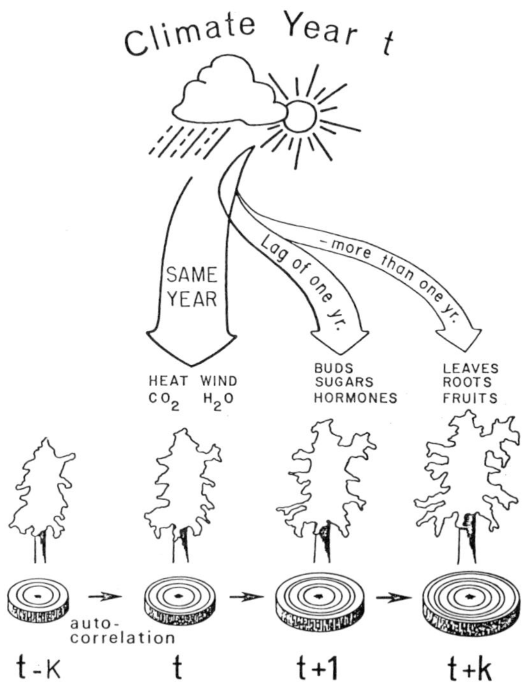

Figure 2. Schematic diagram showing how climate of year t can affect tree growth in year t + k (after Fritts, 1976).

Because of the transformations performed on the climatic data, i.e., the derivation of the amplitudes of the eigenvectors, the climatic variables are orthogonal and fulfill one major assumption - that of the independence of the "independent" variables. Additionally, use of the principal components reduces the number of variables, thereby reducing the dimension of the problem. The use of these transformations, however, somewhat obscures the physical relationship of the effect of climate upon ring width. Fritts et al. (1971) suggest that a solution to this undesirable effect is in the "response function," which transforms the principal components back to the original variables. If the components are expressed in terms of the original variables, x_1, $x_2, \ldots x_n$, Eq. (1) is transformed to a linear equation in x. Each additional component changes the coefficients attached to the several x_i terms, these changes being proportional to the elements of the eigenvector (corresponding to the amplitude) newly added. Thus,

$$y_t = \theta_0 + \theta_1 (a_{11} x_1 + a_{21} x_2 + \ldots) \qquad \text{for } \xi_1 \tag{2}$$

or

$$\begin{aligned} y_t &= \theta_0 + \theta_1 a_{11} x_1 + \theta_1 a_{12} x_2 + \ldots + \theta_2 a_{21} x_1 + \theta_2 a_{22} x_2 + \ldots \qquad (3) \\ &= \theta_0 + x_1(\theta_1 a_{11} + \theta_2 a_{21}) + x_2(\theta_1 a_{12} + \theta_2 a_{22}) + \ldots \qquad \text{for } \xi_1 \text{ and } \xi_2, \end{aligned}$$

where the a's are elements of the respective eigenvectors and the x's are observed values of the climatic variables. Thus, if the variable x_i is factored out of any term, the resultant term is the sum of the regression coefficient times the eigenvector elements. Since these regression coefficients and eigenvector elements are determined in an unbiased manner from the observed values of the variables, the result should be a way to compare the response of the dependent variable y against the respective independent variables x. By plotting these sums of regression coefficients times eigenvector elements for the same independent variables but different dependent variables, one can compare "response functions" for various dependent variables.

Transfer Functions

With the limitations of the response function, it is possible to develop an equation for reconstructing a pattern of past annual runoff from tree-ring indices.

If the tree-ring data are sampled at widely dispersed sites over a moderately large watershed, say 2000 or more square miles, a means is needed to incorporate into the model the spatial distribution of the tree-ring data at any time t. In addition, the persistence with time may vary among tree-ring sites, and a way is needed to compensate for this difference in persistence and at the same time model the generating mechanism within these data so that it can be compensated for in the reconstruction equation. By lagging the matrix of tree-ring series in time up to t - 3 and extracting the eigenvectors from this combined matrix, a space-time distribution of the tree-ring series is accomplished. Thus, a least squares reconstruction equation is obtained:

$$f_t = \beta_0 + \beta_1 \xi_{1t} + \beta_2 \xi_{2t} + \ldots + f_{t-1} + f_{t-2} + f_{t-3} + e_t, \tag{4}$$

where

f_t is runoff at time t

β's are least square regression coefficients

ξ_t's are the amplitudes of the eigenvectors extracted from the correlation matrix of the combined, lagged matrix

f_{t-n} are previous-year runoff values, and

e_t is the error resulting from inadequacy of the model itself

It was found, subsequent to the work of Stockton (1971), that by lagging the values of runoff with respect to the tree-ring series, one can compensate for the generating mechanism differences in the two series. That is, the runoff at time t is a function of the tree-ring data x at times x_{t+1}, x_t, x_{t-1}, x_{t-2}. This provides an expression of the ARMA (autoregressive-moving average) model, established by Stockton (1971) as typical for many tree-ring data series.

In using a transfer function equation like Eq. (4), five basic assumptions are made:

1. The climatic interaction between runoff and tree growth is constant and does not change with time.
2. A linear relationship exists between the tree-ring series and the annual runoff series.
3. The variables are multivariate normal distributed.
4. The residuals are independent (i.e., the cross product of the residuals is zero).
5. Expected value of residuals, e_t, is zero.

EXAMPLES OF HYDROLOGIC RECONSTRUCTIONS

Both the validity and utility of using tree-ring data for hydrologic reconstructions are illustrated by results of two rather diverse projects that have been carried out at the Tree-Ring Laboratory during the past decade. These include: 1) reconstruction of flow for the major streams of the Upper Colorado River Basin back to A.D. 1512; and 2) reconstruction of the occurrence, severity and duration of large-scale drought in the western United States during the past 2½ centuries.

Upper Colorado River Basin

Using the previously outlined techniques, Stockton (1976) and Stockton and Jacoby (1976) investigated the flow properties for various streams in the Upper Colorado River Basin (UCRB). Twelve different watersheds were analyzed and long-term runoff records reconstructed. These included the three main rivers, the Green, Colorado and San Juan.

The UCRB comprises approximately 109,300 square miles and has a wide range of physical characteristics. Much of the region is arid to semi-arid with typical low precipitation and dry-area vegetation. The mountainous areas of the Basin are forested and have much greater precipitation.

Streamflow measurements from the U.S. Geological Survey (USGS) indicate that almost 86 percent of the runoff from the entire UCRB originated on only 15 percent of the area (Figure 3). This area includes all of the high mountains where, due to orographic precipitation, the streamflow per unit area is much greater than in the lower, drier portions of the Basin.

Tree-ring series data were utilized from as many of the major runoff producing areas as possible. Several climatically sensitive series, collected for other projects, were available and these were supplemented by additional samples. Thirty different sites were used for the reconstruction; their location is also shown on Figure 3.

All of the tree-ring series are mean-value functions; that is, at least two series from each tree sampled were averaged to provide the best estimate of the series from a particular tree. At least ten trees were sampled at each site and together formed a site series. This number, 10 trees per site, is based on past experience and represents the sample size necessary to maximize the climatic signal inherent in the tree rings.

A number of subsets were selected from the 30 sites that appear to best reflect tree growth as related to moisture variations. Based on previous work by Stockton and Fritts (1971), it is reasonable to assume that in this region the larger ring-width indices indicate years with above-normal seasonal precipitation coupled with below-normal temperatures. Conversely the narrower rings indicate years during which the opposite was true.

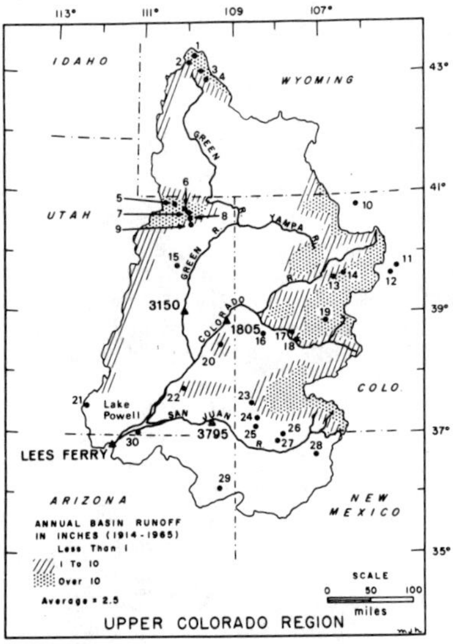

Figure 3. Map of Upper Colorado River Region, showing major runoff-producing areas (shaded), locations of tree-ring data sites (dots) and major gaging sites (triangles).

The individual tree-ring sites used in the reconstructions were not necessarily of equal importance. Consequently, we used a method of spatial and temporal weighting wherein eigenvectors were extracted from a correlation matrix for the suite of tree-ring series to be utilized and for which each was lagged three times. For example, in one of the reconstructions for the Colorado River at Lee Ferry, we used data from 17 tree-ring sites, and when each had been lagged three times, the resulting matrix of data consisted of 68 variables. The resulting eigenvectors were then utilized to weight the original series, and the results were principal components or amplitudes. The resultant weighted value had the desirable property of being orthogonal, and in addition, as long as the variance that the eigenvector accounts for was sufficiently large, the resultant weighting usually was physically reasonable. As the covariance diminished, the eigenvectors were still orthogonal but probably had no physical relationships, as the orthogonality constraint becomes overriding. In all cases, only eigenvectors with corresponding roots greater than 1.00 and accounting for a greater percentage of the variance than would be expected from a matrix of a comparable number of random series were used. In no case were more than 30 eigenvectors used.

The reconstruction equations were established for each model by using least-squares analysis, in which the individual orthogonal variables were evaluated before they were entered into the equation. If the F value did not exceed 3.00, the variable was not used in the calibration equation.

Runoff Reconstructions For Lee Ferry, Arizona

We have reconstructed the virgin, or total, annual runoff at Lee Ferry, Arizona (the 1922 Colorado River Compact accounting point), using different models, different tree-ring data grids, and different flow records for calibration. The models were varied on the basis of percentage variance accounted for in the calibration record, unbiasness in the residuals, and ability to duplicate data not used in the calibration equation. The data used for calibration were from the USGS (Hely, 1969; Table 5, page D-49). The records of actual flow from 1896 through 1945 were published by the Bureau of Reclamation (U.S. Department of the Interior, 1954; pp. 145-146). Records of virgin flow from 1946 through 1966 were furnished by the Upper Colorado River Commission. Another data source was the *Comprehensive Framework Study of the Upper Colorado River Basin* (Upper Colorado Region State-Federal Interagency Group, 1971; Table II) and covered the period from 1914 through 1965. Both sets of data represent the estimated virgin outflow from the Upper Basin. The mean and standard deviation of the data from Hely (1969) are 14.65 and 4.47 maf, respectively, for the period 1914 through 1965 (52 years), whereas data from the *Comprehensive Framework Study* show 14.87 and 4.20 maf, respectively, for the same period. For the 65-year period 1899 through 1963 (data from Hely, 1969), the mean is 15.12 maf and the standard deviation is 4.25 maf. The tree-ring data grids utilized consisted of subsets comprising 13 and 17 tree-ring sites, respectively.

The 65-year calibration period includes a portion of the historical record that was estimated from a longer flow record upstream. There is some question as to whether these data should or should not be used in a calibration equation, especially as our study indicates significant variations in flow for different areas of the UCRB. However, the longer record does include some of the larger flow years which are desirable for inclusion in the calibration equation. To check the reliability of the 65-year calibration equation, we computed another equation using only 50 years of data (1914 through 1963) and compared the reconstructive qualities with the published data covering the period 1896 through 1913. The reconstruction equation for the 65-year calibration period accounted for 75 percent of the variance in the calibration period data, for the 50-year calibration period, 78 percent of the variance is accounted for and for the Comprehensive Framework Study data (50 years), 87 percent of the variance in the calibration period is accounted for by the reconstruction equation.

The yearly respective means for long-term virgin flow are 14.15 maf for the 65-year calibration, 13.94 maf for the 50-year calibration, and 13.06 maf for the Comprehensive Framework Study.

Figure 4 illustrates how the three different reconstruction equations duplicate the estimated total streamflow data for the period 1896 through 1913. As one would expect, the equation calibrated by including this period has a mean that is closer to the mean for this period of 18 years, even though the equation explains less of the variance of the whole calibration period. The extension of the streamflow record back through this 18-year period is based mainly on data from the region where the wet period of the early 1900's began sooner than it did in the rest of the UCRB. Therefore, it is likely that this estimated flow record may be biased toward the high side.

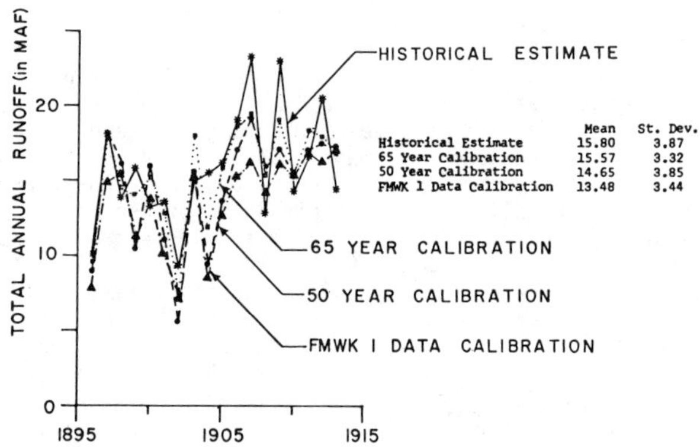

Figure 4. Comparison of the historical estimate of flow at Lee Ferry with that estimated by tree-ring data using a 65-year calibration equation, a 50-year calibration equation, and the Framework I data calibration equation.

The yearly mean of the estimated record for the 18-year period is 15.80 maf and the standard deviation is 3.87 maf. The 65-year calibration gives a yearly mean of 15.6 maf and a standard deviation of 3.3 maf, whereas the 50-year calibration equation gives a yearly mean of 14.65 maf and a standard deviation of 3.85 maf. The overall reconstructions seem to be unbiased, in that for the first equation the reconstructed values both exceed the estimated values nine times and are less than nine times. For the second equation the reconstructed values exceed the estimated values eight times and are less ten times. From the Framework 1 data equation, the resultant yearly mean is 13.5 maf and the standard deviation is 3.4 maf; both are considerably less than are those for the estimated data and for equations (1) and (2). When compared to the independent data for the period from 1896 through 1913, a tendency is seen for slight biasness in underestimation with respect to this particular data set, since for the 18 years there are 11 underestimates and 7 overestimates.

The autocorrelation structure in all three cases appears to be quite similar, with the first-order autocorrelation being approximately 0.33. We have not carefully analyzed the structure of the autocorrelation, but judging from the correlograms (Figure 5), it is more complex than that of a simple autoregressive model.

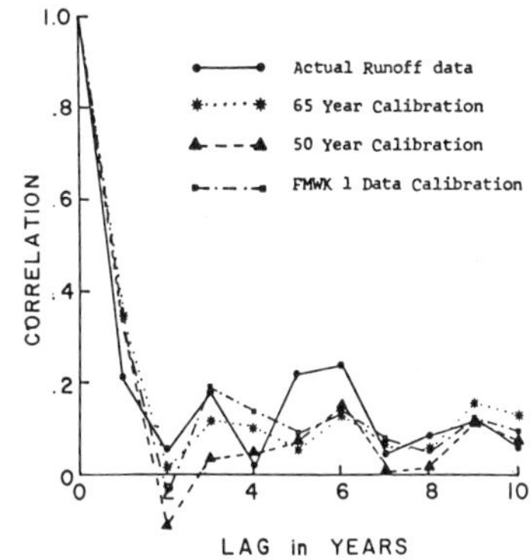

Figure 5. Comparative correlograms for the three reconstructed records of flow of the Colorado River at Lee Ferry and the gaged record.

A critical question is which of the three reconstructions of past runoff at Lee Ferry is the best. Our reasoning is as follows. The reconstruction based on the 65-year calibration record includes data from the 18-year estimated record (1896 through 1913) that are questionable for calibration purposes and for previously stated reasons are probably biased toward the high side. Therefore, we feel that the best estimate of the long-term reconstruction is an average of the results obtained from the two 50-year calibration series resulting in means of 13.94 maf per year and 13.06 maf per year respectively. Consequently, we arrive at an estimated mean annual runoff of 13.5 maf, $\pm$ 0.5 maf, and a reconstructed long-term hydrograph as shown in Figure 6.

The transfer functions do not account for 100 percent of the variance in the calibration period. Therefore, it is desirable to add a random "noise" component to the reconstructed record to procure the total variance. This has been done for the Lee Ferry record by using a pseudo random number generator and the term $\pm t_i S_f (1-R^2)$ where

t_i = a normally distributed random variable with 0 mean and unit variance

S_f = the standard deviation of the original runoff series

R = the correlation coefficient between the tree-ring series and the original runoff record.

Figure 7 shows the resultant hydrograph, and Table 1 compares the statistics of the virgin flow record, the reconstructed record without the noise component added, and the reconstructed record with noise added.

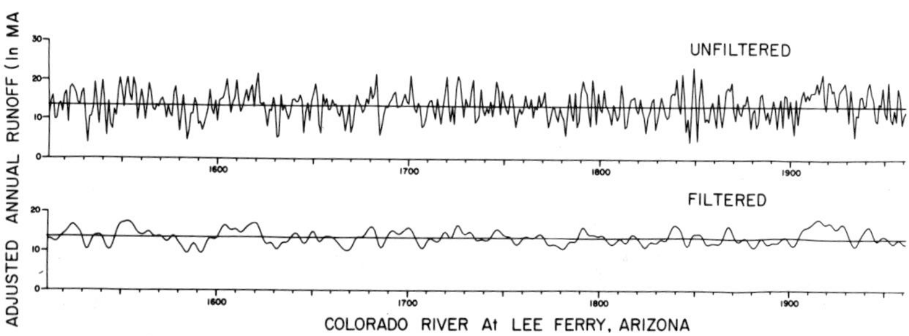

Figure 6. Long-term hydrograph of annual runoff at Lee Ferry, Arizona, based on an average of the results of two reconstructions. (The upper graph shows the actual year-by-year values; the lower graph is the same data, but with the high-frequency components --those with a period less than 10 years--removed).

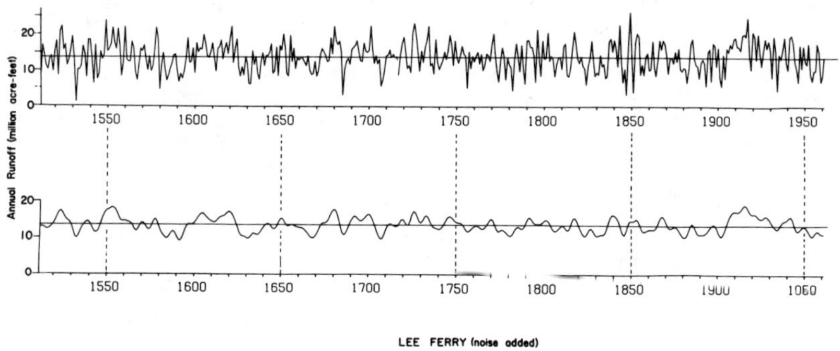

Figure 7. Same as shown in Figure 6 but with random component added.

Table 1.

SUMMARY OF STATISTICS FOR
LEE FERRY RECONSTRUCTIONS AND VIRGIN FLOW FOR VARIOUS PERIODS

	YEARS	N	MEAN[1]	MEDIAN[1]	SD	SKEW	KURT	r_1
Reconstruction	1512-1961	450	13.5	13.4	3.6	-.03	2.73	.32
(with noise)			13.6	13.7	4.1	-.01	2.82	.24
Virgin Flow[2]	1896-1974	79	14.8		4.1	.19	2.44	.16
Virgin Flow[2]	1896-1966	71	14.9	14.8	4.3	.14	2.31	.17
Reconstruction	1896-1961	66	14.6	15.3	3.7	-.34	2.22	.29
(with noise)			14.4	15.3	4.4	-.31	2.25	.21
Virgin Flow[3]	1914-1965	52	14.8	14.6	4.3	.08	2.11	.23
Reconstruction	1914-1961	48	14.8	15.0	3.8	-.22	2.01	.31
(with noise)			14.5	15.0	4.5	-.14	2.17	.20
Virgin Flow[2]	1896-1915	20	15.9	15.2	3.9	.42	2.30	-.17
Reconstruction	1896-1915	20	14.5	15.7	3.6	-.86	2.46	.26
(with noise)			14.8	16.3	4.4	-.85	2.31	.28
Virgin Flow[3]	1914-1933	20	16.9	17.2	4.1	-.19	2.25	.09
Reconstruction	1914-1933	20	16.8	17.5	3.1	-.65	2.54	.07
(with noise)			16.7	17.2	4.0	-.32	2.43	-.04
Virgin Flow[3]	1931-1950	20	13.5	13.7	3.5	-.37	2.25	.04
Reconstruction	1931-1950	20	13.8	13.7	3.6	-.29	2.30	.24
(with noise)			13.8	13.8	4.4	-.35	2.02	-.06
Virgin Flow[3]	1942-1961	20	13.5	13.2	3.8	.36	1.93	.11
Reconstruction	1942-1961	20	13.4	12.4	3.6	.45	1.68	.02
(with noise)			12.5	12.0	4.0	.19	1.59	.10

[1]million acre feet

[2]USGS data

[3]Year by year means of USGS and Framework 1 data

Comparison of the statistics for the gaged data with the tree-ring based reconstructed data shows close agreement. Greatest disagreement is for the period 1896-1915 (20 years) although use of the appropriate distribution function for the estimate of skewness by Wallis, Matalas and Slack (1974) indicates the coefficient of skewness for all cases is not significantly different from zero at the 95 percent level. The remaining statistics have not been tested for significance although all appear to be not significantly different at the 95 percent level.

LARGE SCALE DROUGHT OCCURRENCE

Considerable conjecture has existed for some time that major droughts affecting the Northern Hemisphere, especially the North American continent, have tended to occur with an approximately 22-year periodicity (Willet 1965, Palmer 1964, Thompson 1973, Roberts 1977). This periodicity has been difficult to demonstrate, however, largely because of the relatively short length of the meteorological data base.

Many studies have been directed towards a search for periodicities in tree-ring data (Douglass 1919, 1928, 1936; Siren and Hari, 1971; LaMarche and Fritts, 1972; Bitvinskas, 1974). In fact the origin of dendrochronology in the United States can be traced to Douglass' attempts to find evidence of solar variability in tree-ring series. The sometime existence of an approximately 22-year periodicity in tree-ring data is well documented. Douglass (1946 pp.18) noted that a periodicity of approximately 23 years was prominent in long-lived sequoias [sequoiadendron gigantea (Lindl.) Decne.] and that "Schulman (Edmund) has found it well marked in carefully collected specimens in the intermountain (Utah) area." Douglass also noted that the period length was not constant but varied with time. In a more recent study, LaMarche and Fritts (1972), using modern statistical techniques, found that when tree-ring data were analyzed collectively on an area basis by using empirical orthogonal functions a "pair of peaks corresponding to frequencies of 22.0 and 29.2 years" appeared in the sample auto-spectrum.

The Palmer Drought Severity Index (Palmer 1965) is used as an indicator of the severity and duration of drought periods. This PDSI index, developed by W.C. Palmer of the U.S. Weather Bureau, is based on an empirical water balance approach. The Thornthwaite method (Thornthwaite and Mather 1955) of computing Potential Evapotranspiration from readily available precipitation and temperature data forms a basis for calculating the climatic demand for moisture and the subsequent development of the PDSI. The PDSI for a given month is determined partly by the value of the Index for the preceding month through an autoregressive relationship and partly from moisture received during the month in question. A moisture anomaly is defined as the departure of the actual precipitation received from that expected under "normal" moisture conditions. The "expected" precipitation is calculated by the water balance approach and takes into account accumulated soil moisture levels and estimated evapotranspiration.

The PDSI, for which maps are published monthly by the National Oceanic and Atmospheric Administration in the Weekly Weather and Crop Bulletin, has the desirable property that a given PDSI value means roughly the same degree of drought from one location to another. Also, the PDSI is an integrative index of moisture conditions similar to the response recorded in tree-ring width series. From a hydrologic standpoint, the PDSI is a desirable index for slower response systems such as fluctuations in ground water levels and storage in large reservoirs.

Records of monthly PDSI in 204 climatic divisions in the western United States for the period 1931-1970 were obtained from the National Climatic Data Center in Asheville, North Carolina. The area studied roughly included the states west of the Mississippi River. Since our interest lay mainly with large-scale drought patterns, the 204 divisions were grouped into 40 regions (Figure 8a) using the rainfall classification scheme of Trewartha (1961), and each regional PDSI was then calculated as the arithmetic mean of the PDSI of the divisions comprising the region.

Since the drought was to be reconstructed from annual tree-ring indices, while the regional PDSI series consisted of monthly values, it was necessary to derive some annual measure of drought from the PDSI series that would be expected to yield a strong relationship to tree growth. The July PDSI was chosen for the following reasons: annual tree growth as reflected in the ring widths is usually nearly complete by the end of July; July PDSI reflects to some degree the moisture conditions of the prior spring and late winter -- these conditions are relatively important to most of the species of trees used in this study; major droughts from 1931-1970 tended to peak in intensity during July or August; and July PDSI is particularly important in the growth and yield of agricultural crops.

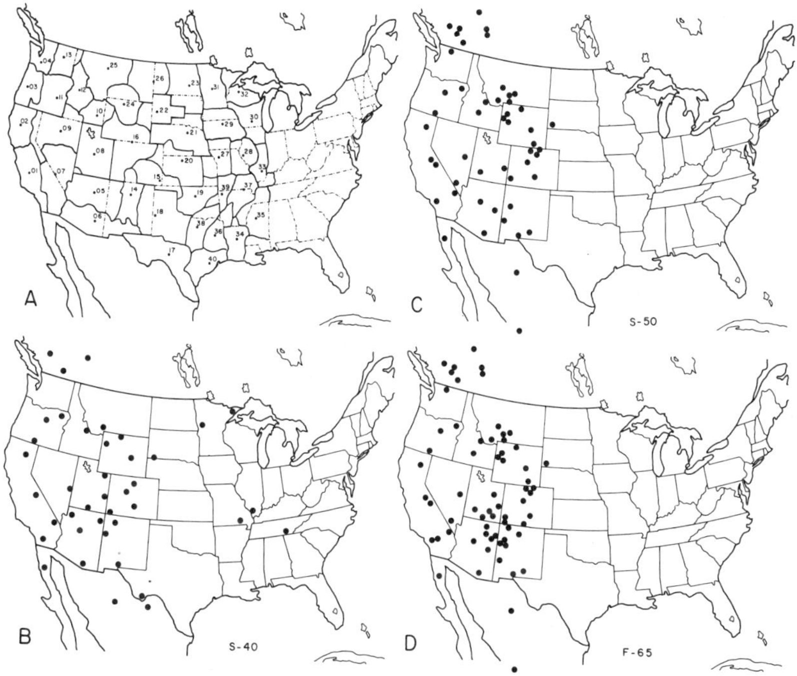

Figure 8. (a) Drought regions: solid lines denote boundaries. (b) Tree-ring grid S-40, extending back to A.D. 1700. (c) Tree-ring grid S-50, extending back to A.D. 1601. (d) Tree-ring grid F-65, extending back to A.D. 1600.

The tree-ring data grids (Figure 8) used in the analysis consisted of 40, 50 and 65 (Fritts and Shatz 1975) tree-ring sites chosen on the basis of replication, location and sensitivity to climatic variation. Species include Pinus spp., Pseudotsuga Menziessi and some Quercus spp. (S-40 Grid). Grid S-40 (Figure 8b) covers the period A.D. 1700-1962 and Grids S-50 and F-65 (Figure 8c and 8d) cover the time period A.D. 1600-1962). These data were not collected specifically for this analysis but were chosen to extend the time span of the drought reconstructions back to A.D. 1600, a period encompassing the Maunder Minimum (Eddy 1976).

Technique of Reconstruction

An empirical orthogonal function analysis was performed on both the PDSI and tree-ring data grids to define a smaller number of variables accounting for most of the spatial and temporal variability in both series (Fritts 1976). Empirical orthogonal function analysis enables fields of highly correlated data to be represented adequately by a smaller number of orthogonal functions (eigenvectors) and corresponding orthogonal amplitudes. Kutzbach (1967) provids a detailed explanation of the mathematical procedure necessary to define the functions and their amplitudes.

Empirical orthogonal functions (eigenvectors) of the PDSI were computed from the 40 observations of July PDSI (1931-1970) in the 40 regions. The analysis indicated the 40 original spatial patterns of PDSI could be represented quite well by the first five eigenvectors (explaining 75% of the variance). The first eigenvector accounted for 38 percnet, the second 17 percent, the third 11 percent, and remaining fourth and fifth about 10 percent.

A similar analysis was performed on the 264-year record (A.D.1700-1963) of tree-ring indices for the 40 sites of Grid S-40 and for the 364-year record (A.D. 1600-1963) for Grids S-50 and F-65. The first 10 eigenvectors accounted for about 60 percent of the variation with the first five of these representing nearly 50 percent of the total variance. The tree-ring record overlapped the PDSI record for the years 1931-1963, and these years were used to derive a transfer function for reconstruction (Stockton, et al., 1978). The approach to reconstructing the long term drought record consisted of deriving relationships to predict PDSI amplitudes from tree-ring amplitudes using canonical analysis(Fritts, Blasing, Hayden, Kutzbach 1971). For Grid S-40 a total of 6 amplitudes were used, accounting for 45 percent of the variance; in the S-50 grid case, only the first two amplitudes were used, but they were entered into the transfer function such that drought at the time t was compared with tree-ring data over the times t + 1, t, and t-1, for a total of 6 variables. The first two amplitudes account for 32% of the total variance in the 50-station tree-ring grid over the period A.D. 1931-1962. The information for all grids is summarized in Table 2.

Canonical analysis was used to derive a matrix of least squares coefficients to predict amplitudes derived from the first five PDSI eigenvectors from the amplitudes of tree-ring series eigenvectors. The resulting transfer function explained about 64 percent of the variance of the PDSI amplitude series during the calibration period (A.D. 1931-1963) for Grid S-40 and about 56 percent for Grid-50 and 65 percent for Grid F-65.

This same transfer function was then applied to the A.D. 1700-1962 and A.D. 1600-1963 tree-ring amplitude series and in this manner the amplitudes of the PDSI were reconstructed back to A.D. 1700 (Grid S-40) and A.D. 1600 (Grid S-50 and F-65). Reconstructed values of the PDSI series were then obtained by multiplying the matrix of predicted PDSI amplitudes by the matrix of appropriate PDSI eigenvectors.

Details of the technique employed here are described in Stockton *et. al*. (1978)

Table 2.

DROUGHT AREA RECONSTRUCTIONS

GRID	(A) TREE-RING INDICES				(B) DROUGHT INDICES (PDSI)			
	Period of Record	Data Points	EVs# Retained	Variance Explained	Period of Record	Data Points	EVs# Retained	Variance Explained
S-40	1700-1962	40	6	45%	1931-1970	40	5	75%
S-50	1600-1962	50	2	31%	1931-1970	40	7	83%
F-65	1600-1962	65	3	41%	1931-1970	40	5	75%

DAI FAMILY	(C) CANONICAL "BRIDGE"[1]		(D) DAI VARIANCE EXPLAINED* PDSI LIMIT			
	Terms Retained	Variance In Common*	<-1	<-2	<-3	<-4
S-40	5	64%	72%	72%	74%	78%
S-50	6	56%	67%	76%	79%	75%
F-65	5	65%	69%	70%	79%	74%

*In overlap period 1931-1962 (32 years)
#EVs = Eigenvectors
PDSI = Palmer Drought Severity
DAI = Drought Area Index

1. The maximum number of "terms retained" can not exceed the smallest number of variables in either set. The variables for the tree-ring set included the amplitudes lagged so that the PDSI amplitudes at time t are compared with those from the tree-ring set at times t and t + 1 (Grids S-40, and F-65) and t , t + 1, and t - 1, (Grid S-50). See Stockton (1975) for justification of the lagged variables.

Results

Reconstructed Drought Area Index

Time series of July PDSI in the 40 regions were reconstructed by the above procedure. Drought area indices (DAI) were then calculated by tabulating the total number of regions in a given drought class in each year. The resulting DAI series are approximate, annual measures of the area extent of a given severity of drought. For example, the DAI series for PDSI $\leq$ -1.0 is simply an annual frequency count of the number of regions reconstructed with PDSI $\leq$ -1.0.

The actual and reconstructed DAI series for PDSI $\leq$ -1.0 and PDSI $\geq$ 1.0 for the calibration period agree reasonably well with the longer term variation especially well duplicated. Since only the first five PDSI eigenvectors were used in the reconstruction, those years when less significant eigenvectors dominated were poorly reconstructed as in 1952 and 1962. A total of 12 reconstructed DAI series were produced -- for PDSI ≤ -1, ≤ -2, ≤ -3, and ≤ 4 from each of the three tree-ring data grids. The percent variance explained -- R^2 x 100 -- by the various DAI reconstructions was generally above 70% and very similar results were obtained for a given drought class from the different tree-ring grids (Table 2).

A time series for the period A.D. 1700-1962 showing the number of regions with PDSI ≥ +1.0 as opposed to those with PDSI ≤ -1.0 is presented in Figure 9. Several points are apparent. First, note the rhythmic nature of the extreme wet and dry periods. Second, note that the wettest three year period prior to 1931 was 1914, 1915, 1916 and the driest was 1846, 1847, 1848. The driest single year in our reconstruction was 1934, the second driest, 1757. In 1934, 19 regions were in extreme drought (PDSI ≤ -4.0) as compared to 10 regions in 1757.

We have attempted to confirm our reconstruction by two methods. First, we have compared reconstructed with actual PDSI computed for single stations for years prior to the calibration period. Second, our maps of reconstructed PDSI were compared with spatial and temporal accounts of climatic extremes taken from diary entries of travelers that crossed the Great Plains during the period A.D. 1845-1849 (Lawson and Stockton 1977). Both approaches confirm the reconstruction, but because of the technique employed to stress the large-scale features, the individual stations often provide confirmation of local rather than regional phenomena.

Spectra of the DAI Series

Spectral analysis of the reconstructed DAI series confirmed the periodicity suggested in Figure 9. The spectrum of the DAI series for PDSI ≤ -1 reconstructed from Grid S-40 shows a peak at 21.3 years that is significant at the 99% confidence level if a simple autoregressive null continuum is assumed (Figure 10). The periodicity tendency diminishes in the more intense drought categories: the peak for the PDSI ≤ -2 series is significant at well above the 95% level, but the peaks for PDSI ≤ -3 and for PDSI ≤ -4 (not shown) are not significant at the 95% level.

Similar results were obtained from spectral analysis of DAI series reconstructed from the F-65 and S-50 tree-ring grids. The F-65 reconstruction had a higher percent variance at the lower frequenices than the S-40 and S-50 reconstructions, but once again a statistically significant (95% level) peak was found at 21.3 years. Cross-spectral analysis between the F-65 and S-40 reconstructions showed an overall coherence of about 0.65, and a coherence of 0.91 at 21.3 years. These results seem to indicate that the periodicity is not unique to a partiular choice of tree-ring sites.

Regional Drought Occurrence Analysis

In the forgoing presentation we reconstructed large-scale drought by deriving transfer functions relating spatial patterns of tree growth. Although this method is useful for reconstructing large-scale contrasts -- for example, dry northern one-half of western United States versus wet southern one-half -- and spatial extents of major droughts, it may be less useful for reconstructing a detailed drought history of a specific region. For practical purposes, like water resources planning,

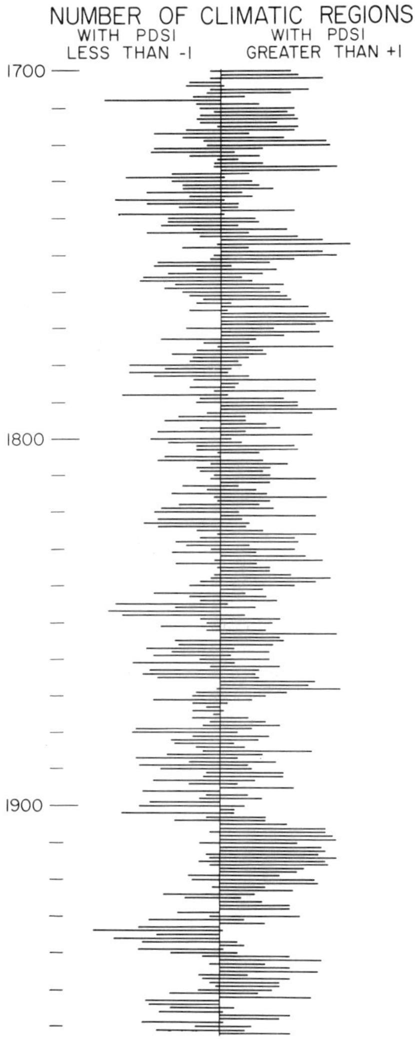

Figure 9. Long-term reconstructed drought-area index for drier than normal (Palmer Index ≤ -1.0) and wetter than normal (Palmer Index ≥ +1.0) categories, A.D. 1700 to A.D. 1962, from grid S-40.

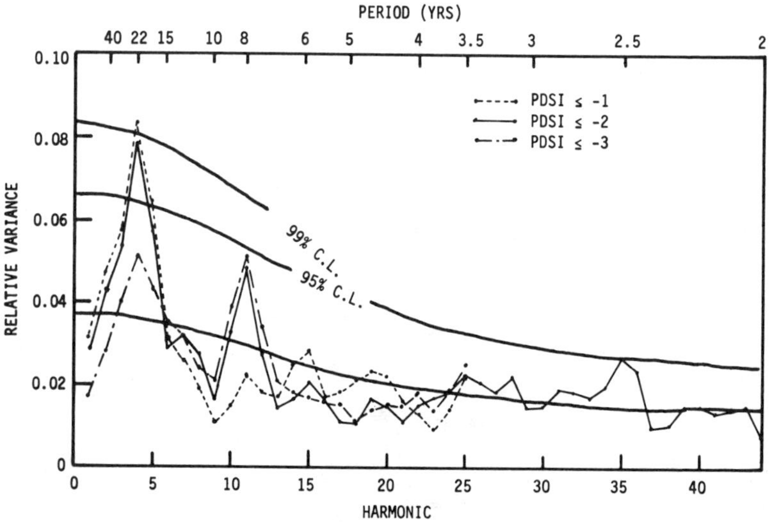

Figure 10. Spectral density functions of drought-areá index for three drought classes reconstructed from grid S-40. Confidence limits based on first-order Markov null continuum.

an accurate reconstruction of drought in a specific region is precisely what is needed. For this reason we have begun investigating the possibility of reconstructing the Palmer Index region-by-region for selected geographic regions in the western United States.

For one such study we chose the southern California region primarily because of the occurrence of severe drought in this area in recent years and we have completed a reconstruction of the Palmer Index for southern California back to A.D. 1700; the results are encouraging. Checks of reconstructed Palmer Index against Palmer Index data not in the calibration period suggest that the reconstructions are highly reliable.

Method

Multiple linear regression (MLR) (Draper and Smith 1966) with stepwise inclusion of variables was used to derive a transfer function to reconstruct a time series of regional PDSI from tree-ring indices at eight sites.

The regional time series of drought is the mean of the July Palmer Index at three stations -- San Diego, Riverside, and Los Angeles, where the PDSI for the individual stations was computed from records of monthly precipitation and monthly mean temperatures. The mean PSDI series, which covers the period from 1882-1975, is shown in Figure 11a.

Tree-ring indices from eight sites (Figure 12) are the predictor variables. These index series were lagged at $\underline{t}$-1 and $\underline{t}$+1 years. The stepwise procedure could thus choose predictors for PDSI in years $\underline{t}$ from 24 potential variables -- tree-ring indices from each of the eight sites at years $\underline{t}$-1, $\underline{t}$, and $\underline{t}$+1.

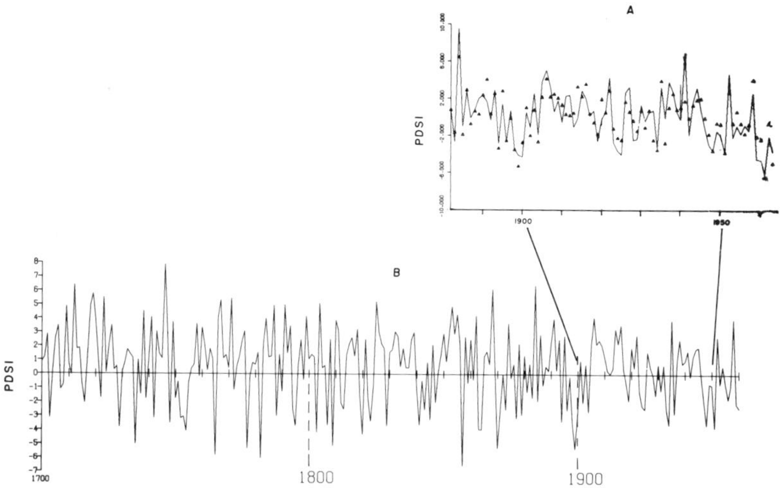

Figure 11. (a) July Palmer Index (PDSI) in Southern California. Mean of PDSI for Riverside, Los Angeles, and San Diego. (b) Long-term reconstruction of Southern California PDSI from tree rings.

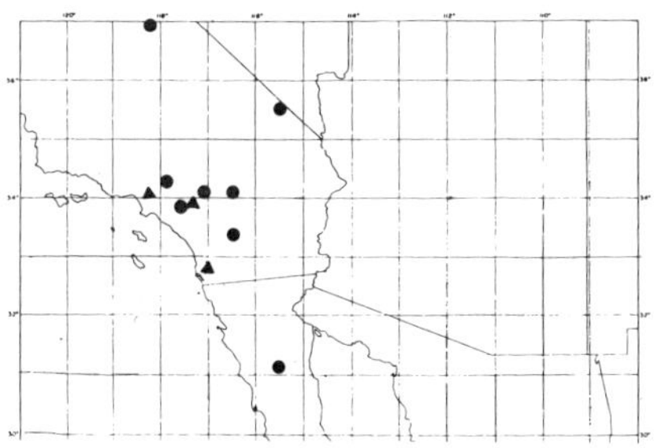

Figure 12. Approximate locations of tree-ring sites and weather stations. Circles denote tree-ring sites. Triangles denote weather stations; from north to south -- Los Angeles, Riverside, San Diego.

Three regression equation were run: 1) the short model, which included only the first three predictors; 2) the maximum adjusted R^2 model, which maximized the quantity

$$R^2 = R^2 - \frac{(K-1)}{(N-K)}(1-R^2)$$

where K is the number of predictors, N is the number of cases, and R^2 is the percent variance explained by the regression equation; this model included 11 predictor variables; and 3) the full model, in which the stepwise procedure was allowed to run its full (F-level of .01) extent.

A comparison of the results (Figure 13) for models 1), 2), and 3) with the actual PDSI for the independent data for 1882-1902 showed that the maximum R^2 model was much better at reconstructing wet years than the short model and better overall than the full model. On this evidence we chose the maximum R^2 model to reconstruct PDSI back to A.D. 1700.

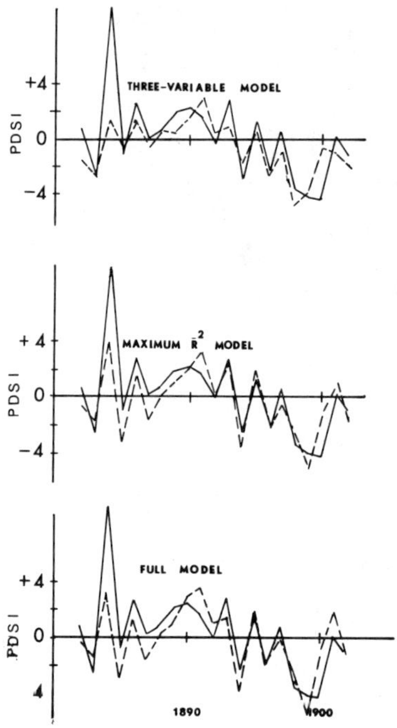

Figure 13. Reconstructed (dashed line) and actual (solid line) July PDSI for the independent verification period (1882-1902). Top: simplest model, based on 3 predictor variables. Middle: maximum-adjusted R^2 model, based on 11 predictor variables. Bottom: model arrived at by letting stepwise procedures run to an F-level for variable inclusion of 0.01.

The most important variables, the first six to enter, were from sites nearest the weather stations whose records were used to compute the mean PDSI series.

The model 2 transfer function accounted for 67% of the variance of PDSI in the calibration period and 70% of the variance in the 21-year independent data. Independent data verification (Figure 13, middle) shows excellent agreement both in droughts and in wet years. The only substantial discrepancies between actual and reconstructed were 1885, 1887, and 1900 on the dry side and 1884 on the wet side. The estimates for these years are greatly improved using model 2 calibrated on the entire record 1882-1962. Figure 14, a plot of the actual PDSI versus the reconstructed based on model 2 calibration for the period 1882-1962, shows the improvement. In this case, 1884 shows a reconstructed PDSI of +7.0 as opposed to a value of +9.5 for the actual.

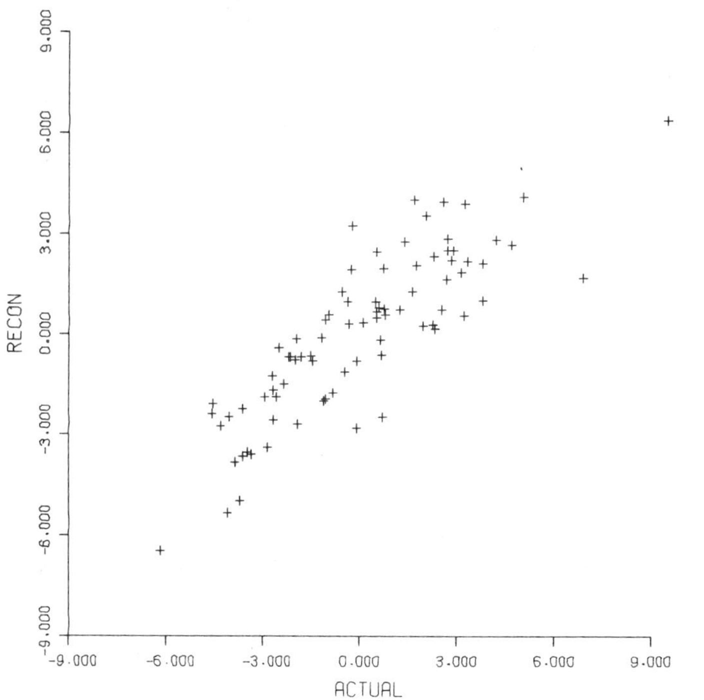

Figure 14. Southern California PDSI as reconstructed from tree-rings versus that computed from nearby weather stations for the period 1882-1962.

It is apparent from this graph that the tree-ring data do a reasonably accurate job of reconstructing drought occurrence in the past on a regional scale, at least in southern California.

A time series of PDSI (Figure 11b) was reconstructed back to A.D. 1700 using the derived transfer function. The results show the first half of this century to have been relatively free from extreme drought (no years with PDSI $\leq$ -4) compared to the 1700's and 1800's, which each had 9 years with reconstructed PDSI $\leq$ -4.0. The driest reconstructed year was 1857, with a Palmer Index of -7.14, compared to -6.45 in 1961, the driest year reconstructed in the calibration period. For the entire reconstruction, 21 years had PDSI $\leq$ -4.0; only two of those -- 1951 and 1961 -- were in this century.

An important question from a planning point of view, is how frequent were long-lasting severe droughts. The only case of back-to-back years in extreme drought (PDSI $\leq$ -4.0) was 1794-1795. Back-to-back years in severe drought (PDSI $\leq$ -3.0) were 1715-1716, 1794-1795, 1819-1820, 1863-1864, 1888-1889, 1959-1960-1961. Interestingly, three consecutive years of severe drought (PDSI $\leq$ -3.0) occurred only once in the reconstruction, 1959-1960-1961. The actual PDSI series (Figure 11a) shows two such three-year droughts -- 1959, 1960, 1961, and 1970, 1971, 1972; the 1970's drought, of course, was outside the time period of reconstruction.

These results indicate that severe droughts lasting several years have been more frequent in the past 20 years than in the previous 250, and that extreme drought has become especially more prevalent in the past 20 years than in the first half of this century.

LITERATURE CITED

Bitvinskas, T. T., 1974. Dendroclimatic investigations. Gidrometeoizdat, Leningrad.

Douglass, A. E., 1919. Climatic cycles and tree growth. Carnegie Institution of Washington, Pub. No. 289, Vol. I.

_____, 1928. Climatic cycles and tree growth. Carnegie Institution of Washington, Pub. No. 289, Vol. II.

_____, 1936. Climatic cycles and tree growth. Carnegie Institution of Washington, Pub. No. 289, Vol. III.

_____, 1946. Researches in dendrochronology. University of Utah Bull., Vol. 37, No. 2. Biological Series Vol. 10, No. 1.

Draper, N. R. and H. Smith, 1966. Applied Regression Analysis. John Wiley and Sons, Inc., New York.

Eddy, John A., 1976. The Maunder Minimum. Science, Vol. 192, 1189-1202.

Fiering, M. B. 1962. On the use of correlation to augment data. J. Am. Stat. Assn. (57) 20-32.

_____, 1963. Use of correlation to improve estimates of the mean and variance. U. S. Geol. Survey Prof. Paper 434-C, 9 pp.

Fritts, H. C., 1976. Tree Rings and Climate. Academic Press, London.

Fritts, H. C., T. J. Blasing, B. P. Hayden and J. E. Kutzbach, 1971. Multivariate techniques for specifying tree growth and climatic relationships and for reconstructing anomalies in paleoclimate. Journal of Applied Meteorology, Vol. 10, No. 5, 845-964.

Fritts, H. C. and D. J. Shatz, 1975. Selecting and characterizing tree-ring chronologies for dendroclimatic analysis. Tree-Ring Bull (35), 31-40.

Hely, A. J., 1969. Lower Colorado River water supply--its magnitude and distribution. U. S. Geol. Survey Prof. Paper 486-D, 54 pp.

Hipel, K. W., 1975. Contemporary Box-Jenkins modeling in hydrology. Ph. D. thesis, Univ. of Waterloo, Waterloo, Ontario, Canada.

Hurst, H. E., R. P. Black and Y. M. Simaika, 1965. Long-term storage: an experimental study. Constable and Co. Ltd., London, 145 pp.

Johnston, J., 1963. Econometric Methods. McGraw-Hill, New York, 300 pp.

Julian, P. R. and H. C. Fritts, 1968. On the possibility of quantitatively extending climatic records by means of dendrochronological analysis. Proc. First Statistical Meteorol. Conf., Hartford, Conn., May 27-29, 1968. pp. 76-82.

Kutzbach, John E., 1967. Empirical eigenvectors of sea-level pressure, surface temperature and precipitation complexes over North America. Journal of Applied Meteorology, 6(5): 791-802.

LaMarche, Valmore C., Jr. and H. C. Fritts, 1972. Tree-rings and sunspot numbers. Tree-Ring Bulletin, Vol. 32, 21-35.

Lawson, Merlin P. and C. W. Stockton, 1977. Exploration of the Great American Desert in the context of climatic reality. Presented at the 73rd annual meeting of the Assn. of American Geographers, Salt Lake City, Utah, April 24-27, 1977.

Lettenmaier, Dennis P. and Stephen J. Burges, 1978. Climatic change: detection and its impact on hydrologic design. Water Res. Research, 14(4): 679-687.

Mandelbrot, B. B. and J. R. Wallis, 1968. Noah, Joseph and operational hydrology. Water Res. Research, 4(5): 904-918.

_____, 1969. Some long-run properties of geophysical records. Water Res. Research, 5(5):967-988.

Matalas, N. C., 1962. Statistical properties of tree-ring data. Int. Assn. of Scientific Hydrology, 7(2):39-47.

Matalas, N. C. and B. Jacobs, 1964. A correlation procedure for augmenting hydrologic data. U. S. Geol. Survey Prof. Paper 434-E, 7 pp.

Mitchell, J. M., Jr., C. W. Stockton and D. M. Meko, 1979. Evidence of a 22-year rhythm of drought in the western United States related to the Hale Solar Cycle since the 17th Century. In Proceedings: Solar-Climate Conference held at Ohio State University, August 24-28, (in print).

Palmer, W. C., 1964. Climatic variability and crop production. Proceedings of a seminar on "Weather and our Food Supply", Center for Agricultural and Economic Development, CAED Report 20, Iowa State University, Ames, Iowa, 173-187.

_____, 1965. Meteorological drought. U. S. Weather Bureau Research Paper 45, U. S. Dept. of Commerce, Washington, D. C.,58 pp.

Roberts, Walter Orr, 1975. Relationships between solar activity and climatic change. In: Possible Relationships Between Solar Activity and Meteorological Phenomena, NASA - sp366, edited by W. R. Bandeen and S. P. Maran, Washington, D. C., 1975.

Rodriguez-Iturbe, Ignacio, 1969. Estimation of statistical parameters for annual river flows. Water Res. Research, 5(6): 1418-1426.

Siren, G., and P. Hari, 1971. Coinciding periodicity in recent tree rings and glacial clay sediments. Rep. KEVO Subarctic Res. Sta. 8, 155-157.

Stockton, C. W., 1971. The feasibility of augmenting hydrologic records using tree-ring data. Ph. D. Dissertation, The University of Arizona, Tucson, 172 pp.

_____, 1975. Long-term streamflow records reconstructed from tree-rings. Papers of the Laboratory of Tree-Ring Research, No. 5, The University of Arizona Press, Tucson.

_____, 1976. Long-term streamflow reconstruction in the Upper Colorado River basin using tree-rings. In: Colorado River Basin Modeling Studies, proceedings of a seminar held at Utah State University, Logan, Utah, July 16-18, 1975. C. G. Clyde, D. H. Falkenborg, and J. P. Riley, eds., Utah Water Res. Lab., College of Engineering, Utah State Univ., Logan Utah, pp. 401-441.

_____, 1977. Interpretations of past climatic variability from paleoenvironmental indicators. In: Climate, Climatic Change and Water Supply, National Academy of Science, Washington, D. C., pp. 34-45.

Stockton, C. W. and H. C. Fritts, 1971. Conditional probability of occurrence for variations in climate based on width of annual tree-rings in Arizona. Tree-Ring Bull. Vol. 31, pp. 3-24.

Stockton, C. W. and G. C. Jacoby, 1976. Long-term surface water supply and streamflow levels in the Upper Colorado River basin. Lake Powell Research Project Bulletin No. 18, Institute of Geophysics and Planetary Physics, U. of California at Los Angeles.

Stockton, C. W., Meko, D. M., and J. M. Mitchell, Jr., 1978. Tree-ring evidence of a 22-year rhythm of drought area in western United States and its relationship to the Hale Solar Cycle. Unpublished paper presented at the meeting of Working Group VIII under the US/USSR Agreement of Protection of the Environment held at the Crimean Astrophysical Observatory and the Kislovosk Mountain Observatory, U. S. S. R., Sept. 13-23, 1978.

Thompson, L. M., 1973. Cyclical weather patterns in the middle latitudes. Journal of Soil and Water Conservation, Vol. 28, 87-89.

Thornthwaite, C. W., and J. R. Mather, 1955. The water balance. Drexel Inst. Tech. Publ. Climatol. Vol. 8(1):1-104, Centerton, New Jersey.

Trewartha, Glenn T., 1961. The Earth's Problem Climates. The University of Wisconsin Press, Ltd., Madison. 334 pp.

United States Bureau of Reclamation, 1954. Colorado River Storage Project: U. S. 83rd Cong., 2nd Sess., H. Doc. 364, 332 pp.

Wallis, J. R., N. C. Matalas and J. R. Slack, 1974. Just a moment! PB-231816, National Technical Information Service, U. S. Dept. of Commerce, Springfield, Va., 197 pp.

Water Resources Council, Upper Colroado Region State-Federal Interagengy Group, 1971. Upper Colorado Region comprehensive framework study, Appendix V, Water Resources: Pacific Southwest Inter-Agency Committee Water Resources Council, 66 pp.

Willett, H. C., 1965. Solar-climatic relationships in the light of standardized climatic data. Journal of Atmospheric Sciences, Vol. 22, 120.

THE CALIFORNIA DROUGHT - ITS EFFECT ON RUNOFF FORECASTING FOR POWER GENERATION[1/]

By
Gary J. Freeman[2/]

INTRODUCTION

The 1976-1977 two year California Drought is now a recorded fact. For those of us involved with runoff forecasting and water planning, the drought in spite of its serious impact on the state's resources, was a significant event for future water studies. Many problems were encountered in the forecasts for these two years as well as for 1978, the wet year which followed the drought. In developing new equations which utilize dry year data, proper lag values and low values have now become better defined.

RUNOFF FORECASTING AT PG&E

Runoff forecasts are an important part of the planning at PG&E for decisions which concern both fuel procurement and water operations. Multiple regression equations are used for seasonal runoff forecasting. Both an April through July and a water year period are forecasted for 64 operational points. A monthly breakdown is made from the seasonal amounts. This is based on rank order comparisons with historic data and base flow trends. The runoff forecasts are updated monthly or as needs require.

The runoff forecast for any given river is done by operational subunits. These subunits are the watershed areas in between major diversions and reservoirs. They range in size from a few square miles to over 2,000 square miles. The forecasted monthly quantities of runoff for a given basin are then routed through the river system by operational subunits. A combination of forecasted subunits for a given river makes up the river in

[1/] Hydrologist, Pacific Gas and Electric Company, San Francisco, California.

its entirety. Based on the forecasted runoff reservoir routing is optimized to minimize spill, meet year-end carryovers, meet fish release and irrigation requirements, and to maximize generation.

Optimum routing of water on a monthly basis for the remainder of the calendar year provides us with a hydro-generation forecast. Planned outages and certain limitations must be included in the generation forecast. The hydro-generation forecast becomes input to another forecast used to plan the company's fossil fuel requirements.

Runoff forecasts are made for an area which extends from the Pit River in northern California to the Kern River near Bakersfield. The geology varies from alluvium and volcanics in the northern watersheds to the more granitic type watersheds in the Stanislaus and Kings Basins. Soils near the headwaters of the rivers are in general deeper in the northern portion of the Sierra and elevation differences within subunits are less noticeable there than in the southern Sierra.

THE 1976-1977 TWO YEAR CALIFORNIA DROUGHT

Prior to the 1976-1977 drought the 1924 and 1928-34 periods were classic dry years. Water planning in California focused on these earlier years for dry year studies. Probably 1931 has become the most commonly used year as a basis for long range resource planning. The California Water Plan, the Central Valley Project and hydro-electric development in the Sierra utilized these dry years as the reasonable minimum events for design and planning purposes.

The 1976-1977 two year drought however, was considerably drier than previous dry years used in studies. The 1977 water year was the driest year in California since records were started 100 years ago. It followed 1976 which was the fourth driest year on record. In the 1976 water year precipitation was 65 percent of average. During the 1977 water year, precipitation was 45 percent of average. Limited rain and snow in 1976 and 1977 reduced runoff to 47 percent and 22 percent respectively.

The impact of the drought on power generation at PG&E was substantial. The amount of energy produced under average year conditions by hydroelectric plants of PG&E and other agencies located in PG&E's northern California electric control area is about 24 billion kilowatt-hours per year. During 1977 the same plants produced about 10 billion kilowatt-hours of hydroelectric energy. This represents a difference of about 14 billion kilowatt-hours which is equivalent to approximately 22 million barrels of oil.

DECISIONS WHICH PRECEDE THE FORECAST

Before a runoff forecast can begin certain basic assumptions or rules must be made. PG&E normally updates the forecast each month for both the water year period and the April through July period, and then breaks the amounts down to monthly quantities. Using an assumption that weather cannot be reliably forecasted beyond 3 to 5 days, median precipitation and temperature conditions are assumed for the weather.

However, beginning in 1977, because of the possible consequences of another year of drought, it was decided that two types of runoff forecasts were needed. The median condition forecast, which utilizes median precipitation for the remainder of the water year, would be made. In addition, a forecast was needed using 1976 precipitation for the remainder of the season. This duel forecast provided management with insight into planning for both situations.

Beginning in 1978, two types of forecasts were again made. This included both a median condition forecast, and a forecast which assumed 1977 precipitation for the remainder of the season. With low storage levels in our reservoirs, this pessimistic forecast and a simulated type operation would present to management the worst situation. Also, a generation forecast requires that a year-end reservoir carryover be assumed for each significant reservoir. This decision during the drought required a judgement in planning between leaving enough water in the reservoirs in case another dry year followed the two year drought, and the threat of no benefit from storage carryover should the year result in a greater than average runoff.

Once the runoff forecast is made for the alternative type weather conditions, a generation forecast follows using the assumed carryover and reservoir draft schedule.

On completion of the hydro-generation forecast, management must then determine how much natural gas is available for power plant use, how much economy energy is available for purchase, and how much fuel oil should be purchased. Oil contracts are usually made several months or even years in advance. If the weather becomes significantly wetter than average, and oil contractual commitments exceed oil storage capacity and use, then penalty charges might come into effect for oil under contract but not taken.

RUNOFF FORECASTS DURING THE 1976-1977 DROUGHT

Multiple regression type forecast equations used along with comparing rank ordered historic runoff to insure the "reasonableness" of the equation prediction is the common method used by PG&E to forecast runoff. However forecasts during the drought and an adverse type forecast done for 1978 were lower than the reliable range for which the regression type equations were developed. A question that inevitably arises during extreme conditions is how well such equations "predict." A limitation exists in using the multiple regression technique with extremes in the minimum range. If the slope intercept is minus, the equation will frequently give a minus answer when the data is below the minimum for which the equation was developed.

Since forecasted flows using the regression technique were frequently on the minus side, decisions had to be made as to how much subjective judgement should be used in producing the forecast. Rank orders of historic natural flows and precipitation were compared with snowpack data from other dry years. A combination of using rank ordered data with a common sense type approach proved very helpful.

It is important in runoff forecasting to keep track of the actual flow that has already occurred. We do this by a continuous updating of preliminary flow numbers with data which in most cases is less than 24 hours old. Natural runoff computations for each of the forecast points are made by the computer each day. This provides us with an updated daily natural runoff to date for each of the forecast points.

Flow trends, especially for the base flows, were observed from the updated daily natural flows and extended forward in time. These projected flows were then compared with rank ordered flow data. Reasonably good results were obtained using this method. The volcanic watersheds on the Pit and North Fork Feather River showed almost no increase in flow during the snowmelt period in 1977. Other watersheds further south showed only a slight increase during the spring snowmelt runoff. In working with extreme minimum type flows there is a definite limit of zero. This fact in some ways makes the forecast easier than for very large flows in which a definite limit doesn't exist. For watersheds without large base flows, the forecast could only reasonably be between the lowest recorded water year flows such as for 1924 and zero flow.

The watersheds we are concerned with vary in geology from the highly granitic watersheds in the southern Sierra to volcanic in the Cascade and northern Sierra. The North Fork Kings River above Wishon Lake has almost no base flow which carries over from the previous year. However in contrast the flow from Fall River tributary to the Pit River derives much of its runoff from the Modoc Lava Flow. This underground storage and consequent base flow which averages about 70,000 acre feet a month is large enough to almost completely smooth over any variations in flow which result from variations in precipitation. During the entire drought the water year flows from Fall River remained close to average. The annual average base flow for Fall River is about 840,000 acre feet. The standard deviation of annual flows is 120,000 acre feet or only 13 percent of its total annual mean flow of 946,000 acre feet. Compared with the average mean flow of 946,000 acre feet, the water year flows for 1976 and 1977 respectively, were 944,000 acre feet and 897,000 acre feet.

Much more difficult to forecast were watersheds such as the Feather and to a lesser extent the Yuba River. On these watersheds, especially on the North Fork Feather River, the base flows are relatively large, and did show a noticeable drop during the drought. The late summer flows for example which occur on the North Fork Feather River at Lake Almanor dropped to about one half the normal flow of 25,000 acre feet in August.

During the drought these flows were watched very closely to see trends in the base flows. This requires relatively accurate flow data. In unusually low flows the gage heights recorded on the rivers are on the low end of the rating curve. In many cases the stage discharge relationships for flows in the very low range must be considered somewhat questionable until rating curves are later verified by actual flow measurement. The problem of using preliminary numbers for low flows during the drought, on several occasions resulted in unreasonable natural flow numbers. In such cases a temporary correction factor applied to the flows was necessary until new flow ratings became available.

A further difficulty in predicting runoff during a period of uncertainty such as during the drought is the inability to define proper confidence limits for the forecasted values. The runoff forecast can normally have a probability range associated with a standard error type approach. Since equations were not a reliable predictor of runoff during the drought, this approach for defining forecast error was not possible.

CARRYOVER EFFECTS FROM DROUGHT INTO 1978-1979

Following two consecutive years of drought, precipitation in 1978 was considerably above average, especially in the southern Sierra. A significant portion of the forecasted water from the Stanislaus River northward did not show up. The regression equations had overforecasted the runoff. The forecasted runoff for the North Fork Feather River at Lake Almanor for example was almost 100,000 acre feet higher than actually occurred. This represented an overforecast of better than 15 percent of the total flow. This overforecast on the northern Sierra watersheds is best accounted for by attributing the losses to replenishment of ground water and soil moisture storage. The combination of these factors creates a "lag effect" which influences flows in future years. This "lag" or memory effect as it could be called can best be illustrated on the Feather River Basin.

The North Fork Feather River Basin is characterized by volcanic rock and soils above Lake Almanor. Using a combination of annual recovery factors and statistical analysis to study the flows, it appears that a large two way dynamic flow occurs into and out from ground water storage. This two way flow of water into ground water storage and outflow of ground water produced a net effect of about 51,000 acre feet returning to surface flow during 1976. However in 1977, a net flow of 4,000 acre feet returned to the ground water storage. In 1978 a net flow of 266,000 acre feet disappeared into ground water storage. If a stabilized mean flow were to theoretically occur it appears that approximately 100,000 acre feet of water enters into soil moisture and ground water storage each year and then returns to surface flow in the form of springs showing up as a lagging base flow.

In 1979, two years after the drought, this lag effect in the watershed above Lake Almanor still appears very prominent. It is estimated that this year about 100,000 acre feet net loss to ground water storage will occur. In developing a regression type equation for this basin following the drought it appears that a three year accumulated lag in precipitation gives the best results. It may take several years of normal or above normal precipitation for watersheds in the volcanics to fully recover from the effects of the drought and reach a new state of stability.

During the summer and fall of 1978 in order to take advantage of drought data, PG&E has redone nearly all of its forecast equations. The actual minimum data obtained as a result of the drought is somewhat analogous to the case of the hydrographer getting a 'point of zero flow' or in other words a low flow measurement. In hydrology such events provide a wealth of important data and information. It is hoped that from such events as the 1976-1977 drought forecasters in the future will have the capability to better forecast these extremes.

FUTURE

In developing new equations emphasis was placed in the following areas:

1. Improvement of the flow data: In order to forecast runoff for power generation it is necessary to predict runoff for small subunits of a larger watershed. Many of the subunits range in size from 25 to 300 square miles. We have the problem of measuring small quantities of water at operational midpoints in the river basin which may have 1,000 square miles of drainage above the entry point into the subunit which we want to measure. These large flows measured through a powerhouse which is used during the day for peaking power present problems in accurately defining the subunit flow. It can be likened to trying to accurately weigh a penny on a bathroom scale; much of the resolution is lost. In water

operations this is a very serious limiting factor. Since nearly all water flow measurements are by nature simply a sample of flow, and not necessarily the true flow, an error usually exists. An excellent way to find the error is to combine the flows on a daily basis into natural flows and produce a daily natural flow hydrograph which includes temperature and precipitation. When data from two or more gages are combined to compute natural flows for subunits, errors can be detected and corrections made.

Most of our forecast points are reservoirs. Since storage changes are used to compute the natural flow, large reservoirs cause the "bathroom scale" effect mentioned earlier - the net effect is a loss of resolution in defining relatively small flows. The storage change problem however is a problem only for the day-to-day readings. Over 3 or 4 weeks time, the inaccuracies of storage tend to cancel out the net error. Staff gages and instrumentation which can be read in increments of 0.01 foot help to solve the resolution problem.

2. Grouping of data and careful selection of variables: This helps to avoid the problem of multicollinearity, or in others words cross correlation among independent variables. After two or three independent variables have been added, further addition of variables usually doesn't improve the actual predictive ability of the equation. Good correlations do not necessarily indicate equations with high predictive ability. Cross correlations and the standard error should be checked closely to insure that the equation has a reasonable formation and its predictive capability has been maximized.

Improvements in runoff forecasting beyond a certain level come slowly. The limiting factor can be lack of good precipitation and/or snow data. Flow data used in equation development usually has a 5% or greater error. The costs for improved runoff forecasting go up enormously once a certain level has been reached with usually only a small improvement in prediction possible. The costs involved should always be in proper proportion to benefits derived. In runoff forecasting for power generation, errors in the runoff forecast are carried through to all levels of fuel procurement planning. The error can directly result in too much or too little purchase of oil. The consequences can cause either contract penalties or shortages in fuel supply to meet our energy demands. The economics of the runoff forecast error in forecasting runoff for power generation generally justifies a continuous improvement program even after a high level of forecast confidence has been achieved.

CONCLUSIONS

The drought's effect on runoff forecasting in general will be that the data now exists for two very dry years. Planning cannot ignore the possibility that there will be other years with similar extremes. Reservoir carryover planning and fuel contracts will have to consider the possibility of another drought of similar magnetude.

Runoff forecasting during the drought provided the input for both fuel procurement and operational planning. The methods used shifted from the rigid equation method to the more subjective "best judgement/common sense approach." Inspection of rank ordered data was very helpful in accomplishing this. It is important to have the flexibility to use alternative approaches during an extreme situation such as occurred during the drought. Reasonable numbers were forecasted as a result of using a subjective approach in this case.

The drought data however, now exists to develop equations with better defined low ends of flow than was previously possible. The reliability of equations to predict during a drought situation has been greatly enhanced as a result of the dry year data now available. It is hoped that we now have the data and knowledge to better cope with such a period when it occurs again.

REFERENCES

Brown, A. J., 1978. The California two year drought. Proceedings, 45th Annual Western Snow Conference, Otter Creek, Oregon. 85-91pp.

California Department of Water Resources, 1957. Bulletin No. 3. The California Water Plan. Sacramento, California. 246p.

_________________, 1978. The 1976-1977 California Drought - a review. Sacramento, California. 228p.

Nie, N. H., C. H. Hull, J. G. Jenkins, K. Steinbrenner, and D. H. Bent, 1975. Statistical Package for the social sciences. Second ed. McGraw-Hill Inc. 675p.

State of California, 1977. Proceedings of the Governor's Drought Conference. Los Angeles, California. 140pp.

BPA Hydrologic Forecasting Considerations and Actions During the 1976-77 Drought. by W. R. Gordon, 1/ C. H. Watkins, 2/ & R. C. Lamb 3/

The Bonneville Power Administration (BPA) is the marketing agency for the electric energy produced from the Federal Columbia River Power System. It is the largest electric energy supply system in the Pacific Northwest (PNW). BPA schedules and controls the power operation of 23 Federal projects and schedules the storage releases from 3 Canadian storage projects.

Hydro power produces about 80 percent of the electric energy resources in the Pacific Northwest. The key point for measuring this hydro potential is the Columbia River at The Dalles, Oregon encompassing a drainage area of 237,000 square miles (614,000 square kilometers) with a mean annual volume of runoff of about 131 million acre-feet (MAF) (162 billion cubic meters). Most of the runoff is spring and early summer snowmelt and is controlled by a large reservoir storage system. Storage for power production above The Dalles totals nearly 40 MAF (49 billion cubic meters) while the entire Coordinated System which also includes reservoirs on western slopes of the Cascade Mountains totals about 45 MAF (56 billion cubic meters). Managing the runoff and reservoir storage is essential to meet the region's electric power needs.

This paper presents the major items addressed by the Bonneville Power Administration's Branch of Power Supply in response to the 1977 drought in the Pacific Northwest. Principal emphasis relates to the BPA Hydrometeorological Section and the operational procedures used to evaluate the 1977 drought and the subsequent actions taken.

Chronological Summary of the 1976-77 Drought

The following summary of the Pacific Northwest's 1977 drought provides a picture of the drought's progress and resultant major events.

1/ Hydraulic Engineer, Hydrometeorology Section, Branch of Power Supply, BPA, Vancouver, WA

2/ Chief, Branch of Power Resources, BPA

3/ Head, Hydrometeorology Section, Branch of Power Supply, BPA

November 1, 1976 - The dry fall and below normal streamflow requires BPA to discontinue secondary energy supplies to industries and private utilities.

December 1, 1976 - The continued dry weather and low streamflows requires BPA to discontinue secondary energy deliveries to public agency customers.

December 8, 1976 - A special December 1 water supply forecast predicts a January-July volume of 85.8 MAF (106 billion cubic meters) (90 percent of median) at The Dalles, Oregon.

December 16, 1976 - BPA meteorologist, Charlie Ferris, states: As of December 15, the state of the Columbia Basin drought apprears at least as severe as the driest autumn and early winter conditions ever recorded. Mountain snowpacks are at the lowest levels ever for so late in the season.

December 20, 1976 - National Weather Service (NWS), River Forecast Center hydrologist Vail Schermerhorn reports a mid-month special water supply estimate of 80 MAF (99 billion cubic meters) (84 percent of median) for the January-July runoff for the Columbia River at The Dalles, Oregon.

December 22, 1976 - A letter to BPA, U.S. Army Corps of Engineers (USCE), and Bureau of Reclamation from Regional Director of the National Marine Fisheries Service expresses concern of fishery agencies for the ". . . welfare of the Columbia River anadromous fish if low conditions prevail in the spring and summer of 1977."

December 23, 1976 - Terrin Pearson of the BPA Hydrometeorology Section reports that based on the 80 MAF (99 billion cubic meters) mid-month estimate, base Energy Content Curves for January 31, 1977 will lower at only Dworshak and Libby and in those cases only 7.4 and 2.5 feet (2.3 and 0.8 meters) respectively. This implies no availability of secondary energy from the Federal Columbia River Power System.

January 1, 1977 - The volume forecasts (75.7 MAF (93 billion cubic meters), 79 percent of median) at The Dalles indicate no additional draft below base rule curves during the January through July period.

February 1, 1977 - The volume Forecasts (62.2 MAF (77 billion cubic meters), 65 percent of median) at The Dalles and studies of the February-July 1977 power system loads and resources indicate that, even if the region's major thermal plants operate up to planned levels, there exists about a 60 percent chance that major reservoirs in the Columbia River basin will not fill.

February 14, 1977 - BPA and the Northwest Power Pool issued a regionwide call for voluntary curtailment of at least 10 percent by all users.

February 15, 1977 - Electro-process industries in PNW start to reduce production and lay off employees.

April 1, 1977 - The April 1 volume forecasts (58 MAF (72 billion cubic meters), 61 percent of median) at The Dalles and power studies indicate that reservoirs in the Pacific Northwest will be 13.8 billion kilowatthours short of refilling on July 31, 1977. Even with this projection, there will be sufficient storage to meet firm power obligations through next fall under the most adverse streamflow conditions. Northwest Governors establish the Interstate Electricity Task Force to address potential power curtailments. Plans for Operation Fish Flow 77 to augment streamflows for enhancement of anadromous fish survival are now complete.

May 1, 1977 - The severe drought of 1976-77 continues to look worse. The estimated January-July runoff of the Columbia River at the Dalles is 53.8 MAF ((66 billion cubic meters), 56 percent of median). The previous lowest runoff was 60.6 MAF (75 billion cubic meters) in 1944.

August 1, 1977 - Drought conditions in the Columbia basin continued through July. The volume runoff of the Columbia River at The Dalles, Oregon for the January through July period was 54.0 MAF (67 billion cubic meters), setting a new low record by 6.6 MAF (8 billion cubic meters). Coordinated System reservoirs, normally at full content on July 31, were 12.7 MAF (16 billion cubic meters) below full. This deficiency is equal to 14.1 billion kilowatthours or about 30 percent of the total storage useable for power production.

August 26, 1977 - Energy curtailment guidelines for the current drought (1) were finalized and accepted by the Pacific Northwest Regional Commission (Governors of Washington, Oregon, and Idaho) and the Governor of Montana at Boise, Idaho meeting.

October 1, 1977 - September ended the driest water year on record across the Columbia basin. The October 1976-September 1977 water year runoff of 78.6 MAF (97 billion cubic meters) for the Columbia River at The Dalles, Oregon exceeded the previous record of 84.9 MAF (105 billion cubic meters) (1925-26) by 6.3 MAF (8 billion cubic meters). The deficiency from normal reservoir levels remains at near 14 billion kilowatthours.

November 1, 1977 - The 1977-78 water year got off to a dry start in October. The total runoff at The Dalles, Oregon from August 1, 1977, the beginning of the drawndown season, through October was 15.2 MAF (19 billion cubic meters), the lowest ever recorded. The reservoirs remain about 14 billion kilowatthours deficient.

January 1, 1978 - Precipitation over the Columbia basin above The Dalles, Oregon was 124 percent of normal for the September-December 1977 period. Most of this precipitation fell in November and December with 124 and 161 percent of normal amounts, respectively. January 1 volume forecast is 120 MAF ((148 billion cubic meters), 126 percent of median) for The Dalles, Oregon.

January 11, 1978 - Power studies based on January 1 volume forecasts and new rule curves indicate sufficient energy resources for BPA to restore secondary energy deliveries to all of its customers.

Thus, unlike the two year drought experienced in the Pacific Southwest, the Pacific Northwest had a one year drought. During this period the prime concern of power managers was what to do if the 1977 drought was followed by another drought year. The Federal System had enough water in reservoirs and snowpack to meet its firm load requirements. Fortunately, November and December rains in 1977 were sufficient to end the critical power situation in the Pacific Northwest.

Procedures Used By BPA

A. Weather

The BPA Weather Office provides data basic to streamflow and load forecasting. The office is staffed by two meteorologists, and a meteorological technician and is supported with standard weather facsimile and teletypes.

The meteorologists prepare original forecasts specifically designed to meet BPA requirements including, quantitative precipitation, temperature, wind and freezing level forecasts for input to the streamflow and load forecasting models. In addition, model output statistic (MOS) forecasts for 94 sites in the Pacific Northwest were developed for BPA by contract with the NWS. These MOS forecasts are probability of precipitation, probability of precipitation amount, and maximum and minimum temperature. Four days of forecasts are transmitted directly from National Weather Service computers in Suitland, Maryland to BPA via the Bureau of Reclamation computer in Denver, Colorado. The MOS forecasts are reviewed and revised as necessary by BPA meteorologists before being transferred to the streamflow or load forecasting models. Specialized weather forecasts are also provided to powerline maintenance crews and power dispatchers. These hazardous weather forecasts are used to assist operations and maintenance functions.

B. Load Forecasts

Two types of load forecasts are used by BPA. The long-range firm load forecast is developed under the rules of the PNW Coordination Agreement and is based on the assumption of median year temperature. These long-range load forecasts are used in long-range energy studies which establish base reservoir rule curves and the amount of firm energy the system can provide.

The short-term operational load forecasting procedure used by BPA is a solar radiation, temperature and wind sensitive model that predicts the hourly and daily loads for 7 days. This computerized model separates the load into four components. These are: (1) industrial load (known); (2) a normal weather load; (3) a load adjustment for deviations from normal weather, and, (4) a component to adjust for recent model errors. The normal weather component incorporates normal load growth, seasonal, day of week, and hourly variations. The third component adjusts the normal temperature load for weather deviation from normal using temperature, wind, and cloud cover data. Finally the fourth component adjusts for observed errors over the last few hours. The sum of these components becomes what we call the BPA system load. This load is then input to BPA's plant scheduling model.

C. Water Supply Forecasts

Seasonal water supply forecasts are generally made monthly beginning in January and run through June or July. The basic period used by power managers in the Northwest is the January-July period in order to draft reservoirs such that they have a 95 percent probability of refill by the end of July which is the month flows are normally receding back to base flow levels. The project or reservoir owners are responsible for developing the forecast procedures and for supplying the monthly volume forecasts. These are standard statistical regression equations using precipitation, temperatures, snow, and in one case, soil moisture as the independent variables. Observed data are used to the date of the forecast with median subsequent data used through the remainder of the forecast period. Subsequent forecasts of volume runoff into and/or effecting the Federal projects of the Federal Columbia River Power System are prepared by the National Weather Service and are coordinated through the Columbia River Forecast Service which is made up of Bonneville Power Administration, National Weather Service and Corps of Engineers. The Soil Conservation Service and National Weather Service also coordinate forecasts at locations common to both agencies. From these reservoir volume forecasts, new 95 percent probability of refill and flood control rule curves are established. Next, power studies are made by BPA to determine the load carrying capability of the system with the new rule curves. These studies determine the marketing conditions, e.g., energy to market in excess of the Pacific Northwest's needs. Policies established from these seasonal studies are then incorporated into BPA's daily and hourly scheduling activity.

D. Daily Streamflow Forecasting SSARR Model

Streamflow is a major variable in the models used to schedule hourly and daily generation. The basic streamflow model used by the Columbia River Forecast Service is the Streamflow Synthesis and Reservoir Regulation (SSARR) model (2). Until November 1976, this model was not operational on an annual basis for the Columbia basin. BPA began development work on the basin runoff logic of the the SSARR model in the early 70's to simulate daily rainfall runoff and snowpack accumulation and depletion. By the fall of 1976, the major portion of the Columbia basin was modelled and BPA began using it operationally. The major modifications BPA made to the basin model (3) were:

1) Ten equal area elevation bands were added to the model to improve the snowpack logic. Each band is treated independently, accounting for precipitation, snow water equivalent, degree day accumulations, soil moisture, and other parameters characterizing the basin runoff model.

2) The temperature lapse rate is computed by subtracting the surface temperature from the 700 millibar free air temperature (this is assumed to be at 10,000 feet) and dividing by the elevation difference, an improvement over using a fixed lapse rate. The zones of snow accumulation and depletion are better defined using the computed lapse rate.

3) An orographic effect was added to distribute precipitation over the zones using an elevation vs. precipitation curve.

4) An accumulated degree day vs. snowmelt rate curve is used to determine the fluctuation in the melt rate in each zone.

The end result is a watershed model that accumulates and depletes snow, produces rainfall runoff, and can be operated on an annual basis. Streamflow forecasts output from this model become input to BPA's plant scheduling model.

E. Plant Scheduling

This is a model of the Columbia River hydro thermal generation system. Thirty-six storage projects and generating facilities are used in the model. Streamflow forecasts, load forecasts, and project operating constraints, e.g., navigation, flood control, and recreation requirements, are the basic inputs. This model contains optimizing logic that attempts to minimize spill and maintain the highest head or best potential energy on the system. The Plant Scheduling program assists in determining how the system will be operated to meet load hour by hour for seven days in the future. Beyond the first few days of hourly output, day and week average output schedules provide a picture of the power situation thirty days into the future.

The models and procedures discussed above were used as the primary tools to determine the power operation of the Federal Columbia River Power System during the drought of 1977. The daily streamflow forecasting, load forecasting, and power planning models are used to solve short-term operating problems while the volume forecasts and seasonal studies are used for longer range operations planning.

Actions Taken in 1976-77

A. Energy Curtailment

Secondary energy deliveries to private utilities and BPA's direct-service industrial customers was curtailed on November 1, 1976, as a result of below normal streamflows. This is not an unusual situation and most of this load was met with advances of energy by drafting below rule curves and by purchases. As dry weather and low streamflows persisted, secondary energy deliveries to public agencies were curtailed on December 1, 1976. Normally the first volume runoff forecasts are made January 1, but due to the unusually dry fall, Columbia River Forecast Service began making water supply estimates in early December. These early estimates indicated that it would be unlikely that the January 1, 1977, water supply forecasts would provide enough runoff to lower rule curves sufficiently to create any secondary energy. This prompted industries and others who rely on secondary energy to begin purchasing energy from outside the region, primarily from B.C. Hydro and Power Authority in Canada. As the season progressed and the drought worsened, energy supplies to meet the industries and other secondary energy requirements diminished. On February 10, 1977, BPA and the Northwest Power Pool called for a 10 percent voluntary curtailment by all electric energy users. During February, BPA's electro-processing industrial customers began reducing production and laid off over 500 people. This situation lasted for nearly one year.

Reservoir storage and volume runoff forecasts contained sufficient water supplies to meet BPA's firm load requirements for 1977. However, studies showed that of the 40 historical water conditions there would be a few years where firm loads would not be met through the winter and early spring of 1978. The Governors of Washington, Oregon, Idaho, and Montana assumed responsibility for setting regional load curtailment guidelines to manage the regional electric energy deficiencies. The Regional Guidelines for Curtailment Planning (1) prepared by the Pacific Northwest Regional Commission's Electricity Task Force was finalized and endorsed by each of the Governors in August 1977. The curtailment guidelines provide a contingency plan to be followed to conserve resources in the PNW when reservoir storage is not sufficient to carry PNW firm loads. The guidelines call for voluntary curtailment if there is a high probability that mandatory curtailments will be necessary. If there is a high probability that the regional firm loads might not be met, then mandatory curtailment will be required. The Northwest Power Pool calculated the probability of insufficient energy which the Electricity Task Force used to counsel their Governors on State curtailment plans. Voluntary curtailments were solicited in February but no mandatory curtailments were requested. A plan such as this is an important tool for future years when needed.

B. Probability of Insufficient Energy to Meet Firm Loads

Many factors are involved in determining the probabilities of insufficient energy to meet the Coordinated Systems firm loads. Techniques for assessing the probability of insufficient energy resource involve computer simulations of the operation of the Northwest Coordinated Power System. The Power Pool, BPA and the Corps of Engineers perform these simulations on a regular basis using forecasted loads, historical streamflows, assumed thermal energy, and other unprecise information as input. These simulations aim at a solution to the power supply equation which can be stated simply as:

Residual Energy = Energy Resources - Firm Energy Requirement

When the residual energy term is positive, the system has energy in excess of the firm needs and when it is negative, the system has insufficient resources to meet firm needs. Because all inputs to solving this equation have uncertainties, the solution yields probabilistic answers. An optimum solution would address the uncertainty of the runoff, distribution of runoff, forecasted loads, performance of thermal resources, and other parameters affecting electric energy requirements. Measuring the uncertainty of any one variable is not too difficult but combining uncertainties of a combination of these variables yields an array of answers which complicate the solution. To keep the solution relatively simple, only the uncertainty in the runoff and its seasonal distribution was addressed during the 1977 drought. BPA's method of combining uncertainties is described in C. H. Watkins memo to L. A. Dean "Uncertainty in Assessment of Power Supply" (4).

As multipurpose uses of PNW waters increase, and as more thermal is developed, modeling uncertainties will become more important and also more complicated. The drought of 1977 made us aware of the need to look at uncertainties on a long term basis but as hydro resources become a smaller percentage of our base load because of new thermal, and as nonpower demands for water increase, modeling the uncertainties will play an increasingly important role in mid-range operating strategies (e.g., scheduling unit outages, operating reserves, or thermal plant refuelings).

C. Water for Anadromous Fish

In December, fishery agencies became concerned that low spring and summer runoff would not be sufficient to move the downstream migration of smolts. The fisheries' request to spill water to move the fish downstream over the dams rather than through turbines and to release water to increase velocities to speed up the migration to sea conflicted with storing water during the spring and early summer. Because the fisheries request would reduce the chances of meeting firm loads in 1978, Governors of the Pacific Northwest assumed responsibility in the water management decisionmaking process. The governors endorsed a plan worked out by water managers and fishery agencies for water releases and spill to be provided to enhance the downstream fish migration. This fish operation titled "Fish Flow 77" involved two distinctly different operations. In the Lower Snake, a system of trapping and hauling downstream migrants eliminated the need to spill or supplement streamflows on the Snake. On the Columbia River, 1.4 MAF (1.7 billion cubic meters) of water for spill and streamflow enhancement was used. This operation through detailed monitoring of the fish migration was able to keep the spill to minimum (150 KAF) (185 million cubic meters). The flow enhancement, part of the program created generation in excess of firm load requirements and BPA and the mid-Columbia River Public Utility Districts had to find places to store this energy for later recall. Energy losses due to spill and transmission amounted to 263,400 megawatthours and operational costs for storage and wheeling amounted to 1.3 million dollars (CRWMG 77p. 108).

LESSONS LEARNED FROM 1977 DROUGHT

1. An energy curtailment plan agreed to by regional political entities and based on the probability of insufficient energy resources is a valuable tool for managing energy supplies during a period of deficiency.

2. PNW residents will respond to pleas for decreased energy consumption during times of drought and threatened energy shortage. Following the February 14 plea, PNW loads dropped by 6-8 percent below forecasted loads which was apparently due to conservation efforts of the consumers of Pacific Northwest electricity.

3. In the Pacific Northwest, forecasts of seasonal runoff are a key input to any effective energy curtailment plan. These forecasts must have verifiable statements of the standard error of the forecast to be useful. It will be necessary to demonstrate an improvement in the standard error of the estimate before any new forecasting procedure will be acceptable by reservoir managers in the PNW.

4. The forecast of seasonal runoff produced by the Columbia River Forecast Service is the best index of the severity of drought conditions that is currently available for the Columbia Basin region. The Palmer Index computed by the National Weather Service and available on their national facsimile circuit during this period was restricted to the United States portion of the Columbia Basin. The Canadian portion of the basin accounts for about 40 percent of the runoff of the Columbia River.

5. Specialized hydrologic and meteorologic forecasting service is a valuable aid to the effective management of water and energy resources during periods of shortage.

6. The 1977 drought demonstrated that there is a need for frequent updating of the forecasts of seasonal runoff for effective reservoir and energy resources management.

7. The distribution of errors in the volume forecasting process needs better definition during extremely low runoff events such as occurred in 1977. Current reservoir management procedures assume that the errors in the forecast process are normally distributed. There is serious question that this assumption of normalcy is valid, especially in extreme runoff events. There is a need for a better estimate of the form of the distribution.

8. There is a need for improved hydrometeorologic "nowcasting" for the Columbia Basin. During 1977 there was frequently insufficient quantitative information about the past weather events over the Columbia Basin to provide for proper updating of the forecasts of seasonal runoff.

Discussion

New streamflow forecasting techniques have not been developed as a direct result of the drought. The drought did confirm the need for current and accurate forecasts of hydrologic conditions. How can current methods be improved? First, accurate long-range weather predictions would be very beneficial in water supply forecasting, but this is not seen as a viable solution for some time. A more effective use of climatology could be used in making better probablistic forecasts of the water supply. This is a viable solution that is being worked on. Climatology is available for the past 30 years or so. Other areas to be addressed are improving accuracy and timeliness of snow and climatological data reporting. The timeliness of these reports is coming with automated hydromet networks and the SCS's SNOTEL system. How accurately these sites represent what is going on in a basin is another question. The use of models, radar, satellite data, etc., should assist in providing more accurate and timely assessments of current hydrologic conditions. Using this along with climatology should enhance the ability to appraise the potential hydrologic condition more accurately and frequently. Many challenges exist in improving the assessment of hydrometeorological conditions and water supply forecasts. Multipurpose uses of water resources are rapidly approaching optimum development and water managers are challenged to use these resources prudently.

Summary

The Pacific Northwest is dependent upon its water supply to meet the region's electric energy requirements. Managing the large amount of reservoir storage of this multipurpose system requires frequent and accurate assessments of the water supply, electric energy supply and electric energy requirements to determine the system's hourly, daily, and seasonal load carrying capability. Weather forecasts, streamflow and load forecasting models, and plant scheduling models provide the tools to plan the electric energy supply. Constant coordination among water managers is maintained to achieve optimum use of the region's water for all interests. The drought of 1977 emphasized the need for this coordination. Conflicts of water use among States and agencies during the drought resulted in the Governors of the States in the Pacific Northwest to endorse certain water management plans. This was the first time an action of this nature was taken and procedures were established that may be used in future years. As non-renewable energy resources diminish and power requirements increase, more demands will be put on the region's water supply. Optimum management of the region's water is an increasing challenge.

References

1. "Regional Guidelines for Curtailment Planning," Pacific Northwest Regional Commission, Vancouver, Washington

2. Rockwood, D.M., "Streamflow Synthesis and Reservoir Regulation," U.S. Army, Division North Pacific, Portland, Oregon. Technical Bulletin No. 22, January 1967.

3. Pearson, T., "Simulating Runoff To The Hungry Horse Reservoir of Western Montana," Bonneville Power Administration, Portland, Oregon. Proceedings of the 42nd Western Snow Conference.

4. Watkins, C. H., "Uncertainty in Assessment of Power Supply," Bonneville Power Administration, Portland, Oregon. Memorandum from C. H. Watkins to L. A. Dean, November 29, 1977.

WATER SUPPLY FORECASTING IN CALIFORNIA - 1976-78

By

Charles H. Howard [1/]

California is particularly rich in natural water resources. Its mountain streams are rarely dry in so-called "normal" years, when these watersheds produce large amounts of surface water in response to precipitation during the fall, winter, and spring. Much of this water supply first takes shape as snow that envelops its highest mountain ranges in winter, and then which melts and flows down streams into reservoirs in the spring and early summer. These streams also maintain a base flow in the usually dry late summer and early fall months.

To make the best possible use of this vast resource, accurate and timely predictions of the amount of runoff that may be available are vitally important for a variety of wide-ranging purposes -- hydroelectric power generation, agriculture, flood control, domestic and industrial uses, recreation, -- among others, all ultimately affecting the life and well-being of every Californian. Water agencies need forecasts of runoff to operate their reservoirs efficiently. Operators of hydroelectric power plants use the forecast information to plan releases of water to generate the greatest possible quantity of power to match their operations to periods of highest power demands. Irrigation districts use them for scheduling surface water deliveries to farmers and for planning ground water pumping needs, and in turn farmers key their planting to probable available water supplies.

The 1976, 1977, and 1978 runoff seasons posed some very unusual and challenging problems in the preparation and use of operational water supply forecasts in California. The 1976 April-July (the normal snowmelt runoff period) and water year unimpaired runoff for Central Valley watersheds, a major source of supply for most of the State, totaled only 34 and 43 percent of average, respectively, near-record low values. 1977 April-July and water year runoff for this area totaled only 25 and 26 percent of average, respectively, and nearly every major Central Valley watershed registered new minimum flows of record. 1977 has been classified as the driest year in over 100 years of record in California, 1976 as the fourth driest, and combined they represent the record minimum two-year runoff. In contrast, the 1978 April-July and water year Central Valley runoff percentages were 149 and 154 percent of average, respectively, with several southern Sierra Nevada basins registering over 200 percent, the second or third highest of record.

1/ Associate Engineer, Snow Surveys Branch, Department of Water Resources, Sacramento, California.

This series of events had its beginning early in November 1975, when an atmospheric high pressure ridge became established over the ocean between Hawaii and California. Month after month, all winter long, it persisted. Most winter storms were shunted northward over Oregon and Washington. Those which did manage to break through into California were so weakened that little moisture was left to deposit. By February 1, 1976, statewide precipitation averaged only about 45 percent of normal for the first four months of the 1976 water year, with about 80 percent of normal near the Oregon border, 45 percent in the Central Valley, and only 10 percent of normal in Southern California. Snowpack water content ranged from about 40 percent of normal in the San Joaquin Valley and Lahontan areas up to about 60 percent in the North Coastal area, with a statewide average of about 45 percent.

Meteorological conditions remained essentially unchanged during February and March 1976, continuing the dry pattern established earlier. April 1 snow surveys indicated that snowpack water content was the lowest of record at about one-third of the State's snow courses, and data from telemetered snow sensors indicated that water content had peaked about mid-March and melt had been taking place since then, several weeks earlier than usual. Forecasts of April-July snowmelt runoff ranged from about 30 percent of normal in the San Joaquin Valley to 60 percent in the North Coast, with a statewide average of about 40 percent. Water year runoff forecast percentages were about 5 percent higher than those for the April-July period. These forecasts indicated the April-July snowmelt runoff from many central and southern Sierra Nevada river basins would be near the record low values set in 1924.

A forecasting problem which arose during the 1976 season was the lack of comparative data from previous dry years. The driest historic years were 1924 and 1931, and since most river basins had no snow records prior to about 1930, there had been little or no dry year data available for use in the development of the forecast procedures. Consequently, there was little basis for verification of the record low forecasts, and there was considerable need for subjective analysis and forecast adjustment.

As the 1976 water year drew to a close, alternative strategies were being developed for water conservation and use should 1977 also prove to be a very dry year. A Department of Water Resources publication "The California Drought - 1976" noted that "it is essential that each water agency develop contingency plans to cope with the eventuality of a dry 1977 and also of a drought of an intensity and duration equal to the worst drought that has occurred in its service area".

During the summer of 1976, reservoirs throughout the State were drawn down to record lows. Although resulting in losses to agriculture, unseasonable rains and cool weather in August and September prevented reservoir drawdown of larger magnitude. Even so, some reservoirs were completely drained by the end of the 1976 water year, with no carryover storage available for the forthcoming year.

Events during the fall and early winter of the 1977 water year did little to reverse the dry pattern established in 1976. Only one storm during the period October 1976 through January 1977 produced significant snowfall. Statewide precipitation during that period was only about 35 percent of normal, a record low. As a result, the February 1, 1977, snowpack was only 25 percent of normal, and April-July and water year forecasts averaged only 40 and 35 percent of normal, respectively. It should be noted here that water supply forecasts made by the California Department of Water Resources are based on known conditions at the date of forecast, plus the assumption of median conditions (precipitation and snow accumulation) occurring thereafter.

February and March 1977 continued the monotonous dry cycle, with precipitation and snow accumulation far below normal for these two months. By April 1, statewide precipitation averaged only 35 percent of normal for the first half of the water year, and snowpack water content was the lowest in 47 years of record for all basins except the Trinity and Feather Rivers.

At that time, forecasts indicated that nearly all major Central Valley and east side Sierra Nevada streams would produce the lowest flows of record for both April-July and water year runoff. A major concern in the preparation of the forecasts was the question of how much of the meager snowpack would be required to replenish the very dry soil moisture conditions, and how much would be available as surface runoff. As in 1976, there was little or no historic data from previous dry years (except 1976) available for comparative analysis. Again, as in 1976, there was considerable need for subjective forecast adjustment, much of it based on so-called "forecaster's feel".

Early in 1977, when it was becoming apparent that the drought was continuing into its second year, many State and local agencies began formulating plans for various methods of water conservation. The urban areas of Northern and Central California were feeling the effects of the drought, and more than 100 communities implemented some form of mandatory water conservation or rationing, and almost every community in the State had placed restrictions on the outdoor use of water.

Despite the conservation measures, reservoir storage continued its downward trend through the 1977 water year, and by October 1, 1977, statewide reservoir storage was only 37 percent of average for that date, and only 57 percent of that recorded a year earlier. That, of course, was a reflection of the very low levels of runoff in 1977, when 15 major Central Valley and east side Sierra Nevada streams had their lowest April-July runoff of record, and 19 streams had their lowest water year runoff of record.

The first two months of the 1978 water year did little to dispel the effects of the drought, as precipitation in October and November 1977 remained at below normal levels, and no relief appeared in sight. Conditions changed dramatically in mid-December, however, when heavy precipitation and snow accumulation began and which continued into late January. By February 1, 1978, precipitation for the October-January period varied from about 135 to 150 percent of normal throughout the Sierra Nevada. Snowpack water content was about 140 percent of normal, nearly equaling the average total seasonal accumulation, with several months of the accumulation period still remaining.

Streamflows during January also increased dramatically in many streams due to the heavy precipitation. This substantially improved the reservoir storage picture, and with February 1 forecasts of April-July and water year runoff averaging 125 percent of normal, prospects were good for most reservoirs to fill to normal levels by June.

The heavy precipitation and snow accumulation pattern established in December and January continued through February and March, and by April 1, 1978, the October-March precipitation ranged from 140 to 185 percent of normal in the Sierra Nevada. Snowpack water content was also well above normal, with values ranging from 120 percent up to 175 percent of normal. Forecasts of April-July runoff at that time ranged from 120 percent of normal in the northern Sierra Nevada to nearly 300 percent in the Kern River at the extreme southern end of the mountains.

Ordinarily, the start of the snowmelt season in California is about April 1. However, in 1978, April precipitation, as in the preceding months, was well above normal throughout most of the State. Cool weather and storm activity during April delayed the start of snowmelt at higher elevations by about a month, although normal melt took place below about 6,500 feet elevation. Snowpack accumulation exceeded snowmelt above that level, and snowpack water content on May 1 was greater than on April 1. As a result, the May 1 forecasts were increased over those of the month before.

There were further delays in snowmelt in early May due to cool weather, although May precipitation was far below normal in most areas. Reevaluation of forecasts following the below normal May precipitation resulted in forecast reductions.

As the snowmelt season progressed into June and July, it became apparent that the observed flows were going to be lower than the forecasted values. This was apparently the result of several possible factors: the forecast procedures probably inadequately took care of dry soil moisture and ground water conditions carried over from the drought; and the delay of snowmelt plus the very large area of snowpack remaining into the summer months resulted in above normal losses of water from the snowpack by evaporation into the atmosphere.

In Tables 1-6 following are tabulated the 1976, 1977, and 1978 April-July and water year forecasts versus the observed flows for major Central Valley and east side Sierra Nevada streams.

TABLE I

APRIL-JULY UNIMPAIRED RUNOFF 1976

Forecasts VS Observed Runoff 1000 Acre-Feet

Stream	Forecasts				Obs. Value
	Feb. 1	Mar. 1	Apr. 1	May 1	
Sacramento above Shasta	195	190	180	220	157
McCloud above Shasta	310	315	295	300	268
Pit above Shasta	850	875	820	775	804
Total inflow to Shasta	1,385	1,410	1,320	1,370	1,135
Sacramento near Red Bluff	1,800	1,860	1,750	1,800	1,503
Feather at Oroville	960	965	700	610	618
Yuba near Smartville	420	500	330	245	279
American at Folsom	420	485	300	245	312
Cosumnes at Michigan Bar	50	45	23	10	15
Mokelumne at Pardee	195	190	140	120	123
Stanislaus at Melones	290	265	185	140	199
Tuolumne at Don Pedro	520	485	400	345	330
Merced at Exchequer	270	250	210	170	168
San Joaquin at Millerton	505	425	370	340	350
Kings at Pine Flat	540	480	400	375	303
Kaweah at Terminus	120	95	85	80	75
Tule at Success	16	12	12	12	13
Kern at Isabella	115	120	110	110	104
Truckee - Tahoe to Farad	120	120	110	100	70
West Carson at Woodfords	16	15	18	15	17
East Carson near Gardnerville	90	85	70	60	64
West Walker near Coleville	80	75	60	55	50
East Walker near Bridgeport	17	17	15	10	7
Trinity at Lewiston	410	440	375	430	370

TABLE 2

WATER YEAR UNIMPAIRED RUNOFF 1976

Forecasts VS Observed Runoff 1000 Acre-Feet

Stream	Forecasts				Obs. Value
	Feb. 1	Mar. 1	Apr. 1	May 1	
Sacramento above Shasta	480	380	350	390	335
McCloud above Shasta	910	800	740	830	856
Pit above Shasta	2,550	2,480	2,440	2,300	2,441
Total inflow to Shasta	4,030	3,750	3,710	3,730	3,613
Sacramento near Red Bluff	5,500	5,100	4,960	4,985	4,845
Feather at Oroville	2,450	2,150	1,865	1,760	1,862
Yuba near Smartville	1,000	910	735	635	690
American at Folsom	1,120	990	755	700	785
Cosumnes at Michigan Bar	130	90	60	57	55
Mokelumne at Pardee	350	305	250	230	236
Stanislaus at Melones	515	440	350	305	377
Tuolumne at Don Pedro	905	790	690	635	624
Merced at Exchequer	440	385	330	290	299
San Joaquin at Millerton	790	660	610	580	629
Kings at Pine Flat	780	690	595	570	536
Kaweah at Terminus	200	155	145	140	147
Tule at Success	50	40	40	40	42
Kern at Isabella	245	235	220	220	239
Truckee - Tahoe to Farad	205	205	185	175	146
West Carson at Woodfords	35	35	30	30	32
East Carson near Gardnerville	150	140	120	105	115
West Walker near Coleville	115	110	90	85	80
East Walker near Bridgeport	60	55	50	45	40
Trinity at Lewiston	745	740	670	725	672

TABLE 3

APRIL-JULY UNIMPAIRED RUNOFF 1977

Forecasts VS Observed Runoff 1000 Acre-Feet

Stream	Forecasts				Obs. Value
	Feb. 1	Mar. 1	Apr. 1	May 1	
Sacramento above Shasta	180	150	145	100	64
McCloud above Shasta	270	240	225	185	184
Pit above Shasta	625	510	465	395	650
Total inflow to Shasta	1,100	900	845	685	798
Sacramento near Red Bluff	1,400	1,120	1,025	850	1,103
Feather at Oroville	830	600	510	335	* 365
Yuba near Smartville	270	200	190	155	* 196
American at Folsom	310	225	210	170	* 229
Cosumnes at Michigan Bar	35	12	10	5	* 10
Mokelumne at Pardee	150	100	90	65	* 106
Stanislaus at Melones	275	200	180	120	* 119
Tuolumne at Don Pedro	470	330	275	205	275
Merced at Exchequer	225	165	135	90	* 118
San Joaquin at Millerton	470	270	230	185	* 261
Kings at Pine Flat	465	230	200	195	* 273
Kaweah at Terminus	115	60	50	40	* 61
Tule at Success	14	5	4	3	5
Kern at Isabella	125	95	80	65	91
Truckee - Tahoe to Farad	90	75	65	50	* 58
West Carson at Woodfords	20	13	12	10	* 12
East Carson near Gardnerville	65	55	45	40	* 43
West Walker near Coleville	55	45	35	30	* 35
East Walker near Bridgeport	12	10	7	5	* 8
Trinity at Lewiston	315	265	240	160	113

* New minimum of record.

TABLE 4

WATER YEAR UNIMPAIRED RUNOFF 1977

Forecasts VS Observed Runoff 1000 Acre-Feet

Stream	Forecasts				Obs. Value
	Feb. 1	Mar. 1	Apr. 1	May 1	
Sacramento above Shasta	305	245	240	190	* 162
McCloud above Shasta	715	625	610	565	* 577
Pit above Shasta	2,055	1,945	1,900	1,815	2,011
Total inflow to Shasta	3,150	2,650	2,550	2,370	2,628
Sacramento near Red Bluff	4,070	3,250	3,100	2,900	3,422
Feather at Oroville	1,700	1,210	1,100	915	* 968
Yuba near Smartville	560	330	325	290	* 339
American at Folsom	580	340	340	300	* 339
Cosumnes at Michigan Bar	70	21	19	14	* 23
Mokelumne at Pardee	200	125	115	90	* 134
Stanislaus at Melones	360	245	225	165	* 161
Tuolumne at Don Pedro	625	400	340	270	* 339
Merced at Exchequer	285	190	160	115	* 128
San Joaquin at Millerton	630	375	330	280	* 362
Kings at Pine Flat	630	345	310	305	* 383
Kaweah at Terminus	180	92	82	71	* 92
Tule at Success	30	17	15	14	* 16
Kern at Isabella	235	190	170	155	186
Truckee - Tahoe to Farad	135	90	75	75	* 85
West Carson at Woodfords	30	20	20	15	* 19
East Carson near Gardnerville	95	75	65	60	* 63
West Walker near Coleville	80	60	50	40	* 47
East Walker near Bridgeport	35	30	25	20	* 27
Trinity at Lewiston	500	350	315	230	* 198

* New minimum of record.

TABLE 5

APRIL-JULY UNIMPAIRED RUNOFF 1978

Forecasts VS Observed Runoff · 1000 Acre-Feet

Stream	Forecasts				Obs. Value
	Feb. 1	Mar. 1	Apr. 1	May 1	
Sacramento above Shasta	390	435	460	570	398
McCloud above Shasta	450	470	550	580	414
Pit above Shasta	1,025	1,025	1,040	1,150	1,099
Total inflow to Shasta	1,920	2,000	2,060	2,500	2,059
Sacramento near Red Bluff	2,625	2,700	2,800	3,500	3,002
Feather at Oroville	2,330	2,520	2,530	2,950	2,034
Yuba near Smartville	1,210	1,300	1,300	1,580	1,341
American at Folsom	1,505	1,650	1,650	2,040	1,641
Cosumnes at Michigan Bar	160	170	175	240	196
Mokelumne at Pardee	560	615	650	760	654
Stanislaus at Melones	860	1,000	1,100	1,340	1,039
Tuolumne at Don Pedro	1,450	1,750	1,930	2,300	1,801
Merced at Exchequer	800	955	1,100	1,320	1,182
San Joaquin at Millerton	1,700	2,070	2,560	2,890	2,332
Kings at Pine Flat	1,600	1,945	2,350	2,600	2,351
Kaweah at Terminus	340	435	550	600	542
Tule at Success	75	95	120	135	128
Kern at Isabella	525	850	1,200	1,240	1,063
Truckee - Tahoe to Farad	330	340	325	380	318
West Carson at Woodfords	60	75	75	85	61
East Carson near Gardnerville	210	260	270	290	312
West Walker near Coleville	180	210	215	230	196
East Walker near Bridgeport	85	110	115	130	94
Trinity at Lewiston	820	900	885	1,010	825

TABLE 6

WATER YEAR UNIMPAIRED RUNOFF 1978

Forecasts VS Observed Runoff 1000 Acre-Feet

Stream	Forecasts				Obs. Value
	Feb. 1	Mar. 1	Apr. 1	May 1	
Sacramento above Shasta	1,235	1,360	1,440	1,555	1,385
McCloud above Shasta	1,340	1,450	1,525	1,615	1,387
Pit above Shasta	3,525	3,350	3,425	3,545	3,500
Total inflow to Shasta	7,120	7,420	7,810	8,260	7.838
Sacramento near Red Bluff	10,400	11,000	11,740	12,490	12,003
Feather at Oroville	5,550	5,580	6,075	6,510	5,628
Yuba near Smartville	2,675	2,680	2,910	3,200	2,971
American at Folsom	3,000	3,040	3,205	3,605	3,205
Cosumnes at Michigan Bar	415	410	465	530	486
Mokelumne at Pardee	815	880	960	1,075	1,002
Stanislaus at Melones	1,215	1,400	1,650	1,845	1,556
Tuolumne at Don Pedro	2,090	2,500	2,790	3,175	2,697
Merced at Exchequer	1,180	1,430	1,620	1,855	1,755
San Joaquin at Millerton	2,330	2,860	3,475	3,815	3,402
Kings at Pine Flat	2,165	2,670	3,140	3,400	3,334
Kaweah at Terminus	465	650	810	865	833
Tule at Success	135	220	260	275	273
Kern at Isabella	745	1,210	1,665	1,715	1,528
Truckee - Tahoe to Farad	440	440	450	505	440
West Carson at Woodfords	75	90	90	100	75
East Carson near Gardnerville	265	320	335	355	275
West Walker near Coleville	205	250	255	270	240
East Walker near Bridgeport	115	140	145	180	143
Trinity at Lewiston	1,900	2,000	2,085	2,215	2,056

CHANGING NEEDS IN URBAN FLOOD FORECASTING

L. Scott Tucker(1)

Urban flood forecasting is an integral part of urban drainage and flood control management. The purpose of this paper is to discuss urban flood forecasting in terms of both present and future needs. The perspective taken is from a management point-of-view.

To forecast, according to Webster, means to plan ahead or to predict some future event or condition as a result of rational study and analysis of pertinent data. Forecasting in the context of urban flooding can be divided into two major categories. One is to forecast the magnitude of a flooding or runoff event of a given frequency in order to plan or design in anticipation of that event. The second is to forecast an impending flood event so potentially affected flood plain occupants can take appropriate action. Both of these categories are consistent with the definition of forecasting even though forecasting is usually considered in the context of the later category, i.e., weather and what is predicted to happen in the next one or two days.

Hydrologic Forecasting for Urban Drainage and Flood Control Problems

Urban flood problems for the purpose of this paper will assume to mean both drainage and flood control. Drainage is considered to relate to the portion of the system that collects rainfall runoff in curbs, gutters, pipes, and swales and transports it to a major conveyance system. Flood control is considered to relate to the major conveyance system that carries collected runoff in natural creeks and rivers, manmade channels, and in some cases, large culverts.

Urban Drainage Considerations. The basic objective in urban drainage is to prevent runoff from relatively frequent storms from causing flooding problems. The problems can be either in terms of nuisance factors such as flooded streets or damage prevention considerations such as flooded basements.

The problems are solved by providing urban drainage systems designed to accommodate given frequencies of runoff typically in the range of 2-years to 10-years. Various hydrologic forecasting techniques are employed to estimate the rate of flow for the desired frequencies. Hydrologic forecasting techniques, or models, that are commonly used in drainage design include the rational method, empirical relationships, and synthetic unit hydrograph procedures. More complex models are generally not used at present in the design of drainage

(1) L. Scott Tucker, Executive Director, Urban Drainage and Flood Control District, Denver, Colorado

collection facilities; however their use is increasing as more engineers become familiar with computer applications.

It is important to be able to forecast with reasonable accuracy the flow for design purposes because of its effect on costs and the need for consistency within a metropolitan area. Drainage facilities are expensive and over design can be costly, thus creating unnecessary expense to the constructing agency, be it public or private. It can also be costly to under design a facility because more frequent flooding will occur. Consistency of design throughout a metropolitan region is important to insure that the level of protection and the cost burden for providing drainage facilities is equally distributed.

It should be understood that it is desirable to be able to forecast with a reasonable degree of confidence because there is a price to pay for inaccuracy. In order to make forecasting techniques more reliable, local rainfall and runoff data is needed to "calibrate" the techniques (models) to the local conditions. A local data base is important and should be developed in all metropolitan areas.

Urban storm drainage historically has been primarily a matter of getting rid of the water as fast as possible; usually through a system of underground conduits. This philosophy is giving way to consideration of a wider range of options such as local and on-site detention, more open swales, porous pavements, etc. It is required in many areas that new development not increase the runoff over historical flows. The consideration of such options and requirements necessitate the use of more sophisticated models that can account for their effects. Urban drainage considerations are becoming more and more complex and the trend will continue. More pressure will be put on local governments to insure that drainage for new developments are being adequately designed. This can only be accomplished through improved hydrologic forecasting, which in turn will depend on sound data bases.

Flood Control Considerations. The basic objective related to urban flood control is to prevent runoff from relatively infrequent runoff events, 10-year or greater, from causing damage along major drainageways. The damage is generally the result of overbank flooding of low lying areas through which the major drainageway flows.

A combination of factors result in the potential of such damage. One is that as urbanization occurs, lands subject to occasional flooding by major drainageways are occupied by residential developments, commercial establishments, and industrial complexes that are not compatible with flooding. The second is that flood peaks and volumes for given frequencies tend to increase as urbanization occurs, because of the reduction in pervious surfaces. For example, the peak flow rate of an 100-year flood may increase by 100% from historical conditions to fully urbanized conditions. A third factor that aggravates urban flooding is that natural flood plain storage is reduced by flood plain occupancy which tends to further increase flood peaks downstream.

The problems can be mitigated by anticipating and planning for the infrequent flooding events that will in time occur. State and

Federal regulations ans legislation require that local governments regulate and control development in the 100-year flood plain. This requires that the 100-year flood plain be accurately defined, which in turn necessitates that the 100-year runoff peaks and hydrographs be reasonably forecasted. Where flooding problems exist, solutions and costs must be identified for various frequencies so local governments may decide what to do about the problems. Again, this requires that runoff peaks and hydrographs for various frequencies be accurately forecasted.

Hydrologic models, or hydrologic forecasting techniques, are used to predict runoff for various frequencies for major drainageways. Many different types of models are available, but in order to obtain reliable results, it is necessary that the models be calibrated for the conditions prevailing in the metro region of interest. As discussed in the previous section on Urban Drainage Considerations, it is essential to have local rainfall and runoff data in order to accomplish such calibration.

In general, there is limited rainfall and runoff data in urban areas that can be used for urban flood control purposes. One reason is that there is little Federal interest in urban flood control, and consequently, long-term national programs of data collection and analysis were never implemented. Long-term flow data on a national basis exists for larger rivers which has aided in the prediction of infrequent events, but not for smaller urban watersheds. A second reason is that urban watersheds are dynamic and in a constant state of change. Data collected in 1973 may be for a different set of basin conditions than data collected in 1969 or 1977 in the same basin. Long periods of historical data for a constant set of conditions are therefore difficult to obtain.

Future Urban Hydrologic Forecasting Needs. Accurate urban hydrologic forecasting will become even more important in the future. Citizens are more often demanding that drainage and flood control problems either be prevented from occurring in the first place, or mitigated if they already exist. At the same time there is an increasing tendency for citizens to go to court if they feel the actions of a governmental entity are improper or causing them damage. Government does not have an obligation to solve drainage and flood control problems, but because of citizen expectations government will be even more involved in the future, and when government does become involved, it must do it right.

There is also the potential issue of the responsibility of government allowing or approving development that causes damage to others. This issue is just beginning to emerge as evidenced by recent court decisions in Nevada and Ohio. The trend is for people to become more conscious of the effects of development on drainage which will also tend to reuqire more accurate hydrologic forecasting.

Another factor so far unmentioned is stormwater quality. Continuing scrutiny of stormwater pollution as it relates to the nation's efforts to meet national water quality goals will force the quality element into hydrologic forecasting. Quality complicates the problem considerably and virtually dictates the need to develop local data bases if the issue is to be addressed intelligently.

The trend for the future is for more emphasis on reliable flood forecasting in terms of planning, design and regulation. Needed at this point in time is a commitment at the local or regional level of government to initiate data collection and analysis efforts that will provide a solid data base to develop more reliable forecasting capabilities in the future.

Hydrologic Forecasting for Early Flood Warning

The forecasting of impending flooding events is also an important aspect of urban flood control. Forecasting is an integral part of an effective flood warning and evacuation plan which has three critical elements, i.e., event forecasting, preparation and dissemination of the warning, and appropriate actions by flood plain occupants. All three elements must operate effectively in order for a warning plan to be successful.

The time available between the first notification of an impending flood and the onset of the flood ("lead time") dictates what can be done in the flood plain. With a very short lead time, there may be only adequate time for flood plain occupants to evacuate the flood plain. With longer lead times, building contents can be removed or elevated, openings in stronger buildings can be sealed shut, and cars and other vehicles can be driven out of the flood plain. On the larger rivers, lead time can be a matter of days allowing adequate time to construct temporary emergency flood protection works, such as increasing the height of levees. On smaller watersheds, particularly those with steep gradients and potential for intense rainfall, the lead time can be as little as one or two hours or even less than one hour in some cases. The design of the event forecasting element depends to a large extent on consideration of lead time.

The National Weather Service (NWS) has a river basin hydrologic forecasting operation. For several years they have been developing a capability for real-time processing and analysis of radar data by automatic computer operation to improve their forecasting activities. Prediction and warning of floods has been a statutory responsibility of the NWS and its predecessors since 1871. The NWS will disseminate two types of flash flood advisories to local and state officials and the news media. The first type is a flash flood watch which means that there is a likely potential for flash flooding in the general area described. The second is a flash flood warning that means flooding is occurring or iminent in the general area described.

For flash flooding situations, it is generally necessary to augment the NWS forecasting service with a local data collection activity. This might be necessary to pinpoint the warning to a specific small urban watershed, to lenghten the lead time, or to increase the reliability of the forecast. Hydrologic data necessary to make a forecast includes rainfall and runoff information in such a form to permit preparation of a forecast of what will happen in the immediate future.

In the Denver area, there are two examples of flood warning systems for specific watersheds that utilize the NWS forecasting capability augmented by additional local data collection and processing. These are briefly discussed to illustrate what is being done in actual situations.

One situation is in the Westerly Creek Basin located in the eastern metro Denver area. Kelly Dam, located on Westerly Creek and built on Lowry Air Force Base in the early 1950's, is subject to spillway and embankment overflow at about the 40-year flood event. The flood plain below the dam is intensely developed with single-family homes, apartments, and commercial establishments. A flood warning plan has been developed that involves the cities of Aurora and Denver, Lowry Air Force Base, and the NWS. The Urban Drainage and Flood Control District of the metro Denver area coordinated development of the plan and coordinates annual warning exercises.

The event forecasting portion of the flood warning plan involves stage gages to measure depth of water in Kelly Pond and another detention facility located upstream of Kelly Dam defined as Pond A-B; two rainfall observers at Aurora Fire Stations and one at Lowry AFB located within the basin; the NWS office and their radar capability and raingage (the NWS office happens to be located within the basin); a central location (Lowry AFB Command Post) to receive the data, evaluate the threat, and report the condition of the flood hazard; and rule curves to be used to determine the likelyhood of potential spillway overflows.

The Westerly Creek flood warning plan is fairly simple, but it does seem to work. The NWS, when conditions are such that heavy rainfall is possible, will issue an "internal alert" (not for public dissemination) to Lowry AFB, Denver, and Aurora. Lowry and Aurora will alert their rainfall observers. The NWS continues to monitor the situation, and if warranted, will issue a flash flood watch. The watch goes to Lowry, Denver, Aurora and the media for broadcast. Aurora sends a radio-equipped vehicle to Pond A-B and Lowry sends one to Kelly Pond to monitor water levels. The NWS monitors the weather conditions with radar and confirms rainfall amounts and intensities with the rainfall observers. The observers at Pond A-B and Kelly Pond report water depths to the Lowry Command Post. The rule curves are observed and Lowry advises Denver and Aurora when conditions are such that spillway and embankment overflows are likely to occur. Denver and Aurora make the decisions on evacuation. It is also possible that the NWS will issue a flash flood warning, and Denver and Aurora would have to decide whether or not to evacuate the flood plain on this basis.

The Urban Drainage and Flood Control District is also working with the City of Boulder and Boulder County on a flash flood warning system for Boulder Creek. The problems on Boulder Creek are serious as evidenced by the fact that the State of Colorado has defined flooding as the most dangerous hazard in Colorado and Boulder Canyon as the most dangerous area with respect to flooding. The Boulder Creek situation is similar in potential to the Big Thompson flash flood of August 1976 that killed 138 people. Boulder Creek is located about 40 miles south of the Big Thompson River, and both basins drain waters from the Continental Divide east through steep, narrow canyons onto flatter plain areas.

A basic problem with the Boulder Creek situation, particularly in the canyon area, is the very small lead times involved. In order to maximize lead time, it was necessary to telemeter remotely sensed hydrological data to a control location so that decisions can be made in the most timely manner possible.

As of this writing (February, 1979), final design of the event forecasting system is underway and the ordering and installation of equipment will be proceeding in the near future. Implementation of some aspects of the system is already underway. The event forecasting element of the Boulder Creek flood warning and evacuation plan consists of eight raingages and four stream gages whose information is radioed to the Boulder Regional Communication Center (BRCC). These gages are located primarily in the lower portions of the basin where time is a critical factor. In addition, there are sixteen raingages that are monitored by visual observations. Ten of these gages are associated with volunteer fire districts and six are with volunteer observers. Radios will be provided to the fire district observers to enable rapid reporting of rainfall amounts to the BRCC. Some fire districts also look at stage gages.

Rainfall and stream stage data will be supplemented by radar imagery through Color Computer Remote Radar (C2R2) receivers located within the Denver metro area including the possibility of one at the BRCC. The C2R2 system will allow real time radar images of the storm from two NWS radar sites to be used to aid in the forecasting. The NWS and a private meteorological service will both be used to advise personnel in the BRCC with regard to flooding potential.

A hydrologic model of the basin will not be immediately available and all forecasts and decisions will be made by personnel in the BRCC based on data being fed into the Center. It is anticipated that simple rule curves will first be developed followed possibly by computer models that can predict future flood crests based on the rainfall and runoff data input.

Other flood warning systems are planned in the Denver region for the future. These systems will be developed on a basin-by-basin basis, but opportunities for transfer of data will be exploited whenever possible.

The present need in terms of hydrologic forecasting is for systems that can serve as a vital element of urban flood warning and evacuation systems. The future will see more such systems developing and the state-of-the-art can be considered as at a relatively young point at this time. Limited "hands on" experience has been documented, although a body of literature does exist that describes and discusses flood warning systems from a "what is needed" or research perspective.

Other Related Urban Water Forecasting Considerations

The urban flood forecasting needs discussed in previous sections relate to problems that are relatively well accepted, i.e., planning, design, flood plain delineation, and flood forecasting. The data needs are thus reasonably well identified and real world applications are fairly common. Operational control for flood mitigation is much less common and automatic operational control is non-existent. Nevertheless, urban water managers should be thinking in terms of what the future may bring.

Consideration has recently been given to the concept of a "Metropolitan Water Intelligence System," a concept that goes beyond that of automatic operational control for flood mitigation. McPherson[1] coined the term to describe the concept "of multi-service automatic operational control, wherein field intelligence on all aspects of urban water might be acquired, including: precipitation; stream stages and flow rates; water and wastewater treatment facilities; water demands and distribution system rates and pressures; settings of regulating structures; quality parameters for water courses and impoundments, and within conveyance systems; and the status of special facilities such as recreational ponds and lakes."[2] While multi-service automatic operational control is not related directly to urban flood forecasting, it is mentioned because establishment of a data network for urban flood forecasting purposes may also serve as a basis or beginning of a more elaborate urban water data network.

Automatic operational control for urban water services is a concept for the future, a concept to keep in mind while designing urban flood forecasting data systems for more immediate needs. Why give automation consideration? McPherson summarizes[2] that increases in urbanization and the concomitant expansion of facilities and services will require improved methods of administration and management. The several urban water service functions will perhaps have to be developed to the point where exchanges in water qualities and quantities can be made between them. Planning strategies and operational requirements will become more complex and more and better field intelligence will be required.

In a sense, operational control of an urban water system whether or not automatic, depends on the capability to forecast reliably future events. For example, several cities have considered utilization of storage to reduce the frequency of combined sewer overflows, a water quality consideration. This requires anticipation of the direction and intensity of storms so that operational decisions can be made. Operation of a water system depends on anticipating or forecasting future water demands or perhaps whether or not it will rain. Thus, field

(1) American Society of Civil Engineers, Basic Information Needs in Urban Hydrology, A Study for the Geological Survey, U.S. Dept. of the Interior, New York, April 1969, p. 81.

(2) McPherson, M.B., "Feasibility of the Metropolitan Water Intellegence System Concept," ASCE.

intelligence for forecasting potential flood events may with adaptation also serve as input for operation of a combined sewer system or a water supply, treatment, and distribution system.

Urban flood control is a quantity consideration. Closely related to quantity is stormwater quality, an issue which has had much recent attention due to Section 208 planning requirements of PL 92-500, the Clean Water Act. Forecasting requirements for quantity are different than for quality, but certainly a compatibility exists. If a data network is established for flood forecasting purposes, can it be modified to accommodate certain water quality needs? Perhaps so, but perhaps not; the point being that such interchange should at least be considered.

Summary

Urban flood forecasting consists of two basic needs. The first is the forecasting of peak flows and hydrographs for the purpose of planning and design of urban drainage and flood control systems. The second is the forecasting of impending flood events for the purpose of warning flood plain occupants of the potential hazard. Both needs represent current activities, but the changing need is for improved or more reliable flood forecasting capabilities.

Some of the reasons for this changing need include the following: the high cost of drainage and flood control facilities make the "accurate" estimation of flows more important. The potential liability of public officials for their actions places a greater need on reliability of designs. The onus of not warning of a flood or issuing warnings when no floods occur can be a serious problem to a public works official. When 100-year flood plains are defined for regulation purposes, the delineation must be defensible in court as being based on the best that the state-of-the-art can offer. The requirement to provide detention of stormwater flows requires that an entire hydrograph be predicted which is more complex than just estimating a peak flow. The concern about receiving water quality and the pollution effect of urban storm runoff complicates the picture significantly. These developments are all contributing to steadily changing needs in urban flood forecasting.

The changing needs in general require more accurate urban flood forecasting. To obtain more accurate forecasting, research based on local rainfall/runoff data is required. Metropolitan areas with foresight are conducting data collection and analysis programs to allow them to better cope with the changing needs in urban flood forecasting.

CONTAMINATION IN THE HYDROLOGIC DATA BASE

Robert J. C. Burnash[1] and R. Larry Ferral[2]
National Weather Service, River Forecast Center
Sacramento, California

Abstract

Published and currently collected precipitation and streamflow records are contaminated by many misleading measurements, errors in measurements and changes in measuring techniques. Analysis of this data without discovering its pitfalls can lead to the design of grossly inadequate or excessively expensive structural or nonstructural responses to anticipated rainfall and streamflow. Careful checking of published gage data in relation to nearby gages and by streamflow simulation can reveal inconsistencies and guide the engineer and planner to safe, adequate and economical designs.

A substantial difficulty in all hydrologic analyses is the problem of properly evaluating the information content in basic historical data. With very few exceptions, the available hydrologic information has not been consistently measured in space or time. As a result, evaluations based on such data must first attempt to reduce the basic information to a homogeneous data set which is appropriate for the task at hand. All too frequently, published data is accepted as positive information without regard for hidden inconsistencies. Long term records of precipitation and discharge, which are the bases of many extremely expensive structural designs, are rarely studied in an attempt to eliminate the misinformation which most such records contain.

Precipitation data can be particularly misleading due to the tendency to continue the use of a station name through many physical moves of the equipment. A more subtle trap occurs when, although the site remains the same, the measuring device is changed (generally under the guise of improvement, modernization or standardization). Other subtle sources of data base contamination involve any change which can alter the approach trajectory of the precipitation or the velocity of the basic airstream in which the precipitation is imbedded. The principal elements which can have such an effect would include changes in nearby

Contamination in the Hydrologic Data Base

[1]Robert J. C. Burnash, Hydrologist in Charge
[2]R. Larry Ferral, Principal Assistant

structures and vegetation, as well as minor changes with respect to the precipitation gage. These would include such things as minor gage movements on the same plot of land, changes in the gage shape or its support, and variations in the height of the gage above the land surface. Unfortunately, many of these factors will take place in the natural course of events. The effects of seasonal and long term changes in vegetation may have differing but important effects on wind conditions around the gage, which influence the precipitation catch. Many precipitation gaging problems are related to the wind conditions at and around the gage orifice - a factor which can be substantially different from time period to time period within a storm, and remarkably different from one storm to another. A relationship between wind speed and gage catch, developed from combining analyses of Koschmieder, Wilson and other investigators, is regularly used by our office for precipitation analysis (Fig. 1). Although such techniques may be necessary to adequately convert precipitation to runoff, there is not a single published record of precipitation, of which we are aware, which includes a correction for wind deficiencies.

Gage data is published with no warnings as to the changes which may influence its effectiveness. Precipitation records are maintained by the National Weather Service for over seven hundred sites in California. Of these, less than 3 percent have a long record with relatively stable measuring conditions. But these gages are not wind-corrected and the data to do so is not available.

What do these deficiencies mean? Are they important enough to worry about? Certainly, if misanalysis of precipitation leads to a dam design which is too small, serious consequences may result. Conversely, if a dam is overdesigned, the costs may be much higher than necessary with little or no increase in benefits.

Let us briefly consider the changes in precipitation measurements on runoff determinations. In relatively wet areas where soil moisture and evapotranspiration requirements are already met, an analysis of streamflow may help to demonstrate the effect of additional precipitation in producing additional runoff. In an area where rainfall averages over 100 inches per year, normal losses from the soil mantle to the atmosphere may average approximately 45 inches, with the remaining 55 inches becoming basin runoff. An analyst who was unaware that a gage change was producing a 5% increase in the recorded catch of a precipitation gage would predicate an annual runoff of 60 inches (100+5-45=60), a 9% increase in runoff. If the analyst had a split population of historic data - which he had not properly evaluated - he would generate an artificial determination of average annual precipitation, return periods, the probability of critical time series and other hydrologically important elements which could effect subsequent design efforts or operating criteria.

The analysis of precipitation errors becomes substantially more striking as the area of concern becomes more arid. Based on streamflow simulation studies which had been fitted to produce a zero bias in terms of reconstituting published discharge volumes, we are able to test the impact of a 5% precipitation shift on runoff. Sespe Creek, a 253

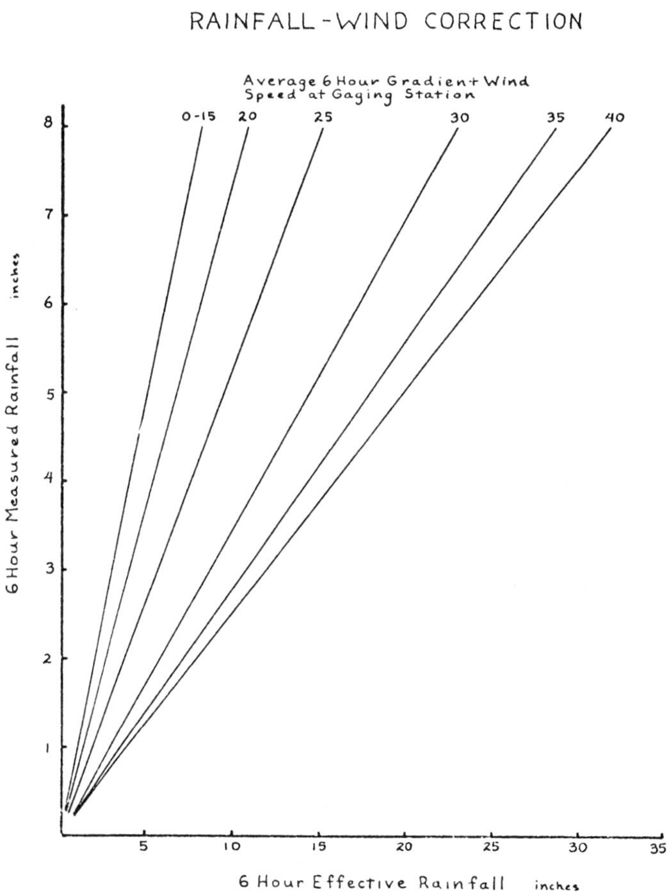
RAINFALL-WIND CORRECTION
Average 6 Hour Gradient Wind Speed at Gaging Station
0-15
20
25
30
35
40
6 Hour Measured Rainfall inches
6 Hour Effective Rainfall inches
1
2
3
4
5
6
7
8
5
10
15
20
25
30
35

Fig. 1

square mile California drainage in Ventura County, was used for this purpose. Precipitation inputs which had produced a zero volumetric bias, were increased by 5% - this increased volume of input resulted in an average increase of simulated streamflow in excess of 17% (Fig. 2). Earlier tests in very dry areas have shown over a 30% shift in simulated annual flows from 5% shift in precipitation. Obviously, nature does not go around altering past precipitation patterns by 5%, but there are very few areas where the precipitation records that man has maintained as his legacy from the past have been maintained with a 5% accuracy level. At a recent AGU meeting attended by the authors, a description of sizing dams based upon long term precipitation records was presented. A record going back into the last century was compared against runoff for a similar type period, but the departures from normal precipitation and runoff did not correspond, and no effective relationship of one parameter could be inferred from the other. The author did not indicate that he had made any effort to verify his data base or to reduce it to a homogeneous data set. In fact, the precipitation he was analyzing as a single time series was actually collected at several different sites with different equipment, but published under a common name.

The point of all this is to caution you to use extreme care when utilizing precipitation data - don't draw conclusions until you have determined:

1) the data is from a single location - specific sites, not just a city name,

2) there has been no change in the recording equipment during the period of interest, and

3) wind influences in the vicinity of the site have not changed air stream characteristics.

There may be a temptation to aschew the use of precipitation data and use streamflow records directly, but similar problems of data base contamination may still exist. Anything which changes the measuring site or the technique used to evaluate flow can produce substantial errors which can contaminate subsequent analyses. As an example, a plot of accumulated discharge from neighboring Sierra drainages is offered (Figs. 3 & 4). For the period 1940-1968, the North Fork of the Kings River at Cliff Camp is plotted against the accumulated flow for Florence Lake on the San Joaquin River. A close, continuing correspondence can be seen. When the record for Cliff Camp is plotted against the Kings at Pine Flat Dam near Piedra, a break in the relationship is suggested. Why do the flows from the two streams in adjoining basins appear to have an invariant relationship, while that between upstream and downstream flows has a time trend? Actually, a change in the downstream measuring techniques due to the construction of Pine Flat Dam took place after 1952. Without comparing the data carefully, it might have been argued that the relative increase of the Kings River flow at higher elevations was due to a cloud seeding operation in that area; however, since the upper Kings and the upper San Joaquin have not changed with respect to one another, a more probable reason for the

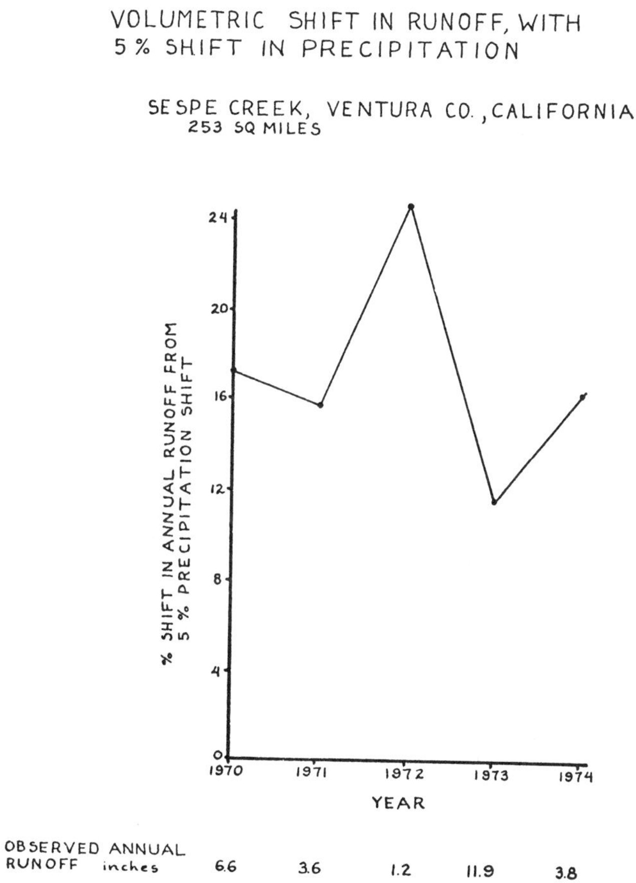
VOLUMETRIC SHIFT IN RUNOFF, WITH
5 % SHIFT IN PRECIPITATION
SESPE CREEK, VENTURA CO., CALIFORNIA
253 SQ MILES
% SHIFT IN ANNUAL RUNOFF FROM
5 % PRECIPITATION SHIFT
24
20
16
12
8
4
0
1970
1971
1972
1973
1974
YEAR
OBSERVED ANNUAL
RUNOFF inches
6.6
3.6
1.2
11.9
3.8

Fig. 2

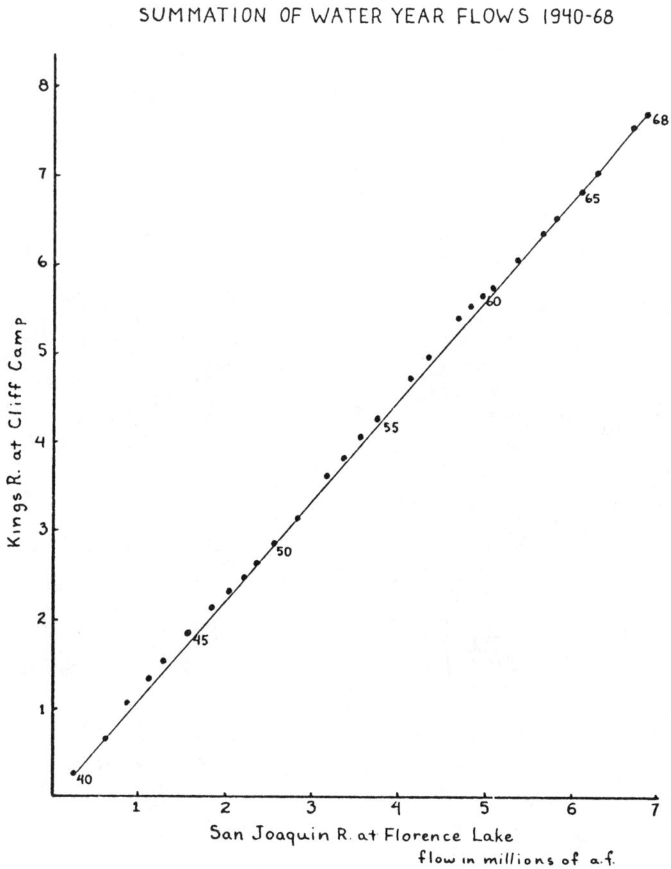
SUMMATION OF WATER YEAR FLOWS 1940-68
Kings R. at Cliff Camp
San Joaquin R. at Florence Lake
flow in millions of a.f.
1
2
3
4
5
6
7
8
40
45
50
55
60
65
68

Fig. 3

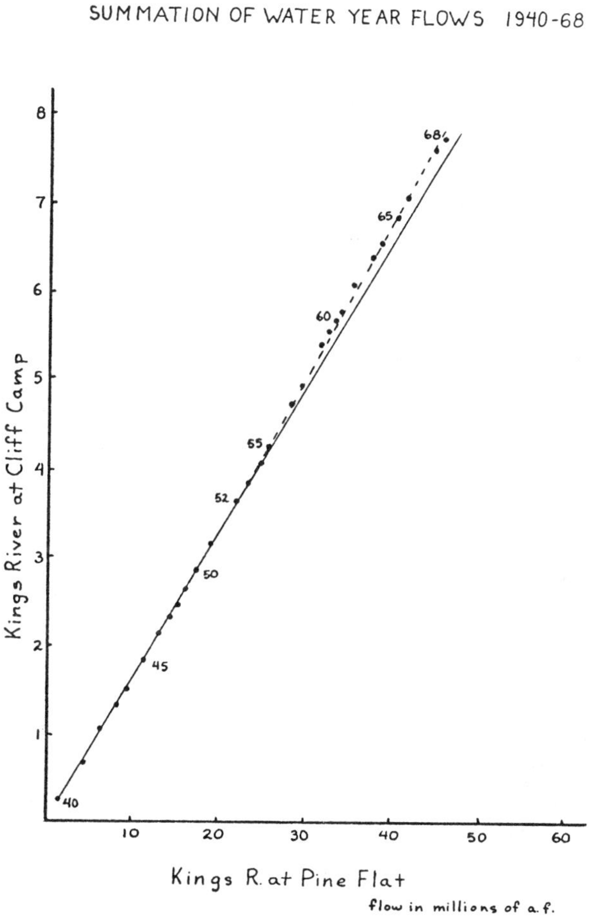
SUMMATION OF WATER YEAR FLOWS 1940-68
Kings River at Cliff Camp
Kings R. at Pine Flat
flow in millions of a.f.
40
45
50
52
55
60
65
68
10
20
30
40
50
60
1
2
3
4
5
6
7
8

Fig. 4

change in the relationship of Kings River measured flows is the change in measuring technique required by completion of Pine Flat reservoir.

In another instance, a major winter flood stirred the desire for improved flood protection in a major eastern city. This city, near the confluence of a major river and a relatively small tributary stream, was the subject of a study based on the discharge records for both the main river and the tributary stream. The engineering group conducting the study for the recent flow of concern, determined little could be done on the main river, but a dam on the tributary would have reduced flood stages in the city by approximately four feet. The dam was built. When one of the authors later reviewed the design information and compared the discharge data used in the design study with corresponding precipitation and adjoining stream records, it was found that the tributary stages, which had been converted directly to discharge for design purposes, were the result of a downstream ice jam. A substantial flood protection structure has been built on the basis of a streamflow determination that was in error by an order of magnitude.

How can an engineer protect himself from data base inconsistencies? There are several techniques, including those mentioned earlier. They include double mass plots of precipitation data and cross basin plots of discharge data. The most effective technique starts with these simple tests and then applies streamflow simulation techniques. Such methodologies have advanced to the point that inconsistencies in the precipitation vs discharge relationship are powerful clues to data base contamination. Such contamination has probably resulted in many billions of dollars being spent on improperly sized structures. Until such time as we can appreciate the importance of a proper data base system and adequately fund it, the linking of cause and effect through simulation techniques is probably our best chance of recognizing the problems of data base contamination which influence the design and operational criteria for increasingly costly water resource projects.

The great value which timely and accurate data has for hydrologic purposes has recently led to the development of a new family of cost-effective precipitation gages which are intended for immediate data acquisition by computer. Such a system will eventually allow the immediate cross checking of data via computer comparison and applied streamflow simulations. It will also provide for the real-time use of data for reservoir regulation and other hydrologic purposes, while simultaneously reducing the cost of data collection and handling for climatologic purposes.

Our efforts in California to automate data collection are focused to a large extent on self-reporting gages. These gages send a radio signal to the base station when a change occurs, and at timed intervals when no change occurs. Thus information on rapid changes in hydrologic data becomes available as it occurs. The field units with self-keying radios need only transmit, not receive, so they are inexpensive and relatively simple and draw very little power. A single battery is adequate for well over a year's data transmissions without servicing of any type or needing any recharging devices. The simplest self-reporting arrangement is a tipping bucket raingage, tipping with each millimeter of rain and sending a radio signal with a station ID and an accumulated value with each tip. Six such gages were installed in Monterey County, California, immediately after the disastrous Marble Cone fire of 1977 to provide immediate information for flash flood warnings on several small rivers in the area.

In those areas where a significant part of the precipitation falls as snow, precipitation is collected in a low-density, self-mixing anti-freeze solution in which it dissolves, mixes and displaces fluid through an overflow into a tipping bucket. At some sites snow water equivalent is measured as it changes by linking the system to snow plates. Several of these installations have been made for use in real-time forecasting of reservoir inflows on the American and Stanislaus Rivers.

Other computer data bases based on DARDC devices, telemark gages and real-time data transfers from other agencies are being linked to provide more effective data systems which can be cross-verified for accuracy.

Data storage, analysis and retrieval are being accomplished by tasks in the same computer that gathers the interagency data base. This program runs continuously, performing different tasks as the needs arise, keeping the latest data continuously updated and readily available for immediate hydrologic forecasts and other analyses.

Thus, data base problems are being attacked in a variety of ways in an attempt to construct a complementary relationship between the collection and verification of real-time data for hydrologic analysis and the hydrologic analysis itself. These approaches, when coupled with streamflow simulation techniques, should substantially assist in determining that a river gage is probably stuck, that an apparently high river discharge is actually backwater from an ice jam, that a raingage is not working or is reporting erroneous data or that precipitation gages are under-sampling because of wind or snow. The data identified as contaminated can be adjusted or excluded and the verified data can be used to produce much more reliable hydrologic analyses and forecasts. All of this will not, however, be completed in the near future, and older data used in many design studies will continue to need very careful handling. Our basic suggestion is this: assume your data is contaminated and apply the necessary checks to discover how badly. You will frequently find that some of your most profitable efforts are those which allow better decisions based upon an intelligent evaluation of data base inconsistencies.

LET'S LOOK AT SOME LEGAL IMPLICATIONS OF STREAMFLOW FORESCASTING

By Thomas M. Stetson*

Clarence Darrow once said "History repeats itself. That's one of the things wrong with history." Hydrologists in planning water resource developments usually lean heavily on history repeating itself. We attempt to determine the dependable annual water supply of a proposed water project by reconstructing the historic annual water supply of the source to be developed. We then forecast that the historic water supply will repeat itself and that the dependable supply will be that quantity available for use during the driest portion of this historic period.

But there is another law. It is called Osborn's Law(1)** and says: "Variables won't; constants aren't." In other words, don't rely solely on what happened historically because historic variations in water supply will not necessarily occur in the future and the variability can easily exceed the historic ranges.

Regrettably, many water supply projects were planned without the advantage of a sufficient, reliable historic record of streamflow. The Colorado River is an outstanding example.

Historical Perspective of the Colorado River

Congressman Frank Mondell of Wyoming was responsible for legislation which became law on August 19, 1921 which authorized the seven states of the Colorado River Basin to enter into an agreement to apportion the waters of the Colorado River. This resulted in the Colorado River Compact, which was signed at Santa Fe, New Mexico, in 1922. Briefly stated, it divided the water supply of the Colorado River between the Upper Colorado River Basin and the Lower Colorado River Basin at what is known as the Compact Point, Lee Ferry, Arizona. It is located on the Colorado River about one mile downstream of the junction of the Paria River with the Colorado River in northern Arizona and about 17 miles downstream of Glen Canyon Dam.

---

* Consulting Civil Engineer and President of Stetson Engineers Inc., San Francisco.

** References are listed at the end of this paper.

The Colorado River Compact divided the Colorado River between the Upper Basin and the Lower Basin as follows(2):

(1) Each basin to receive 7.5 million acre-feet (MAF) annually of beneficial consumptive use of Colorado River water.

(2) The Lower Basin to receive an additional 1.0 MAF annually of beneficial consumptive use.

(3) Mexican Treaty water to be supplied first from surplus over and above (1) and (2) and, if insufficient, the deficiency to be equally borne by Upper Basin and Lower Basin.

(4) States of the Upper Division will not cause the flow at Lee Ferry to be depleted below an aggregate of 75 MAF for any period of 10 consecutive years.

(5) Further apportionment after October 1, 1963 of water unapportioned by (1), (2) and (3).

During the trial of the Arizona v. California case(3), California argued strenuously that the Colorado River Compact divided all the water of the Colorado River system, including all of its tributaries. This would mean, as to Arizona, that all of its uses of water from the Gila River system, for example, would be chargeable to the State of Arizona as part of its use of water allocated to the Lower Basin. At the time of trial (1956 to 1958) Arizona was consumptively using about 2.5 MAF annually from the Gila River system. The Gila River system originates in southwestern New Mexico and flows westerly through southern Arizona to join the Colorado River near Yuma, Arizona, just upstream of the limitrophe section between Mexico and the United States.

Arizona argued against this in two respects. First, that its tributary uses should not be chargeable under the Compact because it is a main-stream compact; and, secondly, that if it is chargeable, the Gila River use should not exceed about 1.2 MAF annually because the virgin flow(4) of the Gila River at its junction with the Colorado River did not exceed that amount. The inflow of the Gila River and its tributaries to the central Arizona area is about 2.1 MAF per year, but it was always a wasting stream and much of its supply was consumed by native vegetation. Thus, its virgin flow entering the Colorado River was about 1.2 MAF per year according to the U.S. Bureau of Reclamation.

California lost this argument before the Special Master who heard the trial and before the U.S. Supreme Court. The Special Master and the Supreme Court ruled that the Colorado River Compact was a mainstream compact insofar as the Lower Basin was concerned and use of water on Lower Basin tributaries was not accountable under the Compact.

When the Colorado River Compact was negotiated, there was very little actual record of the streamflow at Lee Ferry. The stream gage at that station was installed in 1921 and the Compact, as previously stated, was signed in 1922. At that time it was estimated that the virgin flow at Lee Ferry exceeded 18 MAF per year; the Upper Basin would be able to use up to 7.5 MAF per year of Article III(a) water; the Lower Basin would use 8.5 MAF per year of Article III(a) and III(b) water; that there would be sufficient water to satisfy a future Mexican Treaty requirement; and that there would be available under Article III(e) additional water to divide in the future between the Upper and Lower basins.

At that time, however, it was not known how much water would be required for the Mexican Treaty. California assumed it would not exceed about 0.7 MAF. The United States section of the International Water Commission had notified the Mexican section of the Commission on August 30, 1929 that 750,000 acre-feet, representing the highest annual amount used up to that time, would be an equitable division of the Colorado River waters. When the Treaty was negotiated during World War II, it was negotiated by the State Department. The United States wanted to be generous to Mexico, and it was. As a result, the Treaty requires that Mexico receive 1.5 MAF per year.

Thus, if we ignore Article III(b) and III(e) of the Colorado River Compact, there must be at least 16.5 MAF of virgin flow at Lee Ferry to satisfy the Mexican Treaty and something near the Article III(a) provisions of the Compact. And this does not count river losses downstream of Lee Ferry.

The actual virgin flow at Lee Ferry over the now long-term period of record (1921 to 1976) is about 14 MAF. At the time of the Compact negotiations, it was estimated on the basis of correlations with stations(5) many miles upstream to be about 18 MAF. The years prior to 1922 could have been "wet" years and thus led to the assumption that history would repeat itself, as Clarence Darrow said. But the thing that was wrong was not that history repeated itself, it was that history did not repeat itself. The variables did not and the constants weren't--Osborn's law.

Immediately after the Colorado River Compact was signed, strenuous efforts were made to get a large dam and reservoir constructed on the Colorado River at or near Boulder Canyon to provide flood protection and to firm up water supplies for Lower Basin water users. There had been devastating floods in Imperial Valley in the early part of the 20th Century and there were frequent floods in Palo Verde Valley from the spring and summer runoff of the Colorado River.

The Swing-Johnson bill, which finally resulted in the Boulder Canyon Project Act and the construction of the Project, including Hoover Dam and Lake Mead, was a long time in coming. It was carried by California Congressman Phil Swing and California Senator Hiram Johnson.

It was first introduced on April 25, 1922. Then again on December 10, 1923 and then again on December 21, 1925. Two years later the fourth Swing-Johnson bill was introduced and on December 21, 1928 President Coolidge finally signed the Boulder Canyon Project Act.

The effective date of the Boulder Canyon Project Act was June 25, 1929. Hoover Dam, creating Lake Mead, was constructed with a capacity of about 32 MAF. The Secretary of the Interior then entered into water supply contracts to supply water for use in the Lower Basin states.

The assumption in 1930 was that there would be 10.5 MAF of water passing through Lake Mead annually(6). It was assumed that annual diversions to the Lower Basin states and to Mexico could be as follows:

	Acre-feet per year
California	5,500,000
Arizona	3,900,000
Nevada	300,000
Mexico	800,000
	10,500,000

In the short period of time, 1921 to 1929, that there was a stream gage at Lee Ferry prior to 1930, the virgin flow was calculated to exceed 15 MAF in seven of those nine years and to average about 18.2 MAF per year for that 9-year period, according to a 1946 report by the U.S. Bureau of Reclamation.

The quantity of the water supply of the Colorado River was the most important factual question in the Arizona v. California litigation of the 1950's. Records of flow at Lee Ferry through 1956 were available for that trial. Studies presented at the trial covered the 48-year period 1909 through 1956 or shorter periods such as 1922 through 1956 and 1930 through 1956. But the flows from 1909 through 1921 were based on estimates, not recorded flows. The estimated average annual virgin flow for that 13-year period was 18.45 MAF. For the 35-year period 1922 through 1956, the average annual virgin flow was 14 MAF, and for the 27-year period 1930 through 1956 it was only 13.7 MAF.

Four witnesses testified on the long-term water supply of the Colorado River during that trial. The results of all of that testimony spelled future shortages and that neither the Upper nor the Lower basin would receive the full quantities allocated by the Colorado River Compact. Far from it.

Since the Arizona v. California trial much additional information has become available, including some 20 years of additional recorded stream flow data at Lee Ferry. It is now generally considered that the long-term average annual virgin flow at Lee Ferry is about 14 million acre-feet. Although there is presently a surplus of water available from the Colorado River for use in the Lower Basin, it is surplus only because it is not now being used by those entitled to use it. When the Central Arizona Project is completed and comes on line in the mid-1980's it will divert some 1.6 MAF per year and increase Arizona's net diversions to about 2.8 MAF. But there will be shortages to the Central Arizona Project as Upper Basin depletions exceed about 4.3 MAF per year. They are now nearing 4 MAF per year. There will be an almost direct relationship between Upper Basin depletions and reductions in net diversions by Arizona, California and Nevada after Upper Basin depletions reach about 5 MAF per year. Unless the Colorado River supply is augmented in the future, the Upper and Lower basin supplies will total on the order of 12 MAF.

Both California and Arizona, after struggling for more than 50 years in the courts, the Congress and the press, are to be deprived of what they had hoped to receive from the waters of the Colorado River.

Legal Implications

The legal implications can be succinctly stated: law suits. Unfortunately, law suits may end up redistributing water supplies but they seldom develop new water. There have been many law suits over the Colorado River in the past and there are sure to be more in the future.

Historic Litigation

There have been five legal actions brought in the U.S. Supreme Court in attempting to resolve the conflicts in the Lower Colorado River Basin.

The first was a suit by Arizona against the Secretary of the Interior and the six other states of the Basin. The suit sought to enjoin carrying out the Boulder Canyon Project Act and to have the Act declared unconstitutional. Arizona alleged a violation of its quasi-sovereign rights by interference with its appropriations of unappropriated water by building Hoover Dam, partly in Arizona, without complying with Arizona laws and by forcing terms of the Colorado River Compact on Arizona, which at that time was not a signator of the Compact.

The case(7) was dismissed without prejudice to the right of Arizona to defend perfected or future appropriations. The Court found that the Boulder Canyon Project Act was a valid exercise of federal powers under the commerce clause of the Constitution over a river which the Court judicially noticed was navigable; the United States could perform its functions without conforming to the police regulations of a state; and the Act did not abridge the right of Arizona to appropriate waters of the Colorado River.

Soon thereafter Arizona brought another suit(8) against the same defendants. This was for the purpose of perpetuating testimony of the Colorado River Compact negotiators for an action which could be commenced at some future time as a result of the Boulder Canyon Project Act, the California Limitation Act prescribed by the Project Act, the Presidential proclamation declaring the Project Act effective, and the General Regulations of the Secretary of the Interior authorizing water delivery contracts in California.

Leave to file the bill of complaint was denied. The Court held that it had jurisdiction for a proper case involving perpetuation of testimony but it also held that such testimony would not be competent because the real issue was the intent of Congress in 1928 and the material sought to be perpetuated was not embodied in any writing and was not communicated to the governments of the negotiators representing the states nor to Congress.

The third case was brought by the United States against Arizona(9). It sought to enjoin Arizona from a threat of using force to prevent construction of Parker Dam, which is partly on Arizona soil. Arizona had called out its troops--the State Militia.

The Complaint was dismissed because it failed to show that construction of the dam was authorized. An 1899 Act forbids the construction of a dam across a navigable river of the United States until consented to by Congress and the plans have been approved by the Chief of Engineers and the Secretary of War. These approvals had apparently not been obtained.

Thereafter Arizona brought a fourth suit, this one against only California(10). Arizona sought a judicial apportionment of Arizona's equitable share of the water; injunctive relief against California's claims in excess of that allowed by the Project Act and the Limitation Act; a decree by which use by any other state of Arizona's share of the water would not constitute an appropriation; and that any increase in rights for Mexico resulting from works constructed by or for California would be supplied from California's share of the water.

The motion was denied because the United States was an indispensable party.

Finally, or so it was thought at the time, Arizona brought suit in 1952 against the State of California and seven public agencies in California. Nevada, New Mexico, Utah and the United States were added, either voluntarily or on motion. That case was decided by the Supreme Court in 1963. But even then, certain matters were left open for further resolution by the parties. The decree(11) provided in part: "Within two years from the date of this decree, the States of Arizona, California, and Nevada shall furnish to this Court and to the Secretary of the Interior a list of present perfected rights, with their claimed priority dates, in waters of the mainstream within each state...The Secretary of the Interior shall supply similar information, within a similar period of time, with respect to the claims of the United States to present perfected rights within each state. If the parties and the Secretary of the Interior are unable at that time to agree...any party may apply to the Court for the determination of such rights by the Court."

One would think that the determination of present perfected rights would not be that difficult. A perfected right is defined in the Court's decree as a water right acquired under state law, which has been exercised by actual diversion of a specific quantity of water, applied to a defined area of land, etc., and shall include reserved rights for use on federal establishments whether or not applied to beneficial use. A present perfected right is a perfected right existing as of June 25, 1929, the effective date of the Boulder Canyon Project Act.

But it must be recognized that records and maps of water use prior to 1929 are not all that available and if available, not all that reliable. The first vertical aerial photographs along the lower Colorado River were flown in 1930 but did not cover all of the areas of use. Furthermore, some of the federal establishments, namely Indian Reservations, claim they were not adequately represented during the trial and that they are entitled to more water than the decree provided.

To make a long story short, the parties and the Secretary of the Interior have not been able to agree on present perfected rights. On January 9, 1979 the Supreme Court appointed a new Special Master, Judge Elbert P. Tuttle, to consider further pleadings and to take additional evidence on present perfected rights if he deems it necessary. He will be meeting with the parties in Phoenix tomorrow and in San Francisco next month.

Possible Future Litigation

This paper addresses only the basic water supply aspects of the Colorado River. It does not address water quality, environmental matters or other water related problems, such as property boundary changes due to changing courses of the river. These too feed the avaricious appetite of the grist mill of litigation.

The exaggerated water supply quantities which the Colorado River Compact negotiators had before them in 1922 led to the over-apportionment of the Colorado River. As demands for water approach and exceed the limited supply there will undoubtedly be litigation seeking such objectives as reallocation between the Upper and Lower basins, among the states within each basin, among the users within the states, and possibly among the various historic types of use of the water.

The Colorado River Compact apportioned water between the Upper and Lower basins and made provisions for water to supply Mexico. It did not apportion water specifically among the states. It did, however, speak to the questions of priorities among various types of water use. In Article IV(a) it indicated that the use of water for navigation would be subservient to uses for domestic, agriculture and power generation purposes. Article IV(b) provided for the impoundment and use of water for power generation, but such use to be subservient to the use of such water for agricultural and domestic purposes. Domestic use is defined in Article II(h) to include use of water for household, stock, municipal, mining, milling, industrial and other like purposes, but it specifically excludes the generation of power as a domestic use.

Article VII of the Compact says that nothing in the Compact shall be construed as affecting the obligations of the United States to Indian Tribes. There is a similar provision in the Upper Colorado River Basin Compact of 1948.

As future shortages occur, most of these matters will be wrestled with, either in litigation or in negotiations for possible agreements. It is rather obvious that the Colorado River Compact envisioned agriculture and domestic use as the dominant uses in the basin. However, it included industrial use as a part of domestic use. This will lead to many questions such as the use of Colorado River water for the development and/or transportation of mineral resources (coal, oil shale, coal slurry lines) which may then be used in power plants to generate electric power. Is that a domestic use? Or is it a subservient use for electric power generation? The Compact, in Article IV(b), describes the subservient use as water _impounded and used_ for the generation of electric power. So without impoundment is there a question? It is rather obvious that the negotiators of the Compact had in mind the storage required to provide the head for hydro-electric generation at the storage reservoirs.

With impending water shortages there will be serious legal issues with respect to the claims of Indian reservations to use water to develop their resources. This was brought into focus in the _Arizona v. California_ decree(11) wherein it was determined that the Indian reservations had the right to use water to develop their irrigable lands, whether or not the lands had ever before been irrigated, and with a priority as of the date of the establishment of the reservation. These can be very large claims for water and most will predate many of the existing non-Indian uses. Some states will have difficulties in accomodating those uses within their apportioned quantities of Colorado River water without reducing some of the existing non-Indian uses. Uses of water within each state are chargeable to that state, whether or not such uses are on federal reservations.

The foregoing are but a few of the issues which must be faced in the future. It cannot be said that these issues would not exist if the long-range water supply of the Colorado River had been accurately forecast at the time the Colorado River Compact was negotiated (1921 and 1922). However, if the Compact negotiators knew they were dealing with a 14 MAF per year river instead of a 18+ MAF river the apportionments under the Compact would in all likelihood have been less. Also, the Mexican Treaty allocation may have been less.

Outlook for the Colorado

It can be said that the Colorado River has a muddy past and a cloudy future. What can be done to enhance the future water supply of the Colorado River? The U.S. Bureau of Reclamation has undertaken a cloud-seeding project to determine if weather modification can significantly increase the water supplies of the Colorado River. The Bureau believes that weather modification may supply as much as 1 MAF or more annually to Colorado River supplies.

Vegetation management on watersheds is also being studied by the U.S. Forest Service. It indicates a potential increase in water supply of about 1.2 MAF per year in the Upper Basin and about 0.3 MAF per year in the Lower Basin.

The other major alternative, one which could supply several MAF per year, is augmentation of the Colorado River supplies by importation from distant sources that have surplus supplies. But in today's world of environmental concern, all of these possible sources of augmentation face a long and hazardous trail to fruition.

When water projects are built, they are usually built for specific purposes which are not easily susceptible to change. Major projects now exist in the Lower Colorado River Basin which were planned on the basis of a full water supply under valid state water rights and valid federal water delivery contracts. The Colorado River Aqueduct of the Metropolitan Water District of Southern California is a specific example. The Aqueduct was constructed to divert and convey to the south coastal area of California 1,212,000 acre-feet per year. The only losses that were expected were transmission losses of perhaps five percent of the water diverted.

Now, due to over-estimates of the supply available and to the Supreme Court's interpretation of the "law of the river"--the Colorado River Compact, the Boulder Canyon Project Act, the California Limitation Act, etc.--the Metropolitan supply will be reduced to 550,000 acre-feet per year as soon as the Central Arizona Project goes into full operation, estimated to be in the mid-1980's.

But that isn't the full extent of the reduction in supply which may befall the Metropolitan Water District. In one of its recent reports(12) it states: "One other pertinent aspect of the 1964 Supreme Court decree(11) concerns Indian water rights. The decree awarded the Lower Colorado River Indian Tribes certain rights that, in essence, come out of Metropolitan's 550,000 acre-feet. Recently, the tribes have petitioned to have such rights expanded. After allowing for higher priority uses and aqueduct conveyance losses, eventually no more than 450,000 acre-feet and possibly only 400,000 acre-feet of Colorado River water will be available for delivery in the District's service area."

This is, of course, a serious reduction in water supply for an existing project supplying an area of vast economic importance to the nation as well as the State of California. Fortunately, the California Water Project was constructed in time to supplement Metropolitan's needs. But that project is also facing some problems at the present time which must be resolved soon or it will be deficient in supply.

In conclusion, it is unfortunate that planning for an entire river basin for decades was based on the assumption that the total water supply was several million acre-feet per year greater than nature has provided during the past 50 years. You could blame that on the early forecasts, but the problems of the Colorado River can also be attributed to the lack in being able to forecast 50 years ago the various interpretations that would be forthcoming on the law and politics of the river.

REFERENCES

(1) Don Osborn, Associate Director, State of Arizona, Solar Energy Commission.

(2) Article III of the Colorado River Compact

(3) 373 U.S. 546 (1963).

(4) Virgin flow is the estimated flow that would have occurred without interference by any man-made developements upstream--virgin conditions of the watershed.

(5) San Juan River near Bluff, Utah, 1914-1917 record; Colorado River Near Cisco, Utah, 1914-1917; and Green River near Green River, Utah, 1894-1899 and 1904-1922.

(6) Report of Col. William J. Donovan to Secretary of the Interior Ray Lyman Wilbur, February 14, 1930.

(7) 283 U.S. 423, 75 L.Ed. 1154 (1931).

(8) 292 U.S. 341, 78 L.Ed. 1298 (1934).

(9) 295 U.S. 174, 79 L.Ed. 1371 (1935).

(10) 298 U.S. 558, 80 L.Ed. 1331 (1936).

(11) 376 U.S. 340 (1964).

(12) Planning Concepts Used in Determining the Water Supply Available from the State Water Project, Metropolitan Water District of Southern California, Report Number 935, January 1979, p. 3, Overview of Future Water Supply Available to Metropolitan.

FLOOD PLAIN MANAGEMENT NEEDS

G. R. Phippen [1/]

The last 11 years while I have been Chief, Flood Plain Management Services in the Office, Chief of Engineers, have given me a good basis for reflecting on what floodplain management needs are. There are obviously many. But paramount is common understanding of terms and concepts. Misunderstandings are rampant. Ideas are not being communicated consistently. I must be partly to blame for this and many of you are not free from blame either. Recent experience as chairman of a Task Force on Nonstructural Flood Protection Methods under the President's Water Policy Initiatives and especially after participating in "public workshops" on this subject would have convinced me even if I hadn't already come to that conclusion. Expectations based on misunderstandings about the methods and the players in the process, especially perceptions in the environmental community, have led to disappointments about Federal activities, and a reinforcing of the gulf which exists between these concerned individuals and the public water resources agencies.

In the parlance of floods, forecasting connotes what is done at the Weather Services river and hurricane centers, and has a short future time span. I offer this as an example germane to this conference. Imagine the problems introduced when the same word is used, as it may well have been at this conference, to describe translations of probability which in reality have no time dimension. Nothing accumulates, i.e., absence of flood in one year doesn't increase probability in the next.

So that I wont be quite as guilty as I might otherwise be of contributing further to misunderstanding, I offer the following summary in threes as a means of stating where I'm coming from and to offer some hope that those listening will place ideas in the same context. As can be seen from this quick methods of introduction (Fig. 1 & 2), many of the FPM ideas can be grouped in threes. The general scheme comes from the U.S. Army Corps of Engineer Regulations on Alternatives in Flood Related Planning dating from 1970. The Water Resources Council expanded on these notions in developing the report entitled "A Unified National Program for Flood Plain Management," which was finally printed in 1976 after several years of preparation. To illustrate the use of three's--there are three phases of FPM, three sets of tools or adjustments, three illustrative conditions of floodplain development, and three public actors--Federal, State and local.

1/ George R. Phippen, Chief, Flood Plain Management Services Branch, Planning Division, Directorate of Civil Works, Office of the Chief of Engineers, Washington, D.C.

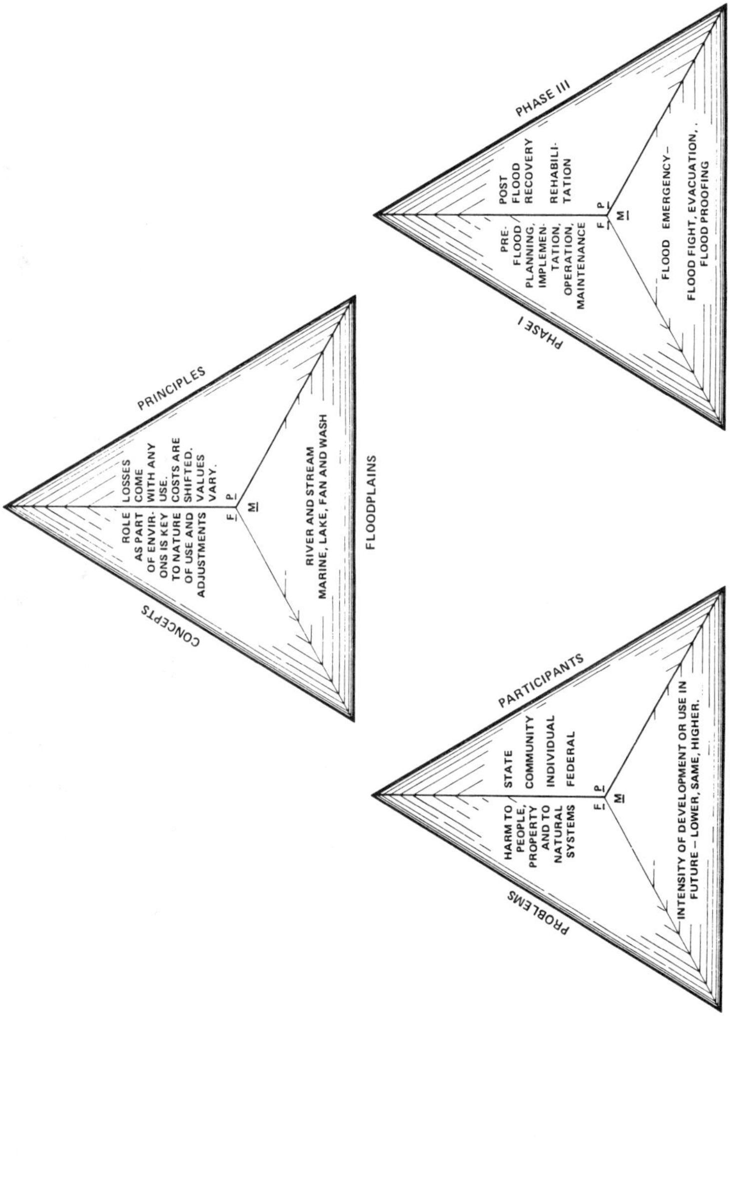
PRINCIPLES
CONCEPTS
FLOODPLAINS
ROLE AS PART OF ENVIR- ONS IS KEY TO NATURE OF USE AND ADJUSTMENTS
LOSSES COME WITH ANY USE. COSTS ARE SHIFTED. VALUES VARY.
RIVER AND STREAM
MARINE, LAKE, FAN AND WASH
F P M
PHASE III
PHASE I
PHASE II
PRE- FLOOD PLANNING, IMPLEMEN- TATION, OPERATION, MAINTENANCE
POST FLOOD RECOVERY
REHABILI- TATION
FLOOD EMERGENCY–
FLOOD FIGHT, EVACUATION, FLOOD PROOFING
F P M
PARTICIPANTS
PROBLEMS
F.P. ROLES
HARM TO PEOPLE, PROPERTY AND TO NATURAL SYSTEMS
STATE
COMMUNITY
INDIVIDUAL
FEDERAL
INTENSITY OF DEVELOPMENT OR USE IN FUTURE – LOWER, SAME, HIGHER.
F P M

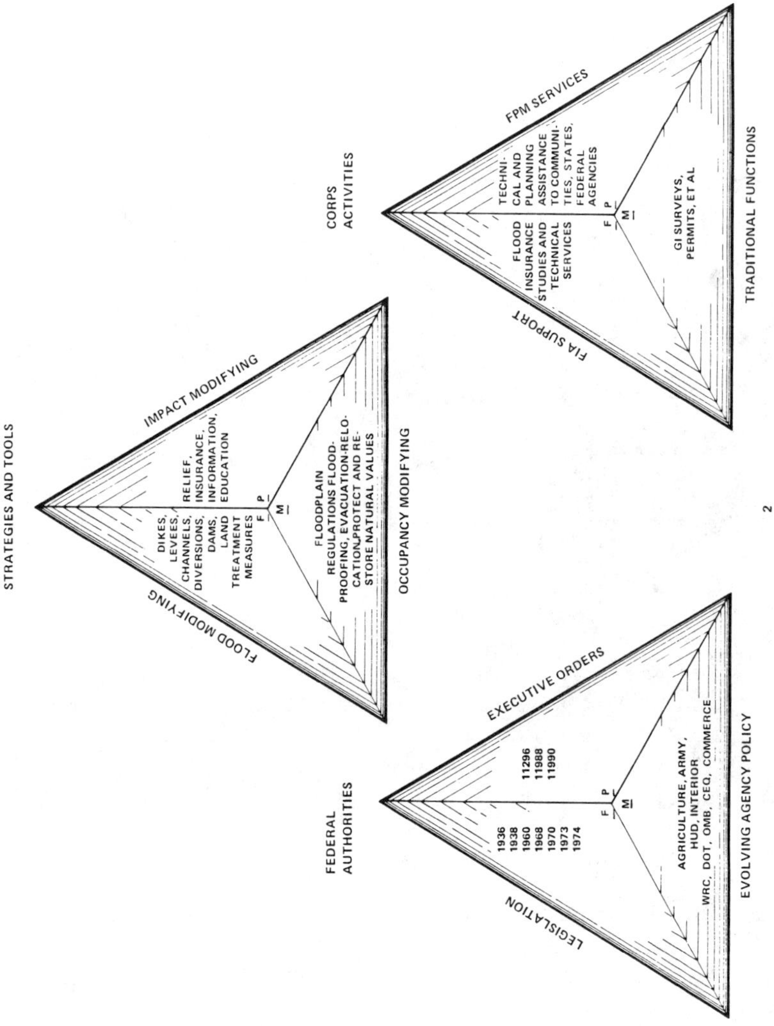
STRATEGIES AND TOOLS
IMPACT MODIFYING
FLOOD MODIFYING
OCCUPANCY MODIFYING
DIKES, LEVEES, CHANNELS, DIVERSIONS, DAMS, LAND TREATMENT MEASURES
RELIEF, INSURANCE, INFORMATION, EDUCATION
FLOODPLAIN REGULATIONS FLOOD-PROOFING, EVACUATION-RELOCATION,PROTECT AND RE-STORE NATURAL VALUES
F P M
CORPS ACTIVITIES
FPM SERVICES
FIA SUPPORT
TRADITIONAL FUNCTIONS
FLOOD INSURANCE STUDIES AND TECHNICAL SERVICES
TECHNICAL AND PLANNING ASSISTANCE TO COMMUNITIES, STATES, FEDERAL AGENCIES
GI SURVEYS, PERMITS, ET AL
FEDERAL AUTHORITIES
EXECUTIVE ORDERS
LEGISLATION
EVOLVING AGENCY POLICY
1936 1938 1960 1968 1970 1973 1974
11296 11988 11990
AGRICULTURE, ARMY, HUD, INTERIOR WRC, DOT, OMB, CEQ, COMMERCE
2

Foundations of Floodplain Management

Three subjects--hydrology and hydraulics (H&H), occupance, and environmental values--form the main foundations for floodplain management. Otherwise, floodplains are governed pretty much by the rules and conditions that apply to other areas of the community and region. In the past there has been a tendency to concentrate more on the H&H aspects than on the others. All are essential.

Hydrology and Hydraulics (H&H)

These topics, H&H, provide the basic data dealing with the nature and behavior of floods and of flood regimen. Of interest to all three phases of floodplain management are topics--quantities, extents, velocities, durations and times of rise of floods. Others might include debris loads and pollution.

Another matter dealt with is H&H terminology in flood description. Subjective terms used to describe floods include small, medium, and large (sometimes, very large) and in a parallel--frequent, rare, and very rare. However, the engineer usually prefers to name floods in terms of probabilities, or in derivations such as percent chance of occurrence (exceedence), and return period. Although the descriptions in the probability mode are less subjective, in theory, than the size-rarity description, the latter may be better definitions for planning. Nothing causes more problems according to what I've observed than the idea of return period, which immediately translates into everyone's idea of time. Further, this concept of time is usually, if not always, anchored at year one.

The decision by many to use the so-called 100 year flood both before and since the National Flood Insurance Program for floodplain regulations was not the source of the problem. The problem was calling it "100 year" in public presentations. (For many years, we in the Corps of Engineers referred to it as the Intermediate Regional Flood in our Flood Plain Management Services (FPMS) reports.) It is important that all users understand that the probability remains the same in theory whether a large flood occurs in a given year or whether it doesn't, and that the next year is the same as the past one in this regard.

Occupance

Floodplain management (FPM) is obviously involved in and concerned with how the floodplain is occupied and used--occupance. Patterns and modes of use are essential information in FPM planning. The amount of existing development and the amount of likely future development, or its reduction, as part of the future of the floodplain's surroundings are key elements in FPM planning for all three phases. Planners need to know if the floodplain is developed, developing, or open, and if the future is likely to be one of less intensive, more intensive or about the same intensity of development. In other words, the floodplain is part of some larger growth system with varying requirements for developed and open spaces. Decisions are therefore as dependent on an understanding of these conditions as they are on the H&H information.

Environmental Values

Although the recognition of environmental values has been part of the floodplain management context since its inception, the emphasis given environmental values has been greatly overshadowed until recently by concerns for harm to people and property. NEPA began the intensification of awareness, at least for Federal agencies, and the whole matter was brought to the forefront in the environmental messages of President Carter in 1977. Environmental values often are much more easily expressed as the reciprocal of lost values, i.e., for flood attenuation, quality improvement, sedimentation reduction, temperature improvement and the like, where environmentally important areas are substantially altered. These are not the most obvious, however. More direct values can be found in the form of fish and game habitat, food chain source (locus of wetlands), and esthetics.

The conclusion to be drawn from this is that since these three basic foundations are interactive, the needs for floodplain management go beyond the sum of the parts.

Problems for the Professional Hydrologic Engineer

The title of this Engineering Foundation Conference ". . . Hydrologic Forecasting," as suggested earlier, gives us a clue as to the kinds of problems that can creep into communications as a result of professional activities. Forecasting for the short term has a morphological basis in the real world of climatic-hydrologic relationships. This is the type of forecasting that the public has a right to expect when they hear the term. But, here and in similar professional meetings the same word has been or will be used in connection with the probability notions of hydrology. The latter is a completely different, statistical system, and while it is used as a basis for decisions concerning the future, it isn't fair to expect the public to understand what is involved in probabilities when they are relying on the common meaning of forecasting.

Definitions and Users

To further illustrate the needs for care in communicating ideas, we can turn to what one would expect is a rather simple concept--floodplain--and see the seeds for confusion. Most people are landscapists and visualize floodplains in terms of their experience, i.e., as bottoms and low terraces; the geomorphologist on the other hand may think in terms of processes--degradation and aggradation--or the extent of water deposited materials; and then there is the hydrologist who thinks of the floodplain as an area which has been or could be flooded. The latter is, of course, the one which has been widely adopted for floodplain management. But, the confusion is present in almost every discussion of floodplains. Areas such as estuaries present a mixture of conditions. Marine floodplains transition into riverine floodplains or vice versa depending on perspective. The probability systems used with one do not readily fit with those of the other.

A major decision of vast importance to the subject of floodplain management, was the selection for National Flood Insurance Program (NFIP) of the "regulatory flood." The variations and differences and the wide variety of floodplain circumstances that floodplain managers must face were not over-looked in the selection process. For although the so-called "100 year" flood is not a good descripter of the flood problem in a given area, it does permit an approach to the regulation - management problem that provides the only basis for "fair treatment" available. The name, i.e. 100 year, is much more the problem than the concept of a broad unifying guideline for a National program dealing as it does with riverine, marine, and other floodplains.

Why is the name bad? The average return idea, of say 100 years, expresses a translation of probability that people almost never are otherwise asked to comprehend. Consequently, they immediately lock into the frame of reference they normally use, i.e., beginning today I have a hundred years before anything will happen to me. Even if they get over the hurdle of the departure point, and recognize that they might be somewhere within the suggested period, they still don't understand what is being said. There is no real event that is a 100 year flood. Percent chance of occurrence (exceedence) improves the likelihood of understanding somewhat and chances that the proper notions come out are further improved when probability is stated. But, even here there is a frame of reference which is largely mystical to the layman. In addition to the problem of probability translation, is the connotation of a specific flood when what is meant is a condition of flooding that is exceeded.

Defense of Subjective

It seems clear to me that the layman would best be served by descriters based on subjectively defined words that (1) avoid probability translation; and (2) don't suggest a discrete flood as do small, medium, large and very large and frequent, fairly frequent, fairly rare, and very rare. I suggest that the flood situation can best be described for planning decisions and for lay understanding by combining a statement about flood conditions with one conveying the relative frequency of exceedence of those conditions. For what has been called the 10 year flood, this would say that the given elevation or stage of flooding and hence land area involved (if any) would be exceeded frequently by flooding. Under this scheme the stage and area now shown as "50 year flood" would be designated as a fairly frequently exceeded flood condition, and so on. (See Table 1.) All the less serious conditions could then be seen for what they probably are--parts of more serious conditions. For communities adopting the NFIP standard this would say to them that the stages and areas shown constitute a flood condition that would be exceeded fairly rarely in the future of flooding in the community. It would also permit

Table 1. Designating Conditions of Flooding

Suggested Terminology:	Alternative Terminology:	Identify With:
Flooding Conditions (Depth and Area)	Risk of Flooding	Flood Designations (in Current Use)
Exceeded		
Frequently	Very High	Up to 10 yr. or 10 percent chance
Fairly Frequently	High	Over 10 to 50 yr. or 2 percent chance
Fairly Rarely	Moderate	Over 50 to 100 yr. or 1 percent chance
Rarely	Low	Over 100 to 500 yr. or .02 percent chance
Very Rarely	Very Low	Over 500*

*Standard project flood (no base in probability) could be substituted here.

Notes on Table 1:

The increasing width of the intervals suggested in Table 1 is fairly consistent with the width of the confidence bands related to H&H. Flooding conditions--exceeded, fairly, rarely--are those depths of flooding and corresponding areas identified in current usage with the so called "100 year" flood. Regulations under the National Flood Insurance Program, then, would apply to the areas with the first three conditions. Also, as an aside, the conditions exceeded frequently has a high degree of correspondence in area to Corps of Engineers delineated floodways, see Fig. 3. Although this suggests the possibility of substitution, care must be taken that the "nuisance" basis, noted earlier, is not abandoned.

recognition that the worst condition normally shown may also be exceeded, and that if these conditions are significantly more threatening to the community than the fairly rarely exceeded condition, they too can be explicitly recognized in the planning. The latter is already being done in the design of works which are part of a floodplain management plan for urban and urbanizing areas. An alternative set of descripters conveying essentially the same idea using risk as a root is also given on Table 1. Either would be useful for planning in all three phases of FPM.

Planner Versus Scientist

The point to all this is that there is a job to do and it is not being done as well as it could at this time. A major reason is that we are not communicating in the right way to the right people. The planner is concerned with the implications of all degrees of flooding to floodplain occupance, but he is especially concerned with the extent of the large flood and the problems with exceedence applicable to any level selected. Precision is not expectable from the hydrologic system not is it needed. Consistent treatment is, however. So too is the need to properly characterize the problems as, for example, the residual loss potential with very large floods.

The researcher on the other hand may be much more interested in confidence bands, or the reverse of this, in striving for greater precision. Ideas of precision are not well used in hydrologically related planning where the layman role is important. Striving for confidence by the researcher illuminates issues which can lead to erosion of confidence in the community. PRESISION BECOMES AN ISSUE. It shouldn't be. The engineering decisions for water yields and structural safety for which this statistical system was originally designed are quite different from the decisions being made for land use adjustments. Protection levels, as used with regulations, constitute a case in point. Regulations shape futures, typically, and in this way protect against future decisions that would lead to uses requiring protection. They are not based on systems of economic efficiency. "Economic man" accountable for externalized costs wouldn't require them. The real world requires them because the evidence is overwhelming that people are not making the right decisions about occupying floodplains, at least from the public standpoint. They are shifting substantial costs to others.

Special Case of Floodway

The regulatory floodway provides an illustrative topic. Many have erroneously assumed it is a wholly hydraulic matter. The portion of a two-zone ordinance or statute which is designated as floodway introduces a special concept of use. The key idea stems from the common law concept of "nuisance." This fact is often overlooked in systems of analysis that are based on economic principles. The definition of floodway for regulatory purpose, and the one used in the National Flood Insurance Program makes the "nuisance" awareness explicit. (Fig. 3.) It says that encroachment will not

THE REGULATORY FLOODWAY

Regulatory floodway policy has been established by the Corps of Engineers and the Federal Insurance Administration Criteria established by both agencies are in agreement. These criteria are depicted in Figure and are listed below.

(1) Maximum rise of 1 foot for the 100-year flood (less if established by state or local regulations). Note: It is not intended that the 1 foot rise be reached at every section.

(2) Floodway is generally based on equal deletion of conveyance from both sides of the stream.

(3) Increases in velocities are not excessive.

(4) The floodway configuration and alignment must be such that it is manageable by the local government.

(5) The floodway alignment should be coordinated with state and/or local officials.

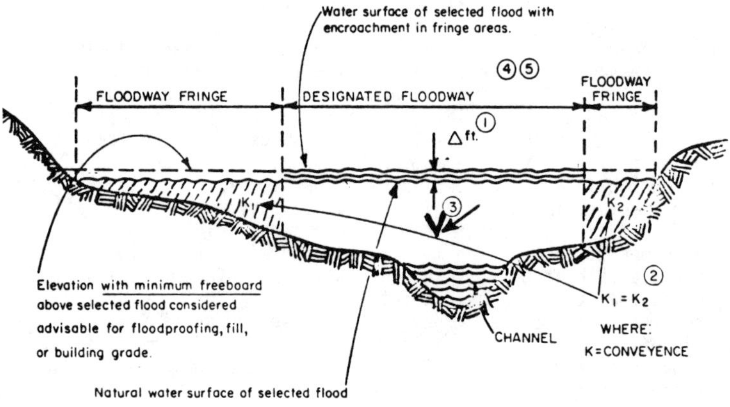

Note: Under the suggested terminology, Table 1, this floodway would be for the flow at the break-point between the fairly rarely and the rarely exceeded flooding conditions.

proceed beyond the state in which the flood surface of a given flood is elevated by a significant amount. This amount was judged by many communities and professionals to be one foot, i.e., if I shift more than one foot of my flood problem to you, you have a basis for suit. In reality, of course, the encroachment is almost never anywhere near as complete as assumed for the floodway problem. Nor is the rise uniform. In order to hold it to a maximum of one foot at any cross-section, for example, the aggregate falls. Corps experience is that the average is about 0.7'. Thus by adopting the two zone ordinance including floodway (where appropriate) the community obviates the basis for individual suits based on nuisance which could follow from shifting a flood (from the same source as the floodway). Once the hydraulic outlines are drawn (trace of toe of assumed fill), lines become the problem. Because the hydraulically defined lines may or may not fit the realities of the planning area, they are typically adjusted as part of the political process. Then it falls back on the hydrologic engineer to say whether the changes negatively impact on the minimum criteria.

The legal basis for a decision is therefore also important in floodplain management. This relates both to the political and engineering aspects. Judicial support for the restrictive application of floodways is present. This raises a further issue related to choices in floodlain management, specifically in the matter of regulations versus acquisition. Once the community sets limits of the floodway and says "no development" in it, the perceived rights to develop disappear. What then is the remaining basis for acquisition? There is none if the intent is to substitute acquisition of rights for regulation of floodway. The basis for acquisition, as specified in Section 73 of the WRD Act of 1974, is in finding a public need for access and management for such things as recreation and fish and wildlife.

Conclusions

Professionals have the responsibility to see that the best possible outcome of their endeavors is made available to the public (not necessarily free of charge). Making it available requires more than publishing in a journal, however. As has been said above, the possibilities for misunderstanding are many in floodplain management. What is of interest for research and improvement to the professional hydrologist in areas of basic hydrology and its application in hydraulics may be confusing to the planner. Almost certainly it is for the layman who is the ultimate decision maker. A basis for working with the concepts of floodplain management has been provided through the Unified National Program of Floodplain Management (WRC 1976, rev.). Improvements in techniques in all areas--physical, economic, social--are both needed and possible. The most obvious need appears to be in the area of understanding by the general public. I've suggested two or three keys to the problems of understanding--notably related to definitions, translations of probabilities, and the special case of the regulatory floodway. Let's try to build on the system, to communicate its ideas in a manner that will eccourage the layman, and to keep the deficiencies under study rather than holding them up as a smoke-screen. Confidence should be found in the transfer of information and not in the certainty of basic computations. If we peg our future to precision where precision is not in the cards, we will fail.

PRACTICES AND PROBLEMS IN FLOOD PLAIN MANAGEMENT

L. Scott Tucker (1)

The purpose of this paper is to set forth and discuss practices and problems related to flood plain management. The perspective taken is from the point of view of local and regional government.

Most flood plain management problems are the result of existing or potential encroachment on nature's prescriptive drainage easements. Also, most of these existing or potential encroachments are related to urbanization and the pressure to develop available open land. The basic objective of flood plain management is to reduce flooding damages in developed flood plains and prevent problems in undeveloped flood plains.

The Practice of Flood Plain Management

Flood plain management has different meanings to different people. For this discussion, flood plain management is considered in a broad sense as those actions individually or collectively that relate to the management of flood plains. Individual practices of flood plain management have been categorized[2] to fall into three basic approaches or strategies as follows:

1) Modifying the susceptibility to flood damage and disruption;

2) Modifying the floods themselves; and

3) Modifying the flood impact on the individual and the community.

Within these three basic strategies are a large variety of measures or "tools" for producing desired uses or changes in uses of the flood plain. The tools have generally been called either "structural" or "non-structural." Non-structural measures have been defined as "...all flood plain, or occupancy, adjustments except those that modify flood behavior..." by the Task Force on Nonstructural Flood Protection Methods, formed pursuant to President Carter's June 6, 1978 Water Policy Initiative. Within the context of this definition of nonstructural, any tool or measure used to implement strategies (1) and (3) would be considered non-structural and measures involved with the second strategy would be considered structural.

The first strategy, modifying the susceptibility to flood damage and disruption, includes several implementing tools or measures. These are land use regulations, development and redevelopment policies,

---

(1) By L. Scott Tucker, Executive Director, Urban Drainage and Flood Control District, Denver, Colorado

(2) "A Unified National Program for Flood Plain Management," Water Resources Council Report, 1976.

floodproofing, disaster preparedness and response plans, flood forecasting and warning systems, and flood plain acquisition and flood plain occupant relocation. This strategy and the associated implementing measures attempt to lessen the hazard from flooding before the floods occur.

The second strategy, modifying the floods themselves, involves structural measures. These include channelization of rivers and streams, construction of levees and dikes, and construction of dams.

Implementing measures for the third strategy, modifying the impact of flooding on individuals and the community, include information dissemination and education, arrangements for spreading the cost of the loss over time (flood insurance), tax adjustments, flood emergency measures, and post-flood recovery. These measures tend to make it easier for flood victims to survive and recover during and after a flooding event.

Problems in Flood Plain Management

The strategies and measures identified above are all involved in practice of flood plain management. The practice of flood plain management, however, is not an exact science and there are many problems associated with the application of the strategies and measures. Some of the problems are discussed in the remainder of this paper.

Bulkanization of Urban Areas. Flood Plain management is generally a multijurisdictional problem. This is because metropolitan areas, where most of the flooding problems exist, generally consist of a central city and several local governments of the outlying suburbs comprised of unincorporated areas of counties, municipalities, and townships. Drainageways of any significant size generally flow through more than one local government. Local government is the level of government that is closest to the people and where most of the interfaces between government and the people take place. Therefore, most of the flood plain management measures must be implemented by local governments or must work through or depend on local governments in some manner.

For example, the Denver metropolitan area consists of 34 units of local government including the City and County of Denver (central city), 27 towns and municipalities ranging in population from 500 to about 140,000, and unincorporated parts of five counties. The total population of the Denver metropolitan area is about 1,550,000 and the size is about 1,200 square miles. Approximately 1,100 miles of major drainageways have been identified in the metropolitan area.

While each metro area is unique, one commonality that exists is the fractionalization or bulkanization of the governmental structure. This is a problem in terms of flood plain management because of the difficulty in obtaining consistency in application of flood plain management tools and measures.

One example is flood plain regulation. State government has delegated land use control powers including regulation of flood plains to local government. Flood plain regulation, therefore, can vary considerably in degree of success and effectiveness throughout a metro region. One community may take it seriously and carefully control development in flood plains. Another community may be trying to maximize their tax base and encourage development regardless of its location with respect to flood plains. Other communities may simply not have the expertise needed to effectively regulate flood plains. Some communities may reject flood plain regulation entirely because of what their citizens perceive as hardships created by such regulation.

With respect to the Denver example, flood plains have been defined in 28 of the 34 units of local governments. All 28 entities with identified flood plains have adopted flood plain regulations, but the degree of effectiveness varies considerably.

Fractionalization of local government also makes it more difficult to implement other flood plain management measures. Development and redevelopment policies will vary from entity to entity, as will disaster preparedness and response plans, flood forecasting and warning systems, flood plain acquisition policies, information dissemination practices, flood emergency measures, and post-flood recovery procedures. It is also more difficult to implement structural measures that involve more than one governmental entity.

Bulkanization may be one of the problems with which the flood plain manager must deal, but it is not insurmountable or necessarily bad. Decentralized local government is argued by some to be inefficient, but little hard data has been developed to support this contention. The other side of the coin is that local government is the most responsive level of government and perhaps some inefficiencies, if they indeed exist, and coordinating problems are the price that must be paid for local control.

Federal Insurance Administration Program. The National Flood Insurance Program (NFIP) administered by the Federal Insurance Administration (FIA) provides both assistance and problems to the flood plain manager. Of most importance is the positive impact of the NFIP in encouraging local governments to adopt and enforce flood plain regulations. In order for a potential buyer of property within a flood hazard area to obtain a loan from a bank having Federally insured accounts, such as FDIC, they must obtain flood insurance. In order for the potential buyer to purchase flood insurance, local governments with identified flood hazards must have adopted flood plain regulations acceptable to FIA. This program provides a significant incentive for local governments to regulate flood plains. Without such incentive, it is doubtful that many local governments would assume the responsibility of flood plain regulation.

Some aspects of the FIA program, however, cause problems. One involves how flood plains are defined. The FIA requires that the hydrology for the regulatory flood be developed for present basin conditions. However, it is an accepted principle that as urbanization occurs in a basin, peak flows increase for given return frequencies. If FIA flood plain delineations for present basin conditions are used, local governments are placed in the position of permitting development to occur in a location that in the future may become part of the regulatory flood plain. The FIA position is that their program is not a regulatory program, but an insurance program and insurance rates must be based on present risks, not potential future risks. At the local level, however, regulatory efforts should be based on future development conditions of the basin. The flood plain administrator at the local level should not be placed in the position where he is allowing development to occur in a known probable future hazard area. The FIA will allow local governments to use flood plain maps more conservative (with a larger flood plain or of greater elevation profile) than those developed by the FIA, but practically speaking, this creates difficulties and confusion. The FIA has indicated it will redefine flood plains in order to account for changes in basin development about every five years. This will be costly and redefinitions from a realistic standpoint will not occur for a long time. A solution to this problem would be for FIA to develop a profile for existing basin conditions for rate setting purposes, and a profile and map for future basin conditions where future basin conditions would result in a rise of the profile of more than a given amount, say, one foot. In basins that are predominantly developed, this is not an issue, but in urbanizing basins it is. Future conditions must be addressed and actions taken now should recognize what will happen in the future.

The FIA has approached the delineation of flood hazard areas on an entity by entity basis. As previously discussed, most urban areas consist of many local government entities which means that as many as three, four, or more different contractors can be involved at different times in the delineation of the flood plain for a given drainageway. While efforts are made by FIA to coordinate the efforts of their many studies, the sheer magnitude of the effort does not permit close coordination between contractors. This results in inconsistencies at jurisdictional boundaries in profiles, flow rates, and flood plain boundaries. This causes difficulties to the flood plain manager in terms of credibility of the information and in trying to determine which is correct.

Another problem with the FIA program is the mapping that is provided to local governments to use to regulate flood plains. From a technical standpoint, the maps contain adequate information for regulation purposes. The problem is, however, that there is resistance to their use because of the lack of detail. To remedy this, local governments can pay for better mapping, but in general, this does not occur. When dealing with real property in the way that regulation does, the flood plain manager wants to be confident the information is correct and to have the most detailed information possible. Right or wrong, this seems to be the way it is.

Flood Hazard Awareness. There is a problem of making flood plain occupants aware of the hazard. Several of the identified measures or tools involve actions by the flood plain occupant such as flood proofing, preparing and responding to flood warnings, learning and understanding the hazard, and purchase of flood insurance. Public meetings to discuss the hazard, public hearings with regard to adoption of flood plain regulations and maps, and notices in the paper do not seem to adequately advise the public. The public still seems to have a great deal of indifference to the potential hazard of flooding. For example, 138 lives were lost and significant property damage was incurred during the August, 1976 flood in the Big Thompson Canyon near Loveland, Colorado. Many lives could have been saved if the people would have known how to react, i.e., climb out of the canyon on foot and to not try to outrun the flood in a car. Unfortunately, only one or two properties were protected by flood insurance and were thus able to recoup rapidly their financial loss.

An approach being taken in the Denver, Colorado area is to annually mail a brochure explaining the hazard to all addresses located in or adjacent to identified 100-year flood plains. Over 20,000 addresses have been identified in about 200 miles of the most heavily developed flood plains in the Denver region. The brochure advises the occupant of the hazard and what actions can be taken including the purchase of flood insurance and routes of escape. This approach seems effective as a research effort undertaken by the Institute of Behavioral Sciences of the University of Colorado[(3)] indicated that flood plain occupants that had received a brochure were more aware of the hazard than those that had not.

Lack of Federal Commitment. There is a lack of a Federal commitment or policy regarding flooding problems related to smaller urban watersheds. The Federal Government has had a commitment to flood control since it was made a national policy issue by Congress in 1936 with passage of the first flood control act. While not explicitly stated in any legislation, Federal policy has been not to become involved in problems associated with smaller urban watersheds where significant flooding problems exist. This situation is exemplified by a recent Corps of Engineers (COE) regulation published in the Federal Register on May 8, 1978, entitled "Rules and Regulations for Flood Damage Reduction Measures in Urban Areas." A criteria is set forth in the Regulation that precludes COE participation upstream from the point where the flood discharge for the ten percent flood for existing conditions is 800 cubic feet per second. Drainage areas of less than 1.5 square miles are assumed to lack adequate discharge to meet this criterion. This policy is of concern to the urban flood plain manager because it arbitrarily limits the extent of Federal involvement.

The COE May 8, 1978 regulation reflects a Federal posture with respect to urban flooding problems. This posture is also reflected in the lack of Federal direction in urban stormwater data collection and analysis, and urban stormwater research. The Urban Water Resources Research Council of the American Society of Civil Engineers has since 1969 recommended to the Federal Government that the USGS develop a comprehensive rainfall/runoff/quality data collection and analysis program. The recommendation was acknowledged but implemented only to a nominal degree. Because of problems with Section 208 planning programs conducted pursuant to PL 92-500 as amended, the USGS and EPA in 1978 entered into an agreement for technical coordination in the area of data collection and analysis regarding urban non-point pollution sources.

(3) "Hazard Mitigation Behavior of Urban Flood Plain Residents," by Marvin Waterstone, Institute of Behavioral Science, Boulder, Colorado, December 1978.

It is still up to each metropolitan area, however, to take the initiative in developing data collection and analysis programs directed toward the problems associated with flood plain management.

Perhaps a lack of Federal involvement should be considered a blessing. This may be the case, but the lack of Federal interest has also created a vacuum in terms of consistency of criteria, approaches, etc. related to urban flooding that are relatively small in magnitude, but relatively large in the number of occurrences and in the damages incurred.

Flood Plain Acquisition. Flood plain acquisition is one of the measures available that can be used to address flood plain management problems. In a remedial sense, a flooding problem can be solved by acquisition and relocation, i.e., by removing the problem. In practice, this approach has had only limited application. Traditionally, Federal cost-sharing formulas have required that local project sponsors provide all lands and easements needed for the project. This has placed a bias on structurally oriented solutions that minimize right-of-way costs. The Water Resource Development Act of 1974 attempted to address this problem. Section 73b of the Act provides a cost-sharing formula, up to 20% non-Federal, for non-structural measures. Still, Federal funding of non-structural measures including acquisition has not been common.

The President also addressed this problem in his June 6, 1978 Water Policy Initiative (WPI). While the President addressed the problem in terms of non-structural measures, flood plain acquisition falls in that category. Specifically, the WPI states:

> "...equalizing cost-sharing for structural and non-structural flood control alternatives. There is existing authority for 80%-20% Federal/non-Federal cost-sharing for non-structural flood control measures (including in-kind contributions such as land and easements). I will begin approving non-structural flood control projects with this funding arrangement and will propose that a parallel cost-sharing requirement (including in-kind contributions) be enacted for structural flood control measures, which currently have a multiplicity of cost-sharing rules."

With this level of Federal interest, perhaps flood plain acquisition will become a more viable alternative in the future.

In the preventive sense, acquisition of undeveloped flood plains is difficult to justify for flood control purposes. Flood plain regulation provides a vehicle at little or no cost to the general public for precluding flood plain development that would be damaged by floods as large as the 100-year event. There is a cost associated with flood plain regulation in terms of limited use of the land or higher development costs, but these costs are borne by the flood plain property owners or developers. Acquisition of undeveloped flood plains must be justified largely on the basis of preserving the values associated with natural flood plains. Within existing funding mechanisms, it is difficult to bring the many programs and interests to bear on a situation that is required because of the multi-purpose nature of acquisition. Motivation for acquisition of undeveloped flood plains will have to come from open space interests with the flood plain management benefit being a consideration.

Maintenance of Flood Plains and Facilities. A significant cost to any operating agency is maintenance of facilities. This is true for any public improvement, property, or facility. The enormous cost to local governments of maintenance of sewage treatment facilities funded in large part with Federal assistance is now being realized. Similarly, there are significant costs associated with maintenance of flood control facilities, including open flood plains. Open grass lined channels in particular are maintenance intensive; trash and debris need to be routinely removed, grass needs to be routinely cut, weeds need to be controlled, drop and check structures need periodic repair, bare areas need seeding, and eroded sections need attention. Other facilities such as lined channels, detention dams, and open flood plains also need continual maintenance attention. Local governments typically do not have the manpower, equipment, expertise, or financial resources to deal with the maintenance required on flood control facilities.

This creates a potential liability problem if anyone is damaged because of the lack of maintenance. Perhaps one solution is to recognize and capitalize maintenance costs and deal with the cost on the front end of a project, particularly when Federal funds are involved. This might require an adjustment in the Federal/non-Federal cost-sharing formula to be palatable to the Federal government.

Federal Project Authorization Procedure. The COE is the major Federal agency that addresses flood control problems. The process by which projects are planned and reviewed by the COE and considered and authroized by Congress is long and tortuous with ten years being an expected time frame for smaller non controversial projects. For large projects involving perhaps $100 million or more, the process can be considered reasonable. For small projects in the $5 million to $10 million range, the process cannot be considered reasonable.

The COE cannot solve all flood control problems; this is understood. The frustration or problem is the time and energy it takes to achieve Federal support for the small number of projects where Federal support is requested. It would be desirable to have an authorization process for smaller urban flood control projects that would have a two or three-year time frame for planning and authorization.

Financing Problems. Financing is a problem with any public works effort, and drainage and flood control is no exception. There are some particular problems, however, related to drainage and flood control financing that bear mention.

One problem area is how to fund major drainageway improvements in developing areas. As development occurs, developers are usually required to provide the local drainage system for their development. The problem is how to fund downstream improvements on the drainageways that serve as collector systems. An approach to this problem is to charge a per-acre development fee that is based on estimated costs for providing the major drainageway system. This is a sound and valid approach, but there is still a problem when it is necessary to provide the facilities before adequate funds have accumulated from the development fees.

Problems also occur when the drainage basin is located in more than one local governmental jurisdiction. Unless the two or more entities can join together and coordinate or pool their efforts, it is difficult to fund improvements through development fees.

Funding of drainage and flood control improvements in developed areas is difficult because the attendant problems generally affect only those that live at the bottom of the hill. Property owners that live on the top of the hill have no interest in committing their money to solving the problems of someone who had the lack of foresight to locate in a low area. In most cases, however, it is recognized that uncontrolled urbanization increases runoff which can either cause a flooding problem or exacerbate an existing problem. Uphill property owners do contribute to downstream flood problems and it can therefore be argued that they have a partial responsibility to help solve those problems. There is a problem, however, in assessing upper land owners for the cost of facilities located some distance away. If monies can be found in the general fund of a local government that is generated from sales taxes or property taxes, then the issue of who benefits and who pays does not arise. However, in many cases other sources of funds must be generated to finance the improvements. One available method is the special improvement district in which benefited property owners are assessed a portion of project costs based on the extent to which their property benefits. Property owners cannot be assessed any more than what they are being benefited. In Colorado, and very likely in other areas, benefits have been narrowly defined by the courts as increases in value of the property, relief from a burden, or adaptation to a superior or more profitable use. Recently in Colorado, the State Legislature has defined the term benefit which is intended to help clarify the situation and broaden the definition of benefit. The legislative definition include "the provision for accepting the burden from specific dominant property for discharging surface water onto servient property in a manner or quantity greater than would naturally flow because the dominant owner made some of his property impermeable." Presumably, an uphill property can be assessed for the service of handling the additional drainage caused by virtue of developments on the property. This definition of benefit has not been court tested yet in Colorado.

Detention. There is a current trend to require detention storage in new developments to reduce the impact of new development on downstream drainage facilities. The objective of this requirement is to hold peak flows at historical rates of flow for a given frequency or frequencies. Detention storage can have very definable effects in terms of reducing peak flows immediately downstream of the detention area, but the effect of small random on-site detention on the larger collector systems or major drainageways is unknown. Even if on-site detention effectively maintained flows in major drainageways at historical rates, there would still be a need to reserve flowage easements along the major drainageway. Also, detention storage in new developments often addresses the more frequent storms, and consequently is most effective in the 2-year to 5-year frequency range.

Flood plain management is generally concerned with the larger or less frequent occurrences, and their effects on the larger drainageways. A problem in terms of flood plain management is the apparent inconsistency in requiring on-site detention of developers while at the same time, still having a need for downstream improvements. The developer argues that he has paid his fair share by providing a local drainage system for his development as well as detention storage to reduce downstream impacts. A fact that is not widely recognized or addressed is that the mere act of urbanization itself necessitates that flowage areas for major drainageways be reserved. The concomitant increase in runoff affects the magnitude of the problem but not the basic need for major drainageway systems. Also, detention of smaller events may not have a significant effect on larger events. Another consideration is that detention has virtually no effect on runoff volume which is also increased by virtue of urbanization. In other words, detention serves a public good and purpose, but it does not solve all problems.

Summary

A purpose of this paper has been to present some of the practices and problems in flood plain management. Problems are emphasized because much time of flood plain managers is spent in addressing and solving problems. The problems enumerated relate to bulkanization of urban areas, the National Flood Insurance Program of the Federal Insurance Administration, lack of flood hazard awareness, lack of Federal commitment, flood plain acquisition, maintenance of flood plains and facilities, Federal project authorization procedures, drainage and flood control financing, and detention.

The discussion of the problems is not intended to be negative, but only to elucidate upon some of the obstacles that must be faced and surmounted. It is only after the problem has been defined, that it can be solved.

ATTITUDE, GUIDELINES, AND TECHNOLOGY IN PLAN FORMULATION

by William K. Johnson[1]

Introduction

Ancient philosphers conceived of the earth as flat, being supported by four elephants one under each corner (Figure 1). Someone asked, "But what do the elephants stand on?" The philosophers tried again and came up with the "answer" that all four elephants stood on one mammoth turtle. But what does the turtle stand on?

FIGURE 1 : Elephants and a turtle.

Today, in plan formulation the plan is analogous to the earth. It is supported by four elephants which represent the principles, standards, procedures, and regulations which are an integral part of our planning process. These elephants are supported by a mammoth technology which allows us to quantify and analyze nearly any aspect of our planning study. But what supports the technology? What is the foundation of plan formulation? It is to meet human needs. Yet today we have the same difficulty recognizing the importance of meeting needs, as the ancient philosophers had in seeing beyond the turtle. It is thought that meeting principles, standards and various other guidelines is the objective of plan formulation. It is thought that using a sophisticated technology in formulating plans is the objective. It is neither

[1]Civil Engineer, The Hydrologic Engineering Center, U. S. Army Corps of Engineers, 609 Second Street, Davis, CA 95616

- it is meeting human needs. This paper is a discussion of this misunderstanding.

Attitude

Attitude is one of those fuzzy things which engineers like not to think too much about and, in fact, do not think too much about because it does not lend itself to being captured in formal procedures or computer code. Yet, professionals in other disciplines see it as a major determinant in the actions we take. Attitude in the context of plan formulation is that perception, that mind-set, that disposition toward the purpose of planning and toward alternatives to achieve that purpose. To illustrate this point consider several attitudes.

Formalist. A Formalist attitude (Figure 2) is to develop a plan in strict conformance with the principles and standards established for planning whether by government agency or professional practice. Adherence to these is the foremost consideration. Measures and means which have been authorized either by traditional use or formal directive are willingly considered, but there is reluctance to adapt to local needs and conditions which do not meet a formal construct of planning.

FIGURE 2 : Formalist

FIGURE 3 : Obstinate

Obstinate. In spite of study information which suggests new opportunities for management of the flood plain, Obstinate (Figure 3) adheres to a previously formulated plan. What needs to be done is decided quickly and ideas which differ from this plan are resented. These ideas come from earlier studies, and the need to change to meet changing attitudes and values in society is difficult to see.

Worldly Wise. Society is a technological society and to be wise in this world is to be wise in technology. Worldly Wise (Figure 4) recognizes this and uses technology to its maximum in plan formulation. An attempt is made to quantify all aspects of the flood plain and to use technology to optimize, systematize, analyze, and mesmerize the data - to make the means become the end.

FIGURE 4 : Worldly Wise

FIGURE 5 : Prejudice

Prejudice. Prejudice (Figure 5) just plain doesn't like extraordinary measures. While a nonstructural plan may be formulated it is only because it is required. From the beginning it is believed that these measures will not work, and the final plan always bears witness to this view. By prejudicing a wide variety of opportunities for flood plain management what needs can be met becomes very limited.

Ignorance. The flood plain as a valuable natural resource is a concept which Ignorance (Figure 6) cannot grasp. Its value as a park or greenbelt, as a wildlife habitat, as an aquifer recharge, as a place of aesthetic quality - none of these values can he see. He sees only flood damage and flood damage reduced. After all he says, "The plan must be economically justified." His overconcern for traditional economic justification causes him to overlook opportunities for management which indeed may be economically justified but not in the traditional sense.

These are five attitudes which are all too common in plan formulation. There are others, but what is the proper attitude? It is one which strives, above all else, to meet human needs. This is the justification for planning itself - to provide information which will result in decisions which better meet our needs. The desire to meet needs must be our perception, our mind-set, our disposition. Our success or failure must be measured on the basis of how effectively we do this. Our guidelines and our technology are but tools to assist us in achieving this end.

FIGURE 6 : Ignorance

Guidelines and Needs

Guidelines. Guidelines, in the context of this paper, means those written procedures and regulations whose purpose it is to describe how plan formulation and evaluation are to be done. At the federal level the principal examples are the Water Resources Council's "Principles and Standards" and soon to come "Planning Manual." These documents together with a host of agency regulations to implement them, state government documents which simulate them, and numerous professional papers which extol them are termed in a broad sense guidelines - guidelines for plan formulation. The fact that they direct planning to be done in a certain way or to meet certain requirements coupled with the implicit authority of the local, state or federal agency requiring their use lead to their ready acceptance. Simply put these guidelines become the ends of planning instead of the means - if in fact it is possible for them to be an effective means at all. Our focus on procedures and regulations turns our attention away from the substance of planning and directs it to the form. Because guidelines offer a specific example of form on how planning should be done and because of their implicit authority they offer a ready vehicle to guarantee a minimum level of acceptability.

Of course planners, the guidelines, and the agency are not the only actors in the scene. Public and political views also influence the formulation process and sometimes use guidelines to direct that process, thus showing the importance attached to guidelines. To illustrate this point consider nonstructural planning at the federal level. While efforts have been made to have nonstructural measures considered in plan formulation for over thirty years it has not been until recently that this effort has been made explicit in the form of procedures and regulations. The 1974 Water Resources Development Act legislated that,

> "In the survey, planning, or design by any Federal agency of any project involving flood protection, consideration shall be given to nonstructural alternatives to prevent or reduce flood damages . . . "

One would think that this general statement would be enough. However, the great degree to which specific guidelines influence planning and their authority to bring about compliance is demonstrated in President Carter's 1978 Water Policy message. In that message the President directs that,

> "The WRC (Water Resources Council) is to make appropriate changes in the P&S (Principles and Standards) to require the preparation and inclusion of a primarily nonstructural plan as one alternative whenever structural project or program options are to be considered."

Thus, to change formulation, political wisdom says, change that which directs formulation - the guidelines - or alternately create guidelines to direct formulation.

Needs. The alternative to focusing - implicitly or explicitly - on guidelines is to focus on meeting needs. Needs includes all social, environmental, and economic aspects of the flood plain resource. In its fullest sense meeting a need encompasses the range of opportunities from preserving the flood plain for its unique environmental, ecological values to protecting property from flood damage. The principal thrust of flood plain management from its earliest inception has been to recognize this unique and valuable resource and to find the best way to manage it. The success of the planning effort and the value of a particular plan cannot be measured by the extent to which it meets guidelines, or the quality of a project report, or by use of a sophisticated technology. These are all means to an end, however, the end itself is not the report but the wise use of the resource.

How then does one focus, in formulation, on meeting needs? The difficulty here is that the answer is that it's a matter of attitude, a personal quality which cannot be written in the form of guidelines or technology. This is not to say that planning is all attitude. To the contrary a well managed planning study, efficient use of time and resources, application of the most appropriate technology, knowledge of means and measures, sensitivity to public views - all of these and more are necessary in formulation. Attitude, however, is the driving quality which pervades, influences, and directs these other aspects. An attitude which actively strives to meet needs is the best assurance that in the formulation they will be met. Emphasizing the importance of meeting needs and circulating examples of studies where needs were met and how they were met can help bring about this attitude.

Technology

The mammoth turtle of today is technology. Various sociological works have been published showing the influence of technology on the way we think and the way we act (1). It is not the purpose of this paper to interpret these works in the context of plan formulation. Suffice it to say that technology's influence, especially computer technology, pervades the planning process and is particularly dominate in formulation where frequently too much time is spent on activities such as preparation of computer input and analysis of computer output.

Properly used, however, technology can assist in formulation. No one wants to make water surface profile computations by hand, or estimate expected annual damage using a planimeter. Rather if we can develop a technology which will not consume us and take our minds off our purpose of meeting needs, such a technology can be useful. To this end it may be worthwhile to briefly review the traditional technology used in plan formulation and then describe recent developments for nonstructural formulation which give the planner a technology which he can use in an interactive mode.

Traditionally, various computer programs have been used to simulate the rainfall-runoff process, to estimate the frequency of flood events, to compute water surface elevations, and to perform economic analyses. They provide numerical estimates of the location, depth, and frequency of flooding under both natural and modified conditions. Also, they provide estimates of expected dollar damage caused by flood events of various magnitudes. Traditionally flood damage has been computed by river reach for different categories of property, and frequency and water surface elevations have been computed for an index location representative of the reach. This type of macro analysis has been adequate for formulation of plans using reservoirs, levees or floodwalls, channel modifications, and diversions.

In recent years another technology has been developed which assists in formulation on a micro as well as macro basis (2, 3). Called spatial data management this technology allows flood plain data to be stored and analyzed on a parcel by parcel or structure by structure basis. A wide variety of data describing the flood plain land or flood plain property can be stored in a computer and analyzed. This technology is especially suited for formulation of nonstructural flood plain management plans, but can also be used for the more traditional measures.

The most micro spatial representation of flood plain data is a point. A point in the flood plain can be identified spatially by an x, y coordinate and data stored and analyzed for that point. Physical features and flood plain property are stored as a point. Coordinate points are appropriate when it is desired and feasible to store data for each structure and where a graphical plot of point data is acceptable.

In flood plains when it is desired to store and analyze flood plain data for a surface area a grid cell is used. A grid network can be overlain on an aerial photo and each cell used to represent property or other features within its boundaries. Each cell is identified spatially by a coordinate system and data characteristic of each cell are stored and analyzed by cell. This type representation is appropriate where it is not desired or feasible to represent each property by a coordinate point or where graphical plots of a land surface are desired.

Data stored. Figure 7 illustrates the type of flood plain data which it may be desired to store to assist in plan formulation using nonstructural measures. The illustration shows the structure being represented by a point, however, the same data can be stored for a grid cell. Briefly, the data may be used as follows.

Structure data. The structure or flood plain property is identified by identification number, name, address, and coordinates. The type of structure and number of levels are necessary for economic analysis and the type construction materials, foundation, and condition are useful for determining the physical feasibility of various measures (4). The ground and first floor elevation are necessary for computing both flood damage, depth of flooding, and level of protection.

Hazard data. The elevation-frequency relationship at the index location is necessary to perform economic analyses and to relate various depths of flooding to frequency of occurrence. This relationship will be transferred to each point location using a relationship of elevation of the water surface at the index location to the elevation of the water surface at the point or structure. In essence this relationship represents the relationship of elevation along a water surface profile for a given flood event.

Economic data. For each structure or property it is necessary to know the relationship between flood elevation and flood damage to compute expected annual damage. Sometimes damage is expressed as a percentage of structure or contents value. In such cases these data should also be stored.

Environmental, social data. Numerous environmental and social data can be stored for each location. The three listed - historic site, aquifer recharge location, explosive storage - are suggestive of the type data it may be desired to store. Whichever parameters are selected they should be necessary for formulation.

Nonstructural measure data. If it is desired to analyze the economic and hydrologic effect of various measures the height of protection or height raised may be stored and the effects computed in the analysis.

Analysis. A variety of analyses are possible using the stored data. With the elevation-frequency relationship transferred from the index

DATA

Structure Data

- I D Number
- Name
- Address
- X,Y Coordinates
- Type
- Levels
- Materials
- Foundation
- Condition
- Ground elevation
- First floor elevation

Hazard Data

- Elevation - frequency relationship at index location
- Elevation index location - Elevation structure relationship

Economic Data

- Elevation - damage relationship at structure
- Value of structure
- Value of contents

Environmental, Social Data

- Historic site
- Aquifer recharge location
- Explosives storage

Nonstructural Measure Data

- Height of protection above first floor
- Height raised

ANALYSIS

- Elevation - frequency relationship at structure
- Frequency of event at first floor
- Depth of flooding above first floor
- Level of protection
- Expected annual flood damage
- Selected structure and flood data

Figure 7

location to each coordinate point the frequency of the event at the first floor, the depth of flooding for specific events, and the level of protection for various measures can be computed. Also, expected annual flood damage or flood damage for specific events can be computed. Also, expected annual flood damage or flood damage for specific events can be computed. Stored data and results of analyses can be selected and tabulated as an aid in formulation. For example, it may be desired to know all structures within the 10 year flood plain or all structures on a raised foundation. By selecting parameters such as these the computer can select structures with those characteristics and tabulate them for further evaluation. Alternately, a specific group of structures can be specified, for example, those in the business district, and output of selected data specified.

In a similar manner other features of the flood plain can be analyzed. For example, the value of the flood plain as a wildlife habitat can be assessed by storing important data on wildlife and habitat either by point or cell, analyzing these data, and using the results in formulation. Likewise technical information on aquifer recharge can be stored, analyzed and used to take proper consideration of the flood plain as a recharge area.

This spatial data management technology is just that - a data management technology. It assists in formulation by storing, for ready access, large amounts of data useful in formulation. These data should be carefully selected to represent data which is necessary to make effective use of the technology.

Work is currently being completed which will make this link between understanding and technology an interactive mode. Earlier research described the physical and economic feasibility of nonstructural measures (4). In that research certain key parameters were identified. Data describing these parameters can be stored for each structure and selectively output on a CRT in tabular form. This interactive mode will provide the opportunity to look at selective structures or data and utilize the computer to quickly process this information. As much, or as little data as desired can be selected and combined with other information to put together an effective plan.

Conclusions

An exceedingly detailed and sophisticated technology has been developed to assist in plan formuation. Guidelines on how to conduct and what to include in the planning process are also available. Though one can be likened to a mammoth turtle and the other to four large elephants, neither are sufficient in themselves to support a flood plain management plan. A plan can only be supported because it meets human needs. The task of plan formulation is to develop such a plan. More important than technology or guidelines is an attitude and desire which strives incessantly, in spite of numerous obstacles, to develop this support.

References

1. Ellul, Jacques, "The Technological Society," Alfred A. Knopf, New York, 1967.

2. Davis, Darryl W., "Comprehensive Flood Plain Studies Using Spatial Data Management Techniques," Water Resources Bulletin, Paper No. 77100, June 1978.

3. Davis, Darryl W., "Flood Damage Assessments Using Spatial Data Management Techniques, The Hydrologic Engineering Center, U. S. Army Corps of Engineers, Technical Paper No. 57, 1978.

4. Johnson, William K., "Physical and Economic Feasibility of Non-structural Flood Plain Management Measures," The Hydrologic Engineering Center, U. S. Army Corps of Engineers, March 1978.

Illustrations

Figures 2 through 6 from John Bunyan's "The Pilgrim's Progress," American Tract Society, 1890.

LOCALLY IMPORTANT ECONOMIC CONSIDERATIONS IN FLOODING AND FLOOD PLAIN MANAGEMENT

D. Earl Jones, Jr.*

Improved hydrologic forecasting, improved availability of hydrologic information, increasing awareness of the economic and social significance of flooding losses and improved understanding of the variety and importance of hydrologic relationships to man and his environment are stimulating rapid development of flood plain management art. Flood plain management practice generally lags development of the art by about ten years.

BACKGROUND

Federal legislation created the National Flood Insurance Program in 1968 and subsequent Nationwide efforts to make flood insurance available in exchange for improved local practices to mitigate future flooding losses. The National Flood Insurance Program has performed a tremendous public service by increasing awareness of flooding problems and their importance, by delineating locations of much flood-prone land in American communities, and by stimulating the first local efforts to achieve flood plain management in thousands of communities. Urban flood plain management for years was the dream of only a few persons; the National Flood Insurance program is making that dream a reality.

Flood plain management necessarily is founded upon hydrologic knowledge and forecasting. In local situations, hydrologic knowledge provides a basis for economic evaluations to identify and quantify potential flooding problems and related alternative benefits and costs. Based upon these evaluations and consonant with statutory enabling authority and applicable case law, local political due process and judgement operate to develop and adopt locally applicable flood plain management regulations. During that process, the National Flood Insurance Program provides local officials with basic hydrologic information, flood mapping and general gudiance regarding the nature of Federally acceptable local regulatory actions.

The National Flood Insurance Program has faced a tremendous mapping challenge that has required much of their effort. Nonetheless, they have supported considerable research to expand the range of land use and construction alternatives available for use in flood plain management. Their efforts necessarily have focused upon development of information that has widespread utility that is not misleading to anyone, and that can be implemented in ways consistent with Federal program authority, structure and operation.

*Consulting Engineer, Arlington, VA

In specific situations, flood plain property owners and local regulating officials have been dissatisfied with the standardized approaches to flood plain management. The Congress anticipated such eventualities and directed establishment of an appeals process whereby a community might seek Federal acceptance of local regulations varying in approach and degree from the standardized regulatory guidance published by the National Flood Insurance Program. Recognizing the tremendous challenges facing the Program, the Congress placed the burden of demonstrating the technical soundness of locally suggested variations squarely upon local communities.

The writer has examined only a few local appeals for Federal acceptance of flood plain management variations. These few appeals suggested: (1) that appellants tend to address perceived problems emotionally, presenting strongly held opinions unsupported by documented technical facts, analyses and comparisons; (2) that few appellants retain broadly informed engineering counsel to prepare appeals documents; (3) that few appellants seem to understand that appeals should identify alternative means of fulfilling flood less mitigation objectives and provide quantified comparisons between suggested alternatives and usual regulatory requirements; and (4) that the general public and many engineers lack optimally broad understanding of hydrologic, economic, legal and administrative aspects of flood plain management.

Inasmuch as individual communities have the burden of demonstrating that alternative local approaches to fulfillment of flood plain management and flood loss mitigation objectives are justifiable, it may be helpful to discuss some considerations that may be meaningful in an appeal. Basically, sound appeals will address either hydrologic forecast accuracy or economic and physical impact analyses, or both. Community appellants should recognize: that appeal documentation should be comprehensive; that persons qualified to develop all aspects of a comprehensive appeal may be locally unavailable or reluctant to become involved; that good local topographic and occupancy inventory information is essential; that development of economic impact and adjustment comparisons must be comprehensive; and, that a properly developed appeal will require financial support.

If affected property owners and local officials think an appeal may be justified, and worth its cost, they would be wise to commission a quick survey to determine whether there actually is a sound basis for appeal. They may discover that application of standard regulatory practices is reasonable in their locale and that costs to develop and document a comprehensive appeal may be unwarranted. When appeal is warranted, elements of the following discussions may be helpful.

ERROR IN HYDROLOGIC RISK FORECASTS

As mentioned previously, the National Flood Insurance Program's mapping activity has been a gigantic task. Their Flood Insurance Study Guidelines and Specifications adequately define sound approaches to flood hazard mapping and their completed maps generally define potential flood-prone areas acceptably. Occasionally, however, a mapping contractor may inadvertently make a mapping error, significantly overestimating or underestimating flood elevations and extents. Correction of such errors is provided for and has proved responsive in the Program's review and appeals processes.

Although a simple hydrologic forecasting error can be simply and effectively dealt with as indicated, there are sources of error whose definition and evaluation is technically complex and more difficult. Such errors can be defined only by analyses of forecasting methods and basic hydrologic and topographic information actually used in derivation of the hydrologic forecasts provided for local flood plain management guidance. Examples of sources of error that may be revealed by study include but are not limited to the following:

* Incorrect transcription of available basic hydrologic data;

* Hydrologic records distorted by a dry period of record or by occurrence of an extremely rare event during the period of record;

* Hydrologic data synthesized without proper recognition of important factors;

* Improper selection of waterway (flood plain) cross-section locations;

* Incorrect bench mark, wrong bench mark elevation or leveling error;

* Failure to recognize hydrologic significance of permanently undevelopable lands (such as parks or cemeteries);

* Incorrect determination of contributory drainage area boundaries or size;

* Failure to identify and consider non-contributory areas within the boundaries of a contributory drainage area (Karst topography, kettles, potholes);

* Insufficient recognition of orographic influences on precipitation;

* Incorrect estimation of energy losses or recoveries at constrictions, or failure to consider updated information about changes in constrictions or other impedances;

* Incorrect determination or consideration of channel and flood plain storage potentials;

* Incorrect selection of hydraulic formulae or of hydraulic coefficients such as Manning's "n", weir coefficients, etc.;

* Failure to consider comparative dependency or interdependency of frequency related variables combined in estimations of probability (such as for the several variables combined in hurricane surge modeling, or in determining probable backwater effects on a tributary stream discharging into a river);

* Use of methodology that predicts the worst event within a long reach of waterway or coast when the exposed frontage is much shorter;

* Failure to consider actual or probable effects of bathymetry;

* Overly simplified inland projections of coastal storm surges (especially failure to consider time/probability relationships involved in filling estuarine storage);

* Failure to correlate (and explain) deviations between historic hydrologic data and hydrologic forecasts (such as disparities between extensive tidal records and projections from surge modeling, or between flow gauge records and forecast profiles, or between individual stream records and regional streamflow patterns);

* Improper evaluation of the significance of hydrologic uncertainties.

* Mathematical errors; and

* Transcription and other human errors.

This partial listing identifies the range of possible forecasting errors and suggests the possible focus of comprehensive technical evaluations of forecasts that seem questionable. The National Flood Insurance Program routinely provides for review of forecasts made by their contractors but it would be unreasonable to expect review process perfection. Local governments and property owners may find it worthwhile to examine the validity of others' hydrologic forecasts if they seem to differ excessively from reasonable expectations. Local funding of studies will be required to evaluate challenged forecasts. Local restudy of hydrologic forecasts, to be meaningful, should reveal significant elevation differences; differences of only a few inches usually have negligible significance because of the many basic uncertainties involved.

FLOODWAY DELINEATION

Even engineers have moved slowly to define or redefine the term "floodway." In 1962, the American Society of Civil Engineers defined it essentially as the portion of a flood plain which is required to carry and discharge flood water. The National Flood Insurance Program does somewhat better in that its definition of floodway applies also to island, coastal and lacrustine overflow areas and bounds all floodway areas by defining them as those portions of flood plains required for passage of the 100-year flood considering that encroachments and constrictions must not increase upstream water surface elevations by more than one foot. Although that definition may serve satisfactorily for insurance purposes, it sometimes proves execssively simplistic and inconsistent with significant safety, flood loss mitigation and legal realities.

Present advanced professional thinking is that floodway delineation properly should integrate six basic considerations: (1) possibility for excessive flood risk, especially potential catastrophic risk; (2) impedance of potential flood discharges; (3) potential legal liability for adverse economic consequences of induced increased backwater elevations upstream from encroachments onto the floodplain; (4) potential adverse environmental consequences of encroachments onto the floodplain; (5) local public needs for open space that might best be fulfilled within the flood plain; and (6) existing public rights within flood plains. Present floodway delineation practices ignore many of these elements. A seventh element, comparative significance of channel and flood plain storage of flood waters, also is important at many locations but not discussed herein. Each of the first six elements is discussed in a succeeding paragraph.

Risk as considered in floodway delineation is essentially measured as the effects of depth of flooding and flood flow velocity upon man and his works. To illustrate comparative risks, let us consider stage-frequency relationships on Thorn Creek at Park Forest, Illinois, and on the Kansas River at Topeka, Kansas. That information is tabulated in Table 1. (Tables and Figures appear at the end of the text herein). From Table 1, it is apparent that at Park Forest the 500-year recurrency frequency event produces flooding only 0.4 foot deeper than the 100-year event and only 1.1 foot deeper than the 10-year event. In contrast, at Topeka, the 500-year event is 7.5 feet deeper than the 100-year event and 30 feet deeper than the 10-year event. A home without a basement encroaching on the Park Forest flood plain, built where the natural ground is at the 10-year recurrence frequency flood elevation, normally would automatically have its first floor almost at the 200-year flood elevation and would be flooded only about 3 inches deep by the 500-year flood; the flow velocity of even the 500-year flood would be so low as to constitute negligible velocity- related risk. In contrast, a similar home built in Topeka with its first floor at the 100-year elevation would be flooded 7.5 feet deep by the 500-year flood, during which it normally would be exposed to the damaging flow velocity associated with movement of more than an 8 foot depth of water. The Park Forest home would be relatively unscathed by a 500-year flood, whereas the Topeka home might very well be destroyed by such an event. Clearly there is a difference in risk between the two locations, and the risk at Topeka must be viewed as potentially catastrophic. Such comparative risks should be considered during the floodway delineation process. They can be rather closely approximated in relation to the National Flood Insurance Program's Flood Hazard Factor [1], and perhaps ever more closely by use of Powell's[2] or McCrory's[3] Stage Index Slope. It is significant that encroachment onto a flood plain to the 10-year frequency elevation at Park Forest, as described, has no significantly adverse flood loss effect and probably offers very little flow impedance. Unless other considerations are overriding, only a narrow floodway may be justifiable in a situation such as that at Park Forest, perhaps only approximating the extent of the area inundated by the 10-year flood. In Topeka, however, the catastrophic potential of infrequent flooding is so great that a protected floodway approximating the extent of the area inundated by the 200-year flood might well be justifiable. Similar considerations are equally apropos for flow in waterways and for evaluation of hurricane surges or tsunamis, but potential wave effects also should be considered at coastal locations. Significantly, where the Flood Hazard Factor or the Stage Index Slope have high values, the potential for catastrophic flooding loss may

TABLE 1

Comparative Flooding Depths at Park Forest, Ill. and at Topeka, Kan.*

Comparative Flooding Frequency in Years	Comparative Flooding Depth in Feet	
	Park Forest, Illinois	Topeka, Kansas
10	0	0
50	0.5	21
100	0.7	23.5
200	0.9	26
500	1.1	30

* Information includes some extrapolation and rounding off of data published in USGS Hydrologic Atlas Maps HA-14 and HA-209.

TABLE 2

Comparative Damageability of Different Building Types

Depth of Flooding in Feet Below or Above First Floor	Damageability Ratio* by Building Type**						
	A	B	C	D	E	F	G
-1.0	1.0	0.0	∞	∞	0.0	∞	0.0
0.1	1.0	0.7	1.75	1.5	2.0	2.5	2.0
1.0	1.0	1.1	1.4	1.6	2.2	2.4	5.0
2.0	1.0	1.25	1.3	1.4	1.9	1.9	4.4
3.0	1.0	1.25	1.3	1.3	1.75	1.85	4.1
4.0	1.0	1.2	1.25	1.25	1.6	1.7	3.6
6.0	1.0	1.1	1.2	1.2	1.5	1.5	3.0
8.0	1.0	1.2	1.2	1.3	1.4	1.4	2.7±
Typical Ratio	1.0	1.2	1.25	1.25+	1.6	1.7	3.6

* Ratio of damage to any other building type to expected damage to a two story building having no basement.

** A = Two story, no basement
B = Split level, no basement
C = Two story with basement
D = Split level with basement
E = One story, no basement
F = One story with basement
G = Mobile home on foundation

be so severe as to justify delineation of a floodway having a considerably greater areal extent than the area inundated by the 100-year flood, the present National Flood Insurance Program boundary of designated special flood hazard areas.

Impedance of flood flows is sufficiently well understood that it will not be discussed here. Its importance centers about the magnitude of upstream backwater effects attributable to the impedance and upon its effects on Manning's "n", the most commonly used hydraulic friction factor.

The creation of legal liability and its potential economic magnitude are of overriding importance for individual and public decisions regarding encroachments upon flood plains. In steep terrain where backwater effects above an encroachment extend only a short distance upstream and where no existing property would be damaged, there usually is virtually no potential for liability for upstream damages caused by an encroachment and no basic hydrologic reason why a backwater effect measured in feet should be unacceptable. In contrast, in flat terrain, backwater effects may extend a mile or more upstream from an encroachment and even a small backwater effect may significantly expand the amount of area flooded and shorten the flood recurrence frequency. Backwater effects above encroachments usually are greatest during periods of greatest flow but usually also are definable and often significant during even frequent flow events; this means that encroachments usually will increase not only upstream flooding depths during extreme flooding events but also will, by increasing flooding depths during frequent events, shorten the recurrence interval between floodings at any given elevation. In the latter situation, a relatively small increase in flooding depth during a one percent probability flooding event can result in significantly increased average annual flooding damages to existing properties; in one recent case, a 7 inch increase in flooding depth during the 100-year flood was calculated to increase average annual damages to a $30,000 existing upstream home by about $200, with the increase largely attributable to an increase in the number of times the property would be flooded. Local public officials and affected property owners must be concerned by the realities of potential liability for damages when they designate floodway boundaries or take construction action within an identified floodplain. Where upstream areas are undeveloped and increased backwater effects would cause no significant damage, and encroachment is practical without excessive exposure to damage, or where either right to damage or settlement for future damages can be purchased, there generally is little justification for prohibiting sound occupancy. As potential liability from backwater effects usually is of paramount local importance, imposition of a "one foot rule" for limiting them will be impractical and unsupportable by established case law precedents. A one foot rule for backwater may have its most severe impact upon local governments trying to provide numerous waterway crossings in flat terrain, as construction costs to avoid cumulative increases in backwater elevations may be very high; communities facing waterway crossing construction needs should approach floodway delineation cautiously.

Potential adverse environmental consequences of encroachments onto flood plains may occur in many ways which will not be discussed here. The worst consequences for biota and sediment transport probably occur where channels are realigned and their cross-sections made more efficient hydraulically. When considering floodway delineation, potential environmental consequences of comparative amounts of encroachment, changing flow velocities, changes in sediment transport potential, and reduction of habitat may locally be important. Quite often, the character of urban waterways will already have deteriorated irreversibly, so that actual environmental impacts caused by contemplated encroachments would be relatively insignificant. In such situations, the only actual environmental benefit from floodway preservation may relate to the aesthetic appearance of an area. The probable environmental significance of alternative floodway delineations may locally be an important consideration, especially in coastal areas[8].

Local public needs for open space quite often may be fulfilled advantageously by linking floodway delineation with environmental considerations and open space planning. The writer[4] and the United States Geological Survey[5] have identified extensive long-term benefits having both private and public importance that derive from wise "Blue-Green" development that may be practical on urban waterways. "Blue-Green" developments are storm water runoff storage areas which may be used for water and shoreline recreation or other purposes. Such development is almost totally inconsistent with a one-foot backwater rule but often provides an excellent opportunity to secure beneficial open space. Wise "Blue-Green" development can increase the marketability and profitability of nearby land, can enhance the desirability of entire neighborhoods and prolong the period of high property value (increasing average tax revenues to the community), can minimize needs for downstream drainage and flood control investments, and can link open space and recreational opportunities with occupancy measures that assuredly will mitigate future flooding losses. Approaches to floodway delineation should not arbitrarily preclude such local integrated benefits.

In some States, the public has fee title to coastal and estuarine lands to elevations defined by State Constitutions, title which usually cannot be negated by prescriptive encroachments. The public similarly has either fee title or other rights along navigable inland waterways in some States, and may have public access rights along waterways in others. There is a growing trend to recognize that lands subject to intermittent occupancy by water (flooding) are encumbered by a "natural" drainage easement, as has been recognized in the Denver area[6]. Such definable public rights, variously applicable below elevations defined by Mean High Tide, High Tide, High Water, etc., can be locally very significant for establishment of proper floodway delineations. It certainly would violate public interests if a community were to establish floodway boundaries at elevations lower than the limits of public ownership or right, which could prove embarrassing to local officials.

It is evident that proper local delineation of floodways involves much more than just consideration of a "one foot backwater" rule, and that any such rule should be used only for guidance, not as a requirement. There are sound arguments for establishing floodway boundaries only after consideration of the many factors that relate to alternative consequences of the decision. From the local public standpoint, broadened approaches to floodway delineation clearly should prove more rewarding. Local officials should recognize that the weights attaching to different considerations probably will differ in different parts of a community, especially where there are significant topographic and geomorphologic differences within the community, so that the bases for floodway delineation in different parts of a community often will be justifiably different. It follows that use of the "one foot rule" should be perfectly acceptable where it will cause no major displacements and will be consistent with the broad range of proper floodway delineation considerations, and that when the "one foot rule" would be inconsistent with those considerations broader and more sensitive evaluations would be proper. This discussion of floodway delineation is presented to emphasize awareness of alternatives that may make floodway delineation more practical and rewarding for all affected parties.

Those interested in additional background on floodway delineation practices will find a recent paper by Goddard[7] helpful. It was prepared for the National Flood Insurance Program and may soon be published by them. It indicates that present floodway delineation practices have evolved essentially from consensus and primarily have considered only generalized effects of alternative boundary selections. Goddard (personal communication) recognizes issues relating to legal questions, private property rights, liability and hydrologic uncertainties, and is not at all certain that one specific universal criterion for floodway delineation is completely proper. Goddard and the writer have discussed needs to consider possibilities of changing stream meanders, and other channel relocation effects due to aggragation and degradation, during floodway delineation, but those subjects would be more properly addressed in a separate paper.

REGULATORY UNIFORMITY

Traditionally, any regulation imposed by a public body has necessitated equal treatment of all regulated parties as an extension of the Constitutional Amendment concept guaranteeing all persons equal protection under the law. During the early evolution of flood plain management practices, a fixed regulatory flood recurrence frequency, universally implemented, was thought to provide uniform and equable treatment of everyone. More recent study by McCrory[3], and stage-damage information published by the National Flood Insurance Program[1], have stimulated realization that equal protection under the law cannot be achieved by universal imposition of one recurrence frequency criterion for all flood plain management decisions.

Today, it is eminently clear that a single recurrent frequency criterion for flood plain management: (1) will not assure a uniform level of flood loss protection or future flood loss mitigation Nationwide; (2) will result in significantly different levels of economic displacement and flood loss expectancy for different types of similarly valued properties at different locations within many cities, even though they may have the same frequency of exposure to flooding (3) will result in significantly different amounts of flood damage to identical properties having the same frequency of exposure to flooding but that may be located in different cities; (4) will not avoid the potential for catastrophic flooding losses in some types of local locations, thereby denying some property owners "equal protection;" and, (5) will cause undue and unnecessary social displacements (especially economic) at some local locations. It should be recognized that a single recurrence frequency basis for flood plain management decisions is fully justifiable in many locales, perhaps in most locales, but that a local appeal for authorization to vary from a generalized exposure frequency rule may be technically and economically supportable when it may be demonstrated that compliance with the rule would create unusual hardhsip or result in non-uniform protection under the law. Any community mounting an appeal upon such a basis should support its contentions with detailed hydrologic and economic facts and comparisons.

When local public officials and property owners find generalized flood plain management guidance difficult or onerous to implement, it may be because unique aspects of their local flood hazard situation warrant alternative approaches to flood hazard exposure regulation. Since the actual flood loss expectancy using a single exposure frequency criterion can vary by a factor of up to ten or more, local appellants seeking uniform treatment may have a good argument. They may find McCrory's[3] approaches useful for documenting their situation and for developing a justifiable alternative regulatory basis. Using McCrory's approaches, regulations would stipulate the maximum average annual damage (expressed as a percentage of property improvements value) acceptable for any property. That approach would be considerably more flexible from a regulatory standpoint because it would establish a quantified flood loss mitigation objective and permit wide latitude in selection of ways to assure that acceptable loss exposure would not be exceeded. McCrory's approach, which points toward identification of maximum acceptable risk as a percentage of value, also could be supplemented by specification of an alternative maximum acceptable risk stated in terms of maximum acceptable average annual flooding loss (constant dollars) per unit of land area, which could provide needed regulatory flexibility to address transitional land uses in an area.

The National Flood Insurance Act of 1968 directed development of regulations that would assure mitigation of future flooding losses, but it provided no specific quantification of that loss mitigation objective. Flood plain management practices to date, based upon regulation at a standard flooding recurrence interval, may produce widely varying amounts of mitigation of future flooding losses. Alternatively, in many situations, specification of acceptable risk stated in economic terms would be more equable and would better assure uniform protection under the law. Succeeding discussions identify some considerations important for risk assessment comparisons.

IMPORTANCE OF BUILDING TYPE

The National Flood Insurance Program[1] published depth-damage information, reproduced as Figure 1, which illustrates expected flooding losses to various types of residential and small business buildings for different depths of flooding. That information reveals that a two story building without a basement is the least damagable by flooding at almost all flooding depths. Taking the expected flooding loss to a two story building without a basement as unity, Table 2 expands the Figure 1 information to show the ratio of expected losses for different building

FEDERAL INSURANCE ADMINISTRATION

SEPTEMBER 1970
Depth Damage Curves
Set A

STRUCTURES-RESIDENTIAL AND SMALL BUSINESS

	CURVE NO.						
	01	03	05	10	13	18	23
Depth in feet			Damage in % of Total Value				
−3.0					.0	.0	.0
−2.0					3.	3.	3.
−1.0	.0	.0	.0	.0	6.	5.	5.
First Floor.0 (0.1)	8.0	4.	3.	8.	10.	7.	6.
1.0	22.	10.	11.	50.	24.	14.	16.
2.0	30.	16.	20.	71.	31.	21.	22.
3.0	35.	20.	25.	82.	37.	26.	26.
4.0	39.	24.	29.	87.	41.	30.	30.
5.0	41.	27.	31.	89.	44.	33.	32.
6.0	44.	30.	33.	91.	46.	35.	35.
7.0	46.	32.	34.	91.	48.	38.	36.
8.0	48.	34.	41.		49.	40.	44.
9.0	50.	39.	46.		50.	44.	48.
10.0		42.	50.			46.	52.
11.1		45.	53.			47.	55.
12.0		47.	55.			48.	57.
13.0		49.	58.			49.	58.
14.0		50.	59.			50.	59.
15.0			60.				60.

Classification	Curve
One story, no basement	01
Two or more stories, no basement	03
Split level, no basement	05
One story with basement	13
Two or more stories with basement	18
Split level with basement	23
Mobile home, on foundation	10

FIGURE 1

types, and identifies comparative damageability ratios for each type. Figure 2 presents the Figure 1 information graphically.

The different building types have quantifiably different comparative vulnerability to damage by flooding. Figure 1 provides information for "two or more story" buildings; Table 2 simplifies that to "two story" because there are few three or more story residences being built today and over one-half of the area of most cities is single family detached residential occupancy.

The local public official or affected property owner seeking equable regulatory treatment may draw the following kinds of conclusions from the Table 2 comparative vulnerability figures:

* Loss expectancies can be either reduced or aggravated by selection of building type.

* For example, where two story dwellings without basements can be substituted for one story dwellings without basements, future flooding losses may be reduced about 37 percent by the substitution.

* Future flooding losses may be significantly limited by permitting construction of only the less damageable types of buildings.

* If a more vulnerable type of new construction is acceptable at the 100-year flood elevation, then equable regulatory treatment should accept less vulnerable construction at quantifiable elevations somewhat below the 100-year flood elevation. If flood loss mitigation is the basis for and objective of flood plain management, uniformity of permitted losses is essential to assure similarity of treatment of individuals.

* If highly damageable types of buildings are typical at the 100-year flood (base flood) level in a city whose flood plain regulations are Federally acceptable, then a lower base flood level should be Federally acceptable in other cities characterized by less damageable construction at the base flood level, provided cities compared have the same Flood Hazard Factor or Stage Index Slope. Acceptable loss probability should be essentially the same for all communities; otherwise, some communities will not be receiving equal treatment.

The importance of building type as a flooding loss variable can be separately considered for established neighborhoods and for developing areas. Separate evaluations may suggest different regulatory approaches in different kinds of neighborhoods.

DIFFERENCES IN FLOODING CHARACTERISTICS

The nature and amount of probable flooding damages to buildings can be forecast with reasonable accuracy. Basic factors that should be considered when estimating probable flooding damages are:

1. Frequency of flooding (the number of times a building can be expected to flood during its useful life).

2. Probable depths to which a building may be flooded during its useful life.
3. Probable velocity of flood waters that may affect a building.

4. Quality of flood waters. This includes suspended sediments ("silt"), solutes (especially salt and alkali), and contaminants (especially oil and other organic compounds).

5. Vulnerability of the particular construction to flooding damage.

6. Duration of flooding

7. Relative scale of flooding.

Vulnerability of construction, Factor 5, has been quantified by the National Flood Insurance Program's stage-damage curves[1]. Grigg[9] has found that quantification reasonable. (See Figures 1 and 2, and Table 2). Somewhat similar adjustment curves for loss-reducing flood proofing measures have yet to be developed. Significance of Factor 5 was discussed earlier herein.

The effects of Factor 6 on flooding losses are largely undefined. Fortunately, most urban floods recede rapidly and have a short crest duration. Only isloated locations, such as along the Red River of the North, are exposed to flooding that may continue for weeks. Duration may be a significant factor in basement flooding losses if floor drains become inoperable, as they may not be pumped out until several days after flooding. The effects of comparative flooding durations will not be further considered herein.

The comparative effects of flood water quality, Factor 4, are complex consideration except at coastal exposures. Such things as season, origins of flood waters, recurrence interval of experienced flooding, existence of debris and kinds of land use practices on the contributory drainage area, erodability of soils, sediment transport and deposition characteristics of the waterway, and others, make point-to-point comparisons of the effects of water quality difficult to estimate with certainty. Although comparative effects of water quality clearly are significant, in most instances they now are too complex and uncertain to be addressed locally when considering the wisdom of a regulatory appeal. Water quality should, however, be considered during the evolution of local flood plain management practices. It will not be further addressed herein.

The relative scale of flooding damages throughout an area, Factor 7, has great economic significance. Yancey[10] has shown that post-disaster repair costs predictably can escalate by as much as 6 times when a large area is impacted by a disaster and total damages in the area greatly exceed local construction industry repair capacity. Effects of the scalar factor are greatest during the rarer flooding events and multiply the influence of infrequent flooding losses upon computed average annual

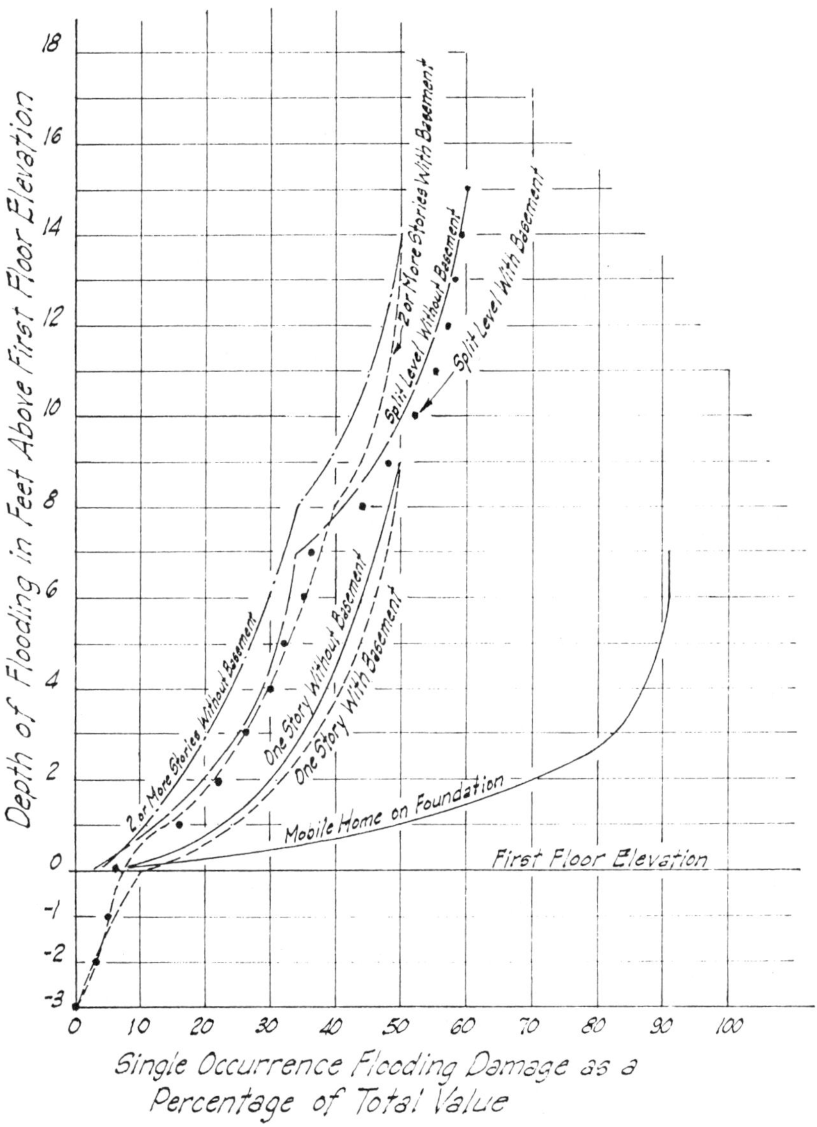
Depth of Flooding in Feet Above First Floor Elevation
18
16
14
12
10
8
6
4
2
0
-1
-2
-3
0
10
20
30
40
50
60
70
80
90
100
Single Occurrence Flooding Damage as a Percentage of Total Value
2 or More Stories With Basement
2 or More Stories Without Basement
Split Level Without Basement
Split Level With Basement
One Story Without Basement
One Story With Basement
Mobile Home on Foundation
First Floor Elevation

FIGURE 2

flooding damages of affected local buildings. Goddard[4] has identified communities in which as much as 81 percent of the developed area (Monroe, La.) would be impacted by a one percent probability flood, as well as several communities that would be 30 to 40 percent impacted. In such areas, average annual damages as usually computed and actual potential losses from extreme floods will both be grossly underestimated. In contrast, the catastrophe at Rapid City, South Dakota affected only six percent of the community and involved no appreciable post-disaster escalation of repair costs, which was at least in part due to local government constraints upon repair in the affected area. Traditional approaches to risk assessment, including Federal approaches, ignore the significance of this scalar factor. Local officials and property owners seeking equal treatment, however, may find that it has significance for them. They may be able to demonstrate that there is no foreseeable scalar effect upon flooding losses in their community as compared to another community where the scalar effect can be shown to be severe but where local flood plain management approaches are Federally-acceptable. If such differences can be demonstrated and were not considered during determinations of locally proper flood plain management practices, there may be justification for appeal on the basis of unequal community treatment.

Frequency of flooding and probable depths of flooding, Factors 1 and 2, combined with standardized information on vulnerability to damage, Factor 5, provide the usual basis for flood damage projections. The variation in loss expectancies attributable to Factors 1 and 2 is significant and predictable, and may be a primary and incontrovertible basis for demonstrating inequalities in regulatory treatment. Figure 3 illustrates stage-frequency curves for several cities, from which differences in flooding depths for various probabilities of flooding can be estimated. The National Flood Insurance Program's Flood Hazard Factor (FHF) is the difference, measured in feet, between the 100-year and 10-year flooding depths. Powell's[2] Stage Index Slope (SIS) is the difference, measured in feet, between the 25-year and 10-year flooding depths. Figure 4 characterizes the Figure 3 Stage-Frequency curves in terms of both SIS and FHF; the FHF characterization is only approximate because FHF does not consider skew. Local officials should be aware that probable depths of flooding may be closely defined by either FHF or SIS and by knowledge of the probability of flooding at any site being considered, much as was discussed earlier herein with regard to Park Forest, Illinois and Topeka, Kansas. Inasmuch as flooding characteristics at those two locations are representative of usually encountered extremes, the differences between flooding effects at the two locations will be considered here.

Standard approaches to computation of average annual flooding damages, such as described by McCrory[3] were applied to two identical one-story dwellings, one in Park Forest, Illinois and one in Topeka, Kansas. Both homes were sited with floors at the 50-year frequency flood elevation; their comparative vulnerability to flooding damages was taken from Figure 3. Average annual flooding damages were found to be $80 for Park Forest dwelling and $212 for the dwelling at Topeka. It is significant that the average annual damages to identical homes exposed to the same frequency of flooding proved 2.65 times (265 percent) greater at Topeka than at Park Forest. Similarly caused inequalities in actual flood loss

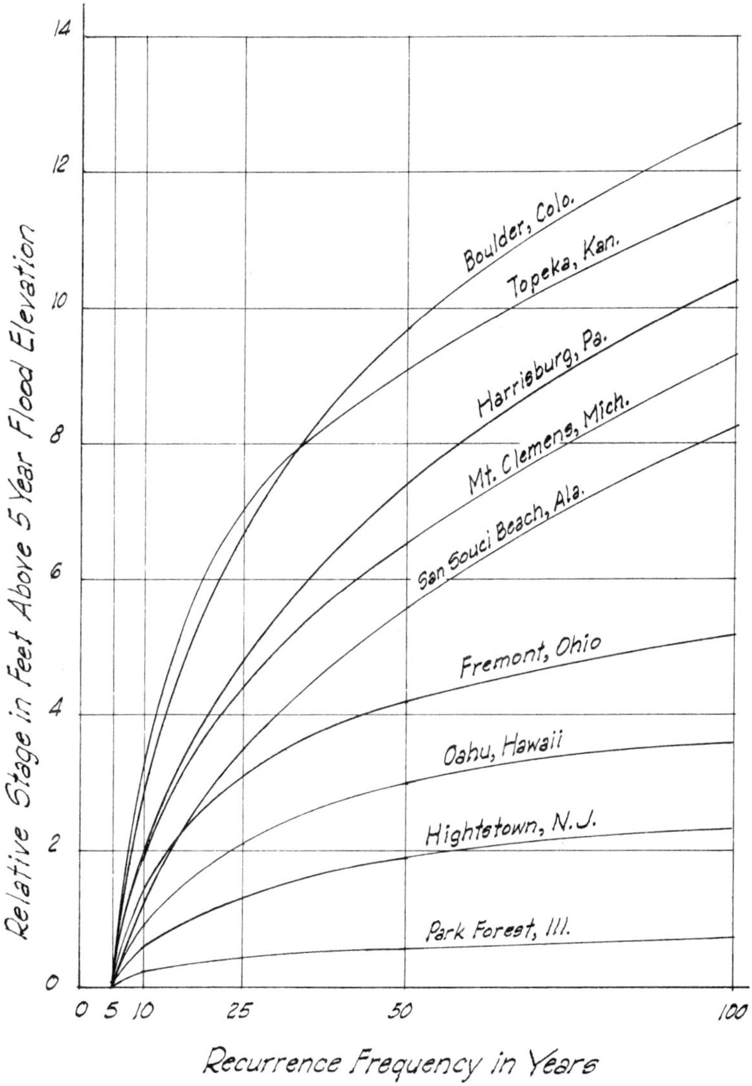

Relative Stage in Feet Above 5 Year Flood Elevation
0
2
4
6
8
10
12
14
Boulder, Colo.
Topeka, Kan.
Harrisburg, Pa.
Mt. Clemens, Mich.
San Souci Beach, Ala.
Fremont, Ohio
Oahu, Hawaii
Hightstown, N.J.
Park Forest, Ill.
0
5
10
25
50
100
Recurrence Frequency in Years

FIGURE 3

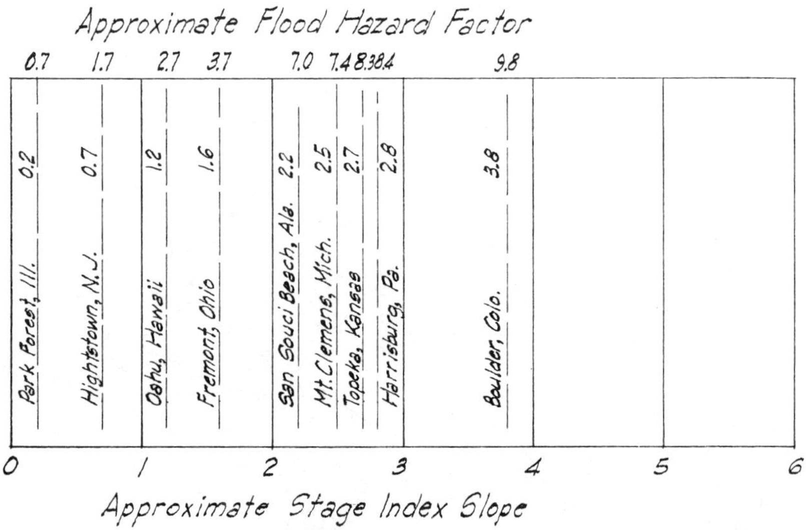

Values for Stage Index Slope and for Flood Hazard Factor are approximate only, taken from raw Stage-Frequency data. Neither has been normalized to reflect adjustments for skew, which will be apparent from examination of the curves illustrated in Figure 3.

FIGURE 4

expectancies (or in flood loss mitigation requirements) can be up to or even greater than 400 percent between identical homes, exposed to the same frequency of flooding, but located in different cities.

The type of inequality illustrated has much significance for local public officials and property owners. Demonstrable inequality is particularly likely to be found in communities where water may cover very large areas but have only shallow depth even during very infrequent floods. Such communities usually will be characterized by low values of SIS or FHF. It often may be demonstrated in such situations that the actual degree of flood hazard mitigation sought by standardized regulations exceeds that accepted elsewhere, where the typical terrain has more relief.

Buildings are affected differently by different depths of flooding and different flood flow velocities. Increased flow velocities, associated with deeper depths of flow, result in increased hydrodynamic forces that can push buildings off their foundations or collapse them. Increased flow velocities also are associated with more severe scour or undermining of building foundations. There has been no comprehensive study of the effects of flood flow velocity upon building damage, but it is generally believed that velocity effects become significant when flow velocities exceed about two feet per second. Where application of the Manning formula for steady flow conditions is proper, it may be shown that the velocity of flow at a 10 foot flow depth may be as much as 4.6 times as great as at a one foot flow depth. Many factors other than depth will affect flow velocity but it is sound to state that flow velocity will always be a function of depth under uniform free flow conditions and that damages to buildings will increase under such flow conditions when depth of flooding increases. Earlier herein, comparisons of flooding depth were made for identical buildings exposed to the same frequency of flooding at locations in Park Forest, Illinois and at Topeka, Kansas; expected flooding damages were noted to be significantly greater at Topeka than at Park Forest, due in part to the greater flow velocity expected at Topeka. Inasmuch as velocity effects generally increase in severity with increasing flooding depth, it is clear that areas subject at worst only to relatively shallow flooding may expect lower flooding damages than areas in which deeper, faster moving flooding may be expected. Although there is no precise basis for comparing flooding loss expectancies attributable to velocity differences, it is clear that such differences are real and that areas subject at worst only to shallow flooding may be asked to deliver a higher level of flood loss mitigation than areas exposed to deeper flooding. Local public officials and affected property owners may find it worthwhile to consider the velocity aspect of potential inequalities in regulatory treatment.

FUNDAMENTAL ECONOMIC RELATIONSHIPS

Actual flooding damages, flood control construction costs, costs for special site development or building construction features to reduce future flooding damages, costs associated with non-use of existing infrastructure or abandonment of existing capital investments, and others, are among those costs whose total must be considered flood-related and that should be considered during economic analyses. Just as all medically-related costs, including those for treatment, prevention, prostheses, etc., are considered "health costs," so all flood-related costs must be viewed as "flooding costs."

All flooding costs, like medical costs, do not occur at one time. An individual home built at the 100-year flood elevation, for example, may be flooded next week or it may not be flooded at all during the next 176 years; statistically, the home should be affected by one flood within the next 100 years, and for cost analysis purposes it normally is assumed that a flood affecting the home will occur in about 50 years. Statistics and statistical probabilities used for risk analyses necessarily are based upon average flood occurrence behavior since there is no way to assuredly forecast just when floods of different sizes will occur. If the concept of statistical averaging is accepted (there is no good substitute concept) actual damage to the home from a 100-year flood would be expected to occur 50 years from now. Recognizing this, the economic analyst asks himself "How much money would it be worth today to avoid damages 50 years from now?" He terms his answer "present value" or "present worth" and arrives at it by using discount equations. Although the economist's terminology and calculations are unfamiliar to most persons, they are familiar with "interest costs" and "interest rates." Interest rates and discount rates are essentially the same thing, but interest rates are used to compute the future value of a current investment whereas discount rates are used to compute the current value of a future investment. Using a 10 percent interest rate, it may be shown that an $85.19 investment made today will be worth $10,000 in 50 years, or that, using a 10 percent discount rate, a $10,000 loss 50 years from today has a present value of $85.19. For comparison, using the same line of statistical reasoning, a $10,000 cost expected 25 years from now (the average time expectancy for occurrence of a 50-year frequency event) would have a present value of $922.96, and a similar cost expected 12.5 years in the future (analagous to a 25-year frequency event) would have a present value of $3,038.03.

The economist's question about the present value of expected future damages is important. His answer can only be given in terms of monetary value, however, as there is no technique for quantifying social losses or benefits, or human misery, which also are worthy of consideration during decision-making. The significance of his answer for local public officials and property owners is that there is no economic justification for an immediate investment to avoid future damages if the immediate cost will exceed the present value of expected future damages. Simple usage of average annual damages as a basis for decision-making distorts the true picture by ignoring foreseeable probable influences of the cost of using money.

Present value or present worth analyses can have the following kinds of implications, among others, for local public officials and affected property owners:

* The analyses are adaptable to both indiviudal properties and to groups of properties (such as neighborhoods or even large parts of cities).

* The analyses place in better perspective the economic significance of alternative approaches to resolving problems.

* The analytic method clearly identifies immediate cost levels and projected amounts of future damages that are mutually cosnistent, justifiable or unjustifiable.

* If the present value of future benefits from an action exceeds the immediate cost of the action, the action is justifiable.

* If the costs of a proposed immediate action exceed the present value of future benefits attributable to the action, the action is economically unjustifiable but may be justifiable when intangible social or human values also are considered.

When considering available alternatives to mitigate future flooding losses to existing buildings, it may be found that provision of flood control works is economically unjustifiable or that basic flood proofing of the buildings is impractical, but that routine remodeling or repair of buildings can incorporate materials and finishes that will significantly reduce future flooding losses without appreciably increasing remodeling or repair costs. Similarly, it may be found that elevating new buildings on fill (where flow velocities will be non-erosive), or on pilings or similar support, will be economically justifiable. Where expected flooding depths will not exceed a few feet or so above the lowest building floor, it may be economically justifiable to construct new buildings so that they will be watertight. In analyzing the costs of such alternative flood loss mitigation measures, present value analyses will provide sound insights regarding the wisdom of each.

When analyzing alternative flood plain managment measures during development of local regulations, economic displacements that would be caused by alternative regulatory approaches may be worth studying from the present value standpoint. Such displacements may include an array of direct or indirect consequences of contemplated actions, such as changes in property values, costs to acquire and occupy alternatve flood-free locations, costs of less than full utilization of existing infrastructure, costs of providing infrastructure to stimulate growth in flood-free directions (this may prove to be a benefit rather than a cost, however), and many others.

Local public officials and property owners should not view present value analyses as mysterious or unfathomable; almost any local banking or savings institution, accounting school or investment house will have staff persons who understand the techniques and who will be glad to help with analyses.

HYDROLOGIC BASIS FOR COMPUTED FLOOD DISCHARGES

Section 1304(a) of the Federal Flood Insurance Act of 1956, as amended to date, is focused upon flooding losses to real or personal property "arising from any flood occurring in the United States." As a practical consideration, present flood risk mapping efforts proceed up tributary streams only to locations where the width of the area inundated by the 100-year flood is 200 feet in developed areas or 400 feet in undeveloped areas. Many areas subject to flooding are upstream from such locations, and probably will be addressed in time by risk mapping activities. In the meantime, local governments have at least a moral responsibility to take reasonable steps to mitigate future flooding losses in such upstream areas, termed "headwater areas" herein.

Headwater areas on tributary streams by their very nature are sensitive to changes in land use on their contributory drainage areas. Present hydrologic practices for forecasting probable flooding elevations generally are based upon actual present drainage area conditions or conditions foreseeable in the near future. Local public officials long ago discovered that failure to anticipate ultimate land uses and land development on contributory drainage areas often disregarded potentials for future flooding. Communities desiring to avoid future flooding problems, therefore, have generally considered post-development potentials for flooding as the basis for their construction and occupancy practices along waterways.

Some communities, in the Denver area and elsewhere, have expressed a desire to have all flood hazard mapping delineate probable ultimate risk areas rather than just current risk areas, because maps delineating only current risk areas may be used to support requests for permission to develop and occupy currently flood-free lands that clearly have a high potential for future flooding. Although the thrust of Federal statutes is to encourage mitigation of all future flooding losses, present standardized mapping practices actually undermine or weaken local efforts to address and mitigate future flooding losses. The current Federal policy of mapping only present conditions but accepting more stringent local regulations actually weakens local efforts to fulfill the Federal statutory objective. Where communities face such a problem, an appeal seems possible focused upon the technical and scientific certainty that upstream occupancies will change and possibly aggravate downstream flooding behavior. This can be very significant locally, as all involved interests should seek to avoid creation of situations such as that along Four Mile Run in Northern Virginia, where the 100-year frequency flood stage forecast in 1953 has been exceeded many times because of the effects of headwater development.

SUMMARY

Federal statute establishes an appeals process whereby local governments can appeal elevations proposed by the Secretary of Housing and Urban Development with respect to an identified area of special flood hazards. Appeals must focus upon the scientific or technical correctness of the elevations established. Unfavorable Secretarial decisions regarding appeals may in turn be appealed to the United States district court for the jurisdiction, whose scope of review shall be as provided by Chapter 7, Title 5 of the United States Code. It seems probable that technical and scientific considerations may be viewed rather broadly in the appeals process, especially since the Statute in Section 1361(c) indicates that local measures for land management and use, flood zoning and flood damage prevention "to the maximum extent feasible" will constrict development of land which is exposed to flooding "where appropriate." Such broad modifying terms would seem to at least encourage equable consideration of the types of unusual conditions and effects discussed herein.

Whether or not appeal of Federal determinations is warranted, State and local governments should strive to evolve regulatory measures that treat property owners and their unique local situations equably. Federal authority and postures are tightly constrained by Federal statutes and related implementing regulations, but local government flood plain management actions are not. Local government's actions may be challenged by their constituencies in many different ways, so local governments must be particularly responsive to inequities and inequalities resultant from their own actions. The implications of the various conditions and factors discussed herein are probably more significant at the local level than at the National level and are believed worthy of careful local consideration.

Foregoing text indicates that, under special local conditions: (1) different kinds of buildings suffer significantly different amounts of damage during flooding; (2) the depth-frequency flooding relationships at different locations can result in significantly different average levels of flooding damage; (3) the relative scale of the potential extent of local flooding can foreseeably increase post-flood repair costs by as much as several hundred percent; and,(4) that flood damage directly attributable to velocity of flood flow also will aggravate flooding losses at some locations. Evaluation of the comparative combined effects of these damage variables indicates that actions taken at one location may coincide in effect with those taken at other places, but it also indicates that comparison of conditions having the greatest contrast may reveal effects of regulation that are up to 25 or more times greater at one location than at another.

It is clear that State and local governments have a basic responsibility to their constituencies to assure that their own regulations have consistent effects throughout their jurisdictions.

REFERENCES

1 "Flood Hazard Factors, Depth Damage Curves, Elevation-Frequency Curves, Standard Rate Tables," Federal Insurance Administration, Washington, D. C., September, 1970.

2 Powell, R. F., James, L. D., and Jones, D. E., "An Approximate Method for Quick Flood Plain Mapping,' a paper delivered at the November, 1975 ASCE National Meeting at Denver, Colorado, Meeting Preprint 2559.

3 McCrory, J. A., James, L. D., and Jones, D. E., "Dealing with Variable Flood Hazard," Journal of the Water Resources Planning and Management Division, ASCE Proceedings Paper 12522, November, 1976.

4 Jones, D. E., "Urban Hydrology—A Redirection," CIVIL ENGINEERING, ASCE, New York, August, 1967.

5 Rickert, D. A. and Spieker, A. M., "Real-Estate Lakes," United States Geological Survey Circular 601-G, Washington, D. C., 1971.

6 Wright, K. R., Drainage Criteria Manual, Paragraph 5.1, Denver Regional Council of Governments, Denver, Colorado, March, 1969.

7 Goddard, J. E., "Origin and Rationale of Criterion Used in Designating Floodways," Report of a special study for the Federal Insurance Administration, Washington, D. C., October, 1978.

8 Jones, D. E., "Housing and Related Coastal Problems--Current Practices Offer Improved Solutions," a paper delivered to the Annual Meeting of the American Shore and Beach Preservation Association, Washington, D. C., October, 1977.

9 Grigg, N. S. and Helweg, O. J., "State of the Art of Estimating Flood Damage in Urban Areas," WATER RESOURCES BULLETIN, Vol 11, 1975.

10 Yancey, T. N., James, L.D., and Jones, D. E., "Disaster-Caused Increses in Unit Repair Cost," Journal of the Water Resources Planning and Management Division, ASCE Proceedings Paper 12564, November, 1976.

11 Goddard, J. E., "An Evaluation of Urban Flood Plains," ASCE Urban Water Resources Research Program, Technical Memorandum No. 19, New York, December, 1973.

Alternatives for Standards and Regulations

Brian R. Mrazik[1]
Richard W. Krimm

Introduction

One of the objectives of this conference is to motivate professionals in water resources management to embrace improved hydrologic forecasting techniques to meet problems and needs enumerated in these presentations. Previous papers in this session focused on the need for forcasting techniques for floodplain management, current problems and practices, public attitude, and local considerations. This paper will present some of the current and future issues facing the National Flood Insurance Program (NFIP) which may require the adaptation of standards and regulations as well as placing new demands on the Engineering Profession to provide the technical data needed to properly administer this Program on a Federal, state and local basis.

Standards and regulations are a part of floodplain management strategy whereby actions are taken to avoid dangerous, uneconomic, or unwise use of the floodplain. In the case of the NFIP, standards and regulations identify the minimum requirements which a community must enforce in order to participate and maintain insurance availability. These requirements are in the form of restrictions on occupancy, alterations, and type of construction within the floodplain.

[1]Hydrologist and Assistant Administrator for Flood Insurance, respectively, Federal Insurance Administration, U.S. Department of Housing and Urban Development, Washington, D. C.

Standards and Regulations, Past and Present

The initial standards and regulations under the NFIP were, of necessity, broad and general. The Federal Government had little experience in the development and administration of a national program for non-structural flood hazard mitigation. The acceleration of the mapping program by the 1973 Flood Disaster Protection Act necessitated the use of uniform standards for floodplain and floodway identification, in order to achieve the 1983 deadline mandated by Congress for the completion of initial rate studies. The use of variable standards and regulations would have made such an accelerated program administratively unfeasible because of the time and expense required to analyze each particular situation. Instead, standards such as the 100-year flood and the one-foot rise for floodway determination were established in an attempt to provide reasonable criteria which would lead to an eventual reduction in annual flood losses while permitting reasonable use of the land.

Experience with the NFIP to date has indicated a number of directions in which the standards and regulations may evolve. With completion of the initial identification of virtually all flood prone areas in the Nation and the completion of rate studies for more than 90 percent of the population at risk within the next two years, the regulatory activities of the Program can be expected to evolve in three general directions. The first of these includes the development of standards and regulations more specifically adapted to the numerous variable hazards under the purview of the Program. The second includes variations of the standards and regulations to better suit the socio-economic situation within the various floodprone areas of the nation. The third includes greater recognition of natural and beneficial floodplain values, particularly those values associated with the natural mitigation of floods.

Recognition of Variable Hazards

Experience with the NFIP has indicated that hazards and risks vary significantly within and among various flooding situations. Floodplain regulations and standards will have to be adapted to meet such situations. Several examples follow.

Wave Height The present standards for coastal flood insurance studies and regulations are based on the 100-year still-water elevation because, in the past, no suitable methodology was available for including wave height. This has created numerous problems for the NFIP, since flood losses in coastal areas rarely involve still water, but rather, reflect the tremendous impacts of wave action. In a recent study,[1] the Program has been criticized for opening mortgage markets in coastal areas which were voluntarily kept closed by lending institutions before flood insurance became available. The availability of flood insurance and the non-inclusion of wave height elevations may be encouraging construction of inadequately protected structures in some areas.

This situation has prompted several special studies by the Federal Insurance Administration (FIA) including: establishment of a methodology for determining wave height;[2] a study of design and construction standards for residential buildings in coastal hazard areas, (presently underway); and a recently completed study[3] determining the effect of the NFIP on lending and construction practices in coastal areas. These studies may lead to revisions in present standards and regulations. Beginning this fiscal year, many flood insurance studies will include wave height determinations on an advisory basis. Mandatory inclusion will be considered following completion of an economic analysis, and review by the U.S. Water Resources Council and members of Congress.

Alluvial Fans Alluvial Fans are an example of a situation where current regulations and standards, as well as technical methodology, are not well suited to the nature of the hazard. These geomorphologic features are extremely hazardous because they are easily developed and appear to be inactive to the short term observer. Conventional hydrologic and hydraulic analyses invariably yield 100-year flood depths which are quite low, 3 feet or less in most instances. Yet the tremendous hazards of erosion, velocity, shifting channels and debris loads, which make elevation of structures a rather ineffective mitigation measure, are not currently addressed by our regulations and standards. Present standards are designed largely for typical riverine situations. Although FIA has recently completed a special study[4] to determine a technical approach for studying such areas, the effective mitigation of flood losses must consider specific regulatory approaches.

Levees The establishment of appropriate standards and regulations for evaluation and management of hazards in areas involving local protection works, such as levees, poses considerable problems. The NFIP was conceived as a non-structural approach to flood hazard mitigation. Yet, the adoption of the 100-year standard for this purpose may be fostering the construction of flood protection works to this standard, for the sole purpose of removing the "protected" area from the special flood hazard designation and the associated floodplain management and insurance requirements under the NFIP. This is a potentially dangerous situation for several reasons. The 100-year standard is generally considered a low degree of protection for structural works which are subject to possible catastrophic failure and overtopping. Upon failure, inundation is often rapid and to considerable depths. Levees designed to provide 100-year flood protection do not provide the equivalent protection and security as the elevation or floodproofing of individual structures and could encourage development in hazardous areas. Therefore, the 100-year standard may not be appropriate for the acceptance of local flood protection works under the NFIP.

The regulatory responsibilities of the program in levee-protected areas are likewise yet to be determined. Levees recognized under the program should have adequate inspection, maintenance and operational requirements to assure that their rated protection levels are maintained. Perhaps these requirements should be specified and mandated under the regulations of the NFIP.

Floodways Floodway regulations and standards have heretofore been viewed in a very narrow context under the NFIP. The one-foot water surface rise criterion was chosen administratively as a reasonable compromise between increased flood hazards and allowing use of floodplain land. This uniform standard provided for a simple technical approach to floodway determination and relatively easy to administer regulations from a Federal viewpoint. Yet, the floodway tool as now applied, does not always conform well with various hazard situations. For example, the floodway addressed in the current standards and regulations considers only the maintenance of floodplain conveyance or hydraulic efficiency as a means of avoiding increased flood hazards due to floodplain encroachment. We know, however, that floodplain storage can have significant effects in attenuating downstream flood peaks in specialized situations. In the future, FIA will be taking a serious look at standards and regulations designed to protect valuable floodplain storage in such areas. Such standards and regulations may take the form of limiting the amount of fill which can be imported into the floodplain.

Another shortcoming of the standards and regulations, with respect to floodways, is that they consider only the increased hazard due to the effect of encroachment on water surface elevation. In many situations, the hazard due to flow velocity can be an equally or more important consideration. The inability to establish standards for hazardous velocities has hampered the development of regulations in this area. A flexible velocity standard may have to be considered because of the tremendous variability in the susceptibility to erosion, of different fill and embankment materials.

Headwater Flooding and Drainage The regulatory role of the NFIP in headwater flood hazard reduction is as nebulous as the transition from what may be considered a flood problem to what may be considered a local drainage problem. It is likely that localized flood damages, from sources with very small drainage areas, inflict millions of dollars of damage annually which are never accounted for because victims absorb the costs. In the past, there has been no direct Federal involvement because such situations never reached disaster proportions. Yet, flood insurance is now available in such areas and the quid-pro-quo philosophy of the NFIP should be somehow applicable.

Because of the tremendous costs involved in studying such areas by conventional engineering methods, and the rapidity with which elevation determinations would become obsolete by drainage alteration, FIA has avoided the detailed study of such flooding sources except in the most critical situations. Instead, the regulations state only that adequate drainage must be provided in new subdivisions to reduce exposure to flood hazards. Within the near future, FIA will have to closely review the role of the NFIP in areas of headwater flooding and drainage, from both a regulatory and insuring role, and establish corresponding standards.

Future Urbanization One final example of the need to meet variable hazard situations, is in the consideration of urbanization effects. The effects of urbanization on flood magnitude and frequency on smaller watersheds have been demonstrated. FIA has been reluctant to consider future urbanization effects in flood insurance studies and regulations for a number of reasons including: the general lack of basin-wide land use planning and forecasting data on which analyses can be based, i.e., the uncertainty in future land use prediction; the rather costly hydrologic approaches required and the unknown accuracies involved in using such methods, i.e., the uncertainty in prediction of future flood elevations; and finally, the legal implication for floodplain regulations and insurance rates based on an uncertain future watershed condition.

Despite these problems, in specific areas of the country development can be predicted in both direction, rapidity and type with a great deal of certainty. In these situations, it makes little sense to base development regulations on a present risk which will be greatly increased with some predictability in the near future. FIA will have to further investigate its "present conditions" standard in such situations and consider regulating for future flood risks where such risks can be accurately assessed.

Recognition of Variable Socio-Economic Situations

According to the U.S. Water Resources Council,[5] "the public interest in flood plain management... includes concern for (1) economic efficiency, (2) environmental quality, (3) individual safety, peace of mind and social well-being in general, and (4) economic and environmental health of regions and localities." Achieving optimium levels for each of these conditions is highly improbable. Indeed, even the determination of what are the optimium levels would be an impossible task in all but the most simple of situations.

The net economic benefits or costs of restricting floodplain development depend largely on the availability of alternatives to the floodplain location and the degree of hazard. Regulations should be equitably applied and sufficient to achieve the goals of flood loss reduction. Yet, they should be flexible, where possible, to provide for reasonable use of the land, given the socio-economic setting of the community.

FIA is currently completing a special study of the social, economic and environmental effects of floodplain regulation. The purpose of this study is to provide a general evaluation of the net public benefit from regulation in terms of these three factors. It is expected that the results of this study will assist FIA in the design of future standards and regulation strategies to minimize adverse socio-economic impacts of the Program.

It is not likely, however, that the primary standards of the Program, such as the use of the 100-year flood, or the one-foot water surface rise criterion for determining the floodway, will be made more flexible or be relaxed. One primary reason is that an important objective of the NFIP, as emphasized in Executive Order 11988, is to avoid direct and indirect support of floodplain development wherever there is a practicable alternative. In a recent observation by the General Accounting Office, FIA was criticized for its emphasis on how floodplains should be developed instead of the avoidance of the floodplain. Any relaxation of our primary standards, or requirements for exceptions from these standards, could constitute a departure from this objective. The responsibility will continue to remain on the community to demonstrate to the Federal Insurance Administrator that an exception from these standards is necessary to avoid severe hardship and gross inequity for the community. The reasons for the request for exception must include sufficient supporting technical data as well as socio-economic and environmental considerations.

Secondary standards and regulations within the program will, however, evolve to provide greater flexibility in the ways that communities can meet the standards for both level of protection and limiting increases in flood hazard by encroachment. In general, the trend in the standards and regulations has been, and will continue to be, away from highly detailed use limitations and construction requirements toward more general performance standards to meet the needs of communities which propose alternative, but equivalent, mitigation measures. Some examples of the direction of evolution of regulations and standards to achieve these flexibilities follow.

Flood Resistant Construction Experience has shown that in some cases the requirement for elevation and/or "dry" floodproofing may impose economic hardships or impractical structural requirements on buildings that must be built in the floodplain. FIA is currently initiating a special study of flood resistant construction, or "wet" floodproofing, which would allow flood waters to enter a structure without destroying its integrity. This study is designed to determine the physical, social and economic conditions under which flood resistant construction is justifiable, and identify construction and design standards which would qualify a structure as flood resistant. It is anticipated that the results of this study may permit the inclusion of flood resistant construction standards in the regulations, for non-residential structures.

Basement Floodproofing Present regulations of the NFIP allow the use of floodproofed residential basements only as an exception, when it can be shown that application of the elevation requirements would create a hardship for the community. In 1977, FIA completed a special study[6] culminating in the publication of a manual for the construction of floodproofed residential basements. As a result of this investigation, FIA is considering a revision to the Program regulations that would permit communities to allow the construction of floodproof residential basements, below the base flood elevation, provided that adequate performance standards are met.

Density Floodways Floodways are presently determined according to an encroachment line concept whereby full development is permitted within a fringe area, while no activity, which would increase flood elevations, is permitted within a floodway area reserved for the passage of flood water. In areas where floodplains and floodways are very wide and flow velocities are not hazardous, floodplain conveyance can be effectively preserved by a limitation on development density. Such standards and regulations might take the form of allowing only a certain amount of floodplain volume to be filled and obstructed per unit area. In this way, sufficient conveyance can be maintained between structures to avoid increases in future flood heights. Such an approach would share the burden of maintaining floodplain conveyance among all floodplain occupants, not just those who happen to be closest to the normal stream channel.

Substantial Improvement - Section 1308(c) of the National Flood Insurance Act of 1968 requires that substantial improvement to properties within the floodplain be treated similarly to new construction as far as insurance premium rates are concerned. Current regulations define substantial improvement, in part, as repair, reconstruction, or improvement which exceeds 50 percent of the market value of the structure. Substantial improvement is also treated the same as new construction for floodplain management purposes under the regulations. A recent study[7] completed for FIA concluded that within 10 years, only 3.8 percent of all housing units located within the special flood hazard area would be affected by the regulation, and in 50 years, only 19 percent would be affected. The study also concluded that substantial improvement was poorly defined and subject to such financial and political constraints to enforcement that the continuation of the existing language in the regulations was highly questionable. A case study[8] of post-disaster reconstruction in Scituate, Massachusetts tended to support this conclusion. As a result of these studies, future changes in the substantial improvement requirements of the regulations are likely. Such changes will likely take the form of redefinition to make the criteria less ambiguous and easier to administer. Total abandonment of the regulation is doubtful because such an action would represent a major movement of the NFIP away from effort to reduce flood problems associated with existing structures.

Recognition of Natural and Beneficial Floodplain Values

The President's Executive Order 11988 on Floodplain Management requires that each agency provide leadership and take action to restore and preserve the natural and beneficial values served by floodplains. The interrelationships between floodplain management, wetland management and coastal zone management are complex and are currently the topic of several public workshops and interagency seminars sponsored by the U.S. Water Resources Council. The standards and regulations of the NFIP can play a significant role in the restoration and preservation of natural and beneficial values served by floodplains. However, the purview of the Program will remain restricted to the floodplain, and the regulatory activity will be directed primarily to the maintenance of the natural flood mitigation abilities of the floodplain. In this regard, there are three potential areas of evolution for the regulations and standards, including the preservation of coastal and inland wetlands, and floodplain aquisition.

Coastal Wetlands It is known that within certain coastal situations, barrier beaches, sand dunes and wetlands have an attenuating effect on storm surge and wave action. The current regulations of the NFIP prohibit the man-made alteration of sand dunes and mangrove stands, within coastal high hazard velocity zones, if such alteration would increase potential flood damage. This is presently a difficult regulation to administer because quantitative data on the attenuating effects of coastal wetlands have been lacking. FIA is presently initiating a special study to provide information indicating the physiographic conditions under which the various wetland types may be considered effective in reducing potential flood damages. It is anticipated that the results of this study will provide standards for the identification of those wetland areas requiring protection.

Inland Wetlands Inland wetlands, particularly those comprising large areas of the floodplain, can have significant and valuable attenuating effects on flood peak flows. The methodology necessary to assess these effects is available. However, its use will require basin-wide analysis, at significant costs. FIA will be initiating a pilot study this fiscal year to investigate the costs, feasibility and alternate regulatory mechanisms for the inclusion of storage preservation criteria in flood insurance studies and regulations. Although not aimed directly at the preservation of wetlands per se, standards and regulations for the protection of floodplain storage will serve as a strong incentive to maintain such areas in natural conditions.

Acquisition Section 1362 of the Flood Insurance Act of 1968 permits the Government to enter into negotiations for the acquisition of flood damaged property. It also provides for subsequent transfer of such property to state or local governments for use as defined by Federal regulation. Funding for this program has not yet been provided. FIA is currently involved in a research study to evaluate alternative means of implementing Section 1362, including an analysis of relevant social, economic, legal and environmental issues inherent to the process. Eventual funding of this section would provide an excellent opportunity to return such lands, or a portion of such lands, to natural uses. FIA will be responsible for the adoption of standards and regulations governing the use of these lands following acquisition and transfer. These regulations will likely emphasize those uses consistent with natural and beneficial floodplain values.

Summary and Conclusions

Early objectives of the National Flood Insurance Program were the rapid identification of flood prone communities, determination of regulatory elevations and risks through rate studies, and conversion of communities the Regular Flood Insurance Program to assure compliance with more comprehensive standards and regulations. Presently, virtually all flood prone communities have been identified and more than 90 percent of the population at risk will have rate studies completed within the next two years. The Program is now being redirected to provide greater community assistance in administration of program requirements at the local level. With greater emphasis on community assistance, the Federal Insurance Administration will be better prepared to adapt the standards and regulations of the Program to meet local needs and situations. Evolution of the Program standards and regulations will likely occur in three primary directions, including greater recognition of variable types of hazards, local socio-economic situations, and natural and beneficial floodplain values.

References

(1) Miller, H. Crane. "Coastal Flood Plain Management and the National Flood Insurance Program: A Case Study of Three Rhode Island Communities". Environmental Comment. Urban Land Institute. Washington, D.C. November, 1975. pp. 2-15.

(2) "Methodology for Calculating Wave Action Effects with Storm Surges". National Academy of Sciences, National Research Council. Washington, D.C. 1977. 22 pp.

(3) Miller, H. Crane. "Coastal Flood Hazards and the National Flood Insurance Program". Federal Insurance Administration. Department of Housing and Urban Development. Washington, D.C. June, 1977. 50 pp.

(4) Dawdy, David R. "Alluvial Fan Flood Mapping: A Uniform Method". Unpublished Report. Federal Insurance Administration. Department of Housing and Urban Development. Washington, D.C. October, 1978. 6 pp.

(5) " A Unified National Program for Flood Plain Management". U.S. Water Resources Council. Washington, D.C. July, 1976.

(6) "Manual for the Construction of Residential Basements in Non-Coastal Flood Environs". NAHB Research Foundation, Inc. Federal Insurance Administration. U.S. Department of Housing and Urban Development. Washington, D.C. March, 1977. 266 pp.

(7) "Alternatives for Implementing Substantial Improvement Definitions." Shaeffer and Roland, Inc. Federal Insurance Administration. Department of Housing and Urban Development. Washington, D.C. January, 1978. 75 pp.

(8) Miller, H. Crane. "Enforcement of 'Substantial Improvement' Regulations: Scituate, Massachusetts." Unpublished Report. Federal Insurance Administration. Department of Housing and Urban Development. Washington, D.C. January, 1979. 9 pp.

Hydrology and Environmental Assessments*

Arliss D. Ray, William P. Wenstrom, and Dan D. Taylor*

It is certain that the future will see the development of improved hydrologic forecasting techniques, particularly since the field of hydrology is still evolving technically and because there are social needs for new techniques.

Hydrology is relatively young and still imperfect and inexact. For centuries, major misconceptions were based on what we now know as the hydrologic cycle which was developed only three to five hundred years ago. Major developments in pulling together diverse information and data on natural phenomena and observations have been made within the last hundred years. Most current hydrologic concepts have become well established only within the past fifty years, and present sophisticated hydrologic techniques have been formulated only within the past twenty five years. The more recent efforts, drawing heavily upon results of precipitation runoff and other relevant data acquisition programs, have utilized mathematical, statistical and computer science approaches.

If a nominal span of some thirty years is considered to represent a human generation, this brief recitation indicates that hydrology, as it presently is known, has evolved within the past one or two generations. Certainly most hydrology has been developed within the past three generations; probably many of the people attending this conference have participated in many of the major and significant developments in the field. The limited history of hydrology, coupled with the relatively rapid rate of change in the field, suggests that the future of hydrology is very much dependent upon the attitudes and abilities of current and future hydrologists as they attempt to respond to the needs of society.

Most hydrology has been related to the physical aspects of water. For example, dams have been justified or built on the basis of flood control, recreation, irrigation and water supply. More recently, the use of reservoir capacity for low flow augmentation of a stream has been recognized, but even this has been considered as a water quantity problem by hydrologists. Power, as conventionally considered by the hydrologist, generally refers to hydroelectric power. Thus, the hydrologist seemingly has restricted his consideration to the quantity of water with little or no regard for water quality; the hydrologist has failed, perhaps deliberately, to recognize that not all waters are the same.

*Hydrology & Environmental Assessments, Arliss D. Ray, William P. Wenstrom, Dan D. Taylor, EMANCO Inc., Houston, Texas.

Based upon papers presented at this conference, most of the current concerns of the hydrologist continue to be related to water quantity, power demand, irrigation water demands, floods and their prevention, recreation and related activities. Attention necessarily has been given to the economics of a given project, but the study of alternatives to a given project traditionally has been based generally on technical and economic considerations only. While this conference gives confidence that hydrologic forecasting will continue to improve with the research, data acquisition and other efforts that additional time will make available, there does not appear to be a desire among hydrologists to broaden their scope of activities. Indeed, there appears to be a determination to get better in traditional activities at the expense of adapting existing knowledge and techniques to new problem areas. However, it is important to note that hydrology, as all other technical fields, must adapt to the challenges posed by our changing social world.

The history of the environmental control efforts of this country, through the U.S. Environmental Protection Agency, its predecessors and analogous state and local agencies, is not that different from the history of the development of hydrology. These efforts have been based traditionally upon a control by permit, with various technical requirements and limitations being prescribed. Necessarily, control of the water environment has required joint consideration of water quantity and quality. Efforts have been made to prescribe both the quality of the receiving waters and limitations on water effluents.

However, many instances of environmental abuse were condoned by the public for a number of years before the intense desire to clean up the environment began to emerge. Material wealth is no longer recognized as the sole measure of success; quality of life has become important to a public with an ability to pay for a better environment. The psychological importance of the photographs of earth, as taken from our space vehicles, should not be unnoticed; the concept of Spaceship Earth has had profound effect on efforts to control our environment.

A major piece of Federal legislation responding to the needs of the public, as perceived by those drafting the legislation, was the National Environmental Policy Act (NEPA) of 1969. NEPA created the Council on Environmental Quality; more importantly, a detailed statement of the environmental impact became a requirement for those major Federal actions significantly affecting the quality of the human environment. An action, as presented in NEPA, is not restricted to facilities or construction; it also covers permitting or other similar actions on the part of Federal agencies. Inherrent in NEPA and its requirement for the preparation of environmental impact statements was the specification that a systematic, interdisciplinary approach be utilized to insure an integrated use of the natural and social sciences in planning and decision making as they pertain to the environment.

There also has been an increasing public awareness of all aspects of our society in recent years; this movement has resulted in demands of the public to be both informed and involved in decision making of social relevance. NEPA also required that methods and procedures be identified and developed such that presently unquantified environmental amenities and values may be given appropriate consideration in decision making. This means, of course, that water-related projects can no longer be built solely on economic and technical considerations; broader environmental aspects also must be considered.

The guidelines governing the organization of materials making up an environmental impact statement for a proposed project were arranged in the following sequence: a description of the proposed action, a description of the existing environment and the environmental impact of the proposed action on that environment. There were additional aspects, including a recitation of various alternatives presumably considered in deciding upon the proposed action, but the completed environmental impact statement under these guidelines is generally a justification of previous decisions.

The general types of problems which may be encountered within an environmental impact statement are those of omissions, technical errors and general inadequacy. In order to minimize these problem areas, many environmental impact statements have become extremely lengthy and boring documents.

To many casual observers, the environmental impact statement approach has occasionally appeared to be ludicrous and a waste of both time and money. However, those closer to proposed actions and the associated environmental impact statements, feel these documents are important. The reason that an environmental impact statement should be treated in a serious fashion is that it becomes a focal point for opponents to the proposed action, and the statement must be defensible in public meetings, hearings and in formal litigation.

Opposition to a project need not be inspired by genuine environmental concern, and the environmental impact statement is a document which can be attacked by anyone; an indefensible statement can be the cause for defeat of a needed project. In this country it appears unlikely that any major project in the future will be constructed without significant opposition. This observation results in the need for competent and professional preparation of environmental impact statements.

Although the environmental impact statement process appears to be responsive to the desire for public information and involvement, problem areas (in the content and focus of the statement) do exist. Recently, the Council on Environmental Quality proposed a new format for environmental impact statements which may have interesting results. Among the proposed new requriements are those of a presentation and discussion of alternatives leading to the proposed major action at the beginning of the statement, wherein environmental, economic and technical aspects are to be considered. Only significant environmental issues are to be stressed and the resulting statement will emphasize brevity and clarity. In the preparation of such statements, most of the meaningful work will be in the study of alternatives; these alternatives presumably, should be considered at the same level of information. Obviously an interdisciplinary effort will be required in such a study.

It is sometimes difficult for individuals, trained and educated extensively in a single discipline, to participate effectively in an interdisciplinary team effort. The reasons for such difficulty are complex and rooted to varying degrees in personal and professional ego, communication difficulties and lack of interest.

There are significant problems with the proposed guidelines for environmental impact statements as they may affect projects in the private sector. The major problems with the environmental impact statement process in the private sector relate to the study of alternatives and the time frame required for statement preparation and acceptance. Many companies in the private sector will feel that their company decisions are made with considerations which cannot be divulged without losing a competitive edge in the marketplace and this concern may pose problems. Probably the larger problem, however, is associated with time requirements. A company in the private sector is profit motivated and, once a decison has been made and monies allocated for a new facility or project, a delay in bringing a new project into full operation represents an increased cost. This subsequently becomes an additional cost borne by the public as a result of its concern for environmental quality.

To the hydrologist faced with participation in an interdisciplinary decision making team, there will appear a need for different forecasting techniques than those available now. Perhaps the need will be for more approximate techniques utilizing less data, which may be in the opposite direction of ongoing efforts within the field of hydrology. Participation in a team effort charged with analyzing alternatives may result in recommending a proposed action which is not necessarily the best technical solution. However, these types of activities in the future may lead to more meaningful environmental impact statements, result in better decisions and provide a sense of social relevancy based upon professional applications of hydrologic technical skills.

NATURAL FLUCTUATIONS OF WATER QUALITY IN A HEADWATER STREAM

by

Joe F. Nix*

Introduction:

Predictions of water quality in streams requires a fundamental understanding of the natural processes occurring in the stream as well as the contribution of dissolved and suspended materials from anthropogenic activities. In highly developed areas, many of the natural variations in water quality may become lost in fluctuations caused by various activities or become so complex that interpreting data from the system would be difficult if not impossible. This is particularly true in the downstream sections of many rivers.

Johnson, et al (1), Cleaves, et al (2), Bilby and Likens (3), and others have described various aspects of the transport of chemical species from relatively undisturbed watersheds. Some investigators have based their conclusions on observations of stream chemistry at relatively lengthy intervals while others have made observations at short intervals. An analysis of short term data indicate that concentrations of dissolved as well as supsended materials in streams are highly variable, particularly in response to storm events.

Variations in the dissolved and suspended materials present in streams are important in determining the fundamental nature of a stream. Such variations also have implications in water quality of the receiving reservoir as well as land use practices throughout the watershed.

The Caddo River of southwest Arkansas drains a section of the Ouachita Mountains which is heavily forested. The predominant rock types are shale and sandstone. Slopes are generally steep and, with the exception of narrow zones along stream banks, soils are thin. Population density is low and only one small town (population approximately 1500) discharges treated domestic sewage into the stream system. In the downstream section of the watershed, the U. S. Corps of Engineers operate a reservoir for flood control, power generation, recreation, and other purposes (DeGray Lake). This reservoir has a surface area of 5,427 hectares at normal pool elevation and has a theoretical exchange time of approximately one year.

Observations of short term variations in both dissolved and suspended materials have been made on the Caddo River at a point near the headwaters of the reservoir. Similar observations have been made in DeGray Lake. The results of portions of these observations are used

*Professor of Chemistry
Ouachita Baptist University
Arkadelphia, Arkansas 71923

to illustrate the dynamic nature of this stream system and the importance of such variations in determining the fundamental nature of the stream. The impact of variations in stream chemistry on DeGray Lake is also discussed.

Methods:

The Caddo River was sampled at intervals ranging from one hour to several hours for three to four day periods at a point two miles upstream from the headwater of DeGray Lake. Samples were taken at mid-stream from a bridge using a horizontal van Dorn type sampler. Sample preservation and analytical methods were those described in Standard Methods (4) and/or EPA approved methods (5). Organic carbon was determined using wet oxidation and an Oceanographic International organic carbon analysis system. Observations of temperature, dissolved oxygen, specific conductance, alkalinity and pH were made in the field.

The same procedures were used on samples taken from the reservoir with the execption that the presence of suspended material was estimated *in situ* using a transmissiometer with a 10 cm light path.

Results and Discussion:

The results of the analysis of selected parameters from samples taken from the Caddo River between October 24, 1976 and October 27, 1976 are shown in Figures 1 and 2. The increased discharge observed during this period was the result of a storm event which occurred on October 24, 1976.

Similar data for a storm event which occurred on June 16, 1977 are shown in Figures 3 and 4. The data for individual water quality parameters determined was subjected to regression analysis to test the relation between each parameter and river discharge. Correlation coefficients for these relationships are summarized in Table 1.

Paustian and Bechta (6) have recently described their observations of suspended sediment as a function of stream discharge for an Oregon Coast Range stream. Their observations show a higher yield of suspended sediment on the rising side of the hydrograph as compared to the falling side. If one uses turbidity as an indicator of suspended sediment, similar relationships were observed in the Caddo River for the event which occurred in June 1977. The relationship between turbidity and stream discharge for the storm which occurred in October 1976 is not as clear, however the initial "wash off" effect is obvious.

In both sets of data, sodium and calcium behave as somewhat conservative parameters in that they respond to dilution from elevated flows. Calculations of the total mass transport of these constituents (discharge x concentration) indicate that larger quantities of these substances are being transported during elevated flows than during low flows; thus they are not truely "conservative" parameters.

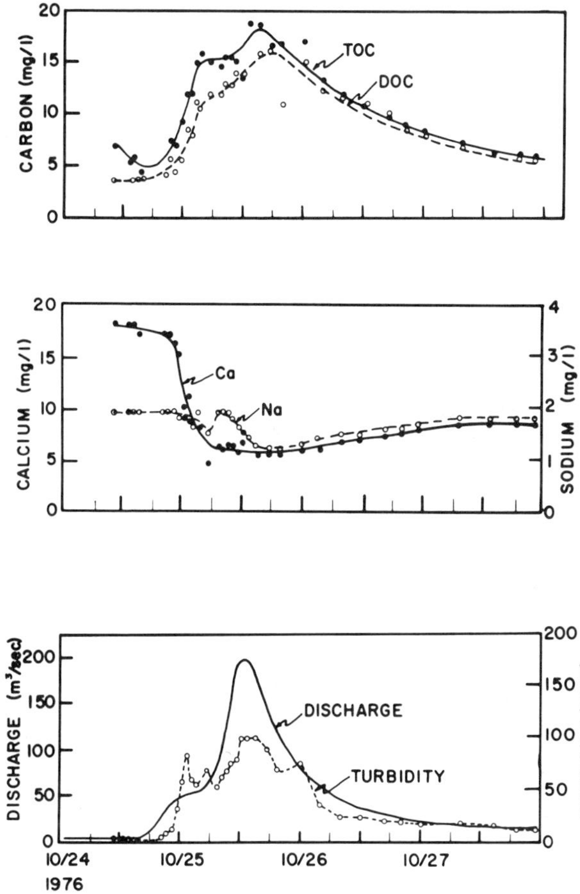
CARBON (mg/l)
TOC
DOC
CALCIUM (mg/l)
SODIUM (mg/l)
Ca
Na
DISCHARGE (m³/sec)
TURBIDITY (N.T.U.)
DISCHARGE
TURBIDITY
10/24
1976
10/25
10/26
10/27

Figure 1

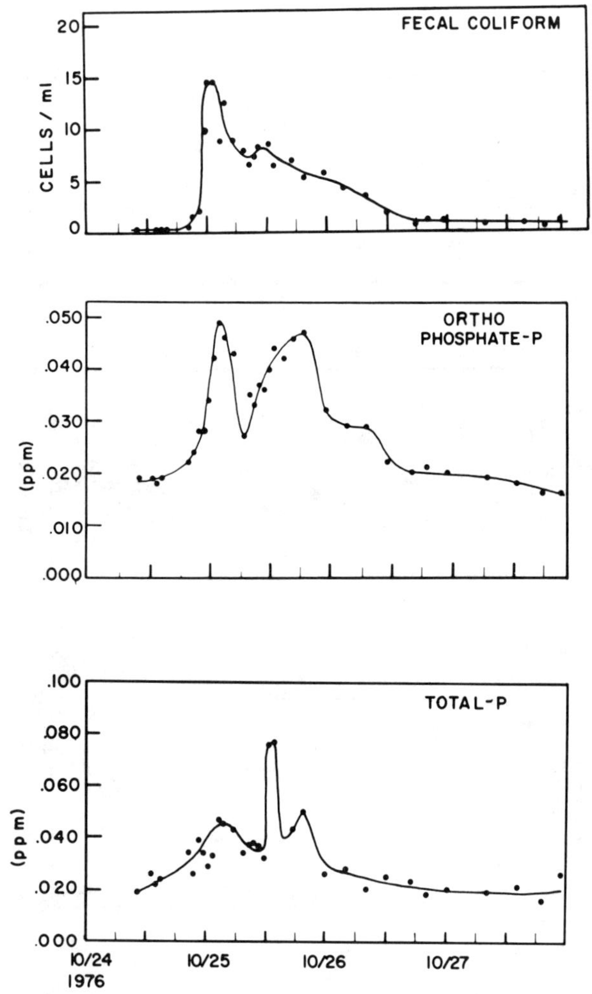
FECAL COLIFORM
CELLS / ml
20
15
10
5
0
ORTHO PHOSPHATE-P
(ppm)
.050
.040
.030
.020
.010
.000
TOTAL-P
(ppm)
.100
.080
.060
.040
.020
.000
10/24 1976
10/25
10/26
10/27

Figure 2

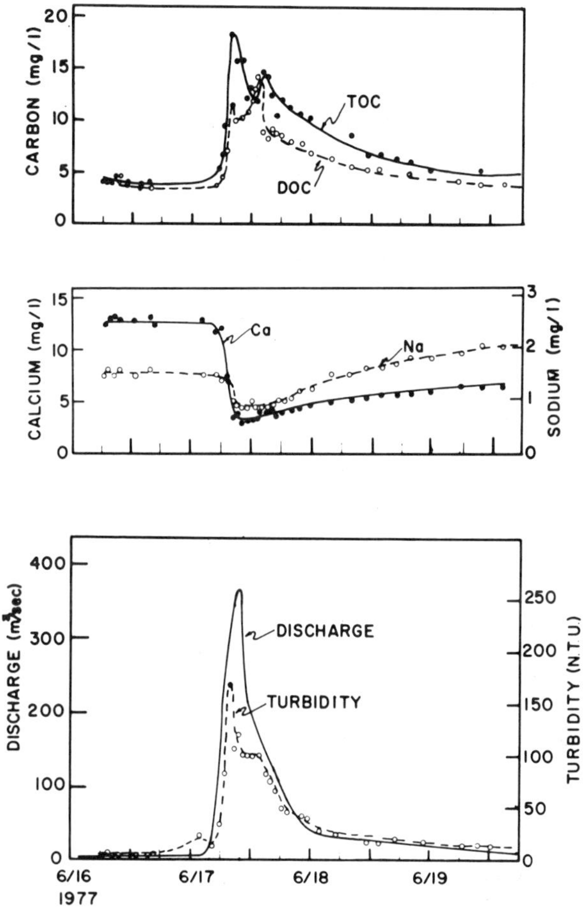
CARBON (mg/l)
20
15
10
5
0
TOC
DOC
CALCIUM (mg/l)
15
10
5
0
Ca
Na
SODIUM (mg/l)
3
2
1
0
DISCHARGE (m³/sec)
400
300
200
100
0
DISCHARGE
TURBIDITY
TURBIDITY (N.T.U.)
250
200
150
100
50
0
6/16
1977
6/17
6/18
6/19

Figure 3

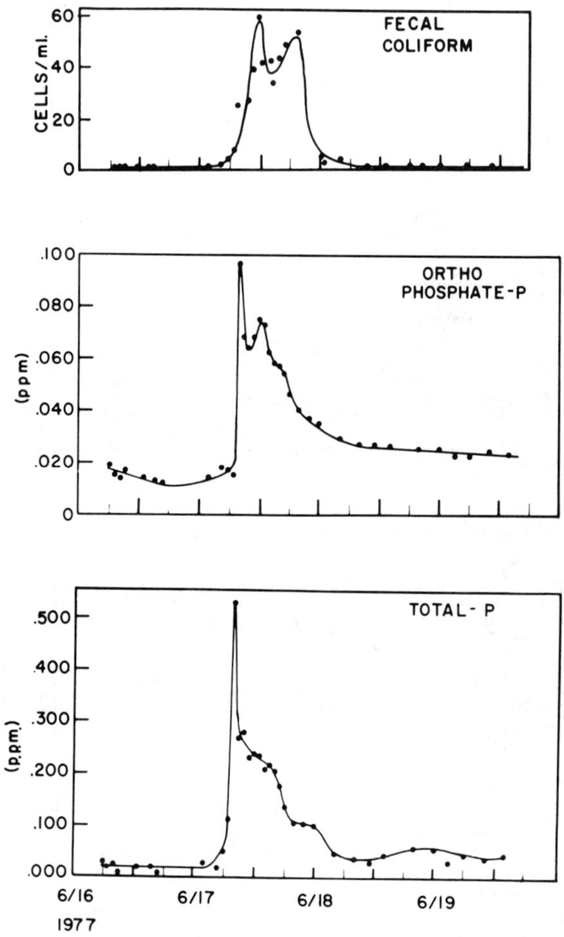
FECAL COLIFORM
CELLS/ml.
60
40
20
0
ORTHO PHOSPHATE-P
(ppm)
.100
.080
.060
.040
.020
0
TOTAL-P
(p.p.m.)
.500
.400
.300
.200
.100
.000
6/16 1977
6/17
6/18
6/19

Figure 4

TABLE 1

Correlation coefficients for specific parameters with river discharge

storm event beginning	10/24/76	6/16/77
Turbidity	0.877	0.883
Alkalinity	-0.628	-0.653
Calcium	-0.606	-0.706
Sodium	-0.528	-0.854
Potassium	0.734	0.824
Silica	-0.669	-0.862
Nitrate-N	0.616	0.361
Ammonia-N	-0.025	-0.041
Total Kjeldahl-N	0.793	0.729
Total-P	0.713	0.824
Ortho phosphate-P	0.694	0.905
Total Organic Carbon	0.817	0.851
Total Coliform	0.724	0.574
Fecal Coliform	0.464	0.769
Fecal Streptococcus	0.340	0.257

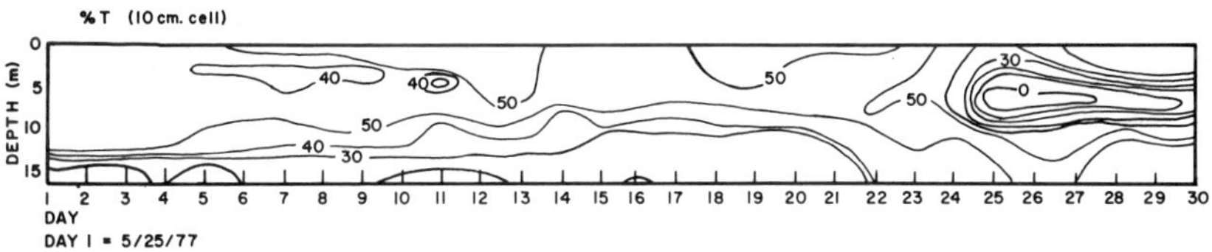

Total organic carbon (TOC) and dissolved organic carbon (DOC) respond in a manner very different to the conservative paramters. Although peaking in discharge and concentrations of both TOC and DOC do not coincide, higher concentrations are definitely associated with periods of elevated flows. The difference between TOC and DOC, which indicates particulate orgainc matter, is greatest during the early part of the rising side of the hydrograph, suggesting that particulate organic matter transport occurs early in the storm event. This is in agreement with Bilby and Likens (3) who observed that fine particulate organic matter concentrations reached a peak prior to the peak in the hydrograph.

Observations of the concentration of fecal coliform in the October 1976 storm show that the higher values occurred prior to the peak in the hydrograph. The fecal coliform data from the storm of June 1977 show the highest concentrations almost coincidental with the peak in discharge, with a second peak following very soon after the first. In this latter case, the bimodal distribution of fecal coliform suggest two overlapping events or inputs from two different sections of the watershed.

The Caddo River, like many of the streams in the Ouachita Mountains and Ozark Mountains have strips of grazing land on alluvial deposits boardering the streams. Grazing throughout the watershed is heavy in this zone. If it is assumed that the origin of the fecal coliform is principally cattle in this zone, the presence of these indicator organisms may serve as a tracer for material originating from grazing lands. For example, the fact that the highest fecal coliform concentrations do not coincide with the highest concentrations of organic carbon suggest different origins. For the same reason, it would appear that the origin of phosphorus in the rivers is not limited to the grazing environment, even though the application of phosphate fertilizers is a common practice on the grazing lands.

Comparison of the total phosphorus concentration and turbidity for the June 1977 storm demonstrates that phosphorus is being transported on sediment. The extremely large quantities of phosphorus being transported in a relatively short period of time is well illustrated by comparing the hourly transport of total phosphorus in the peak of the storm of June 1977 with the hourly transport observed during low flow conditions prior to the storm. During the peak of the June 1977 storm approximately 6×10^6 grams per hour were passing the sampling station while during low flow conditions only 2×10^2 grams per hour were passing the same station. The short term nature of high loading rates suggest that loading calculations based on periodic observations may not adequately reflect true transport unless the sampling interval is very short.

Precise loading rates become extremely important in predicting the trophic state of reservoirs (Lee, et al) (7). The introduction of turbid storm water into the upstream section of DeGray Lake is shown in Figure 5. A storm event occurred on the 23rd day of this study and the resulting imposition of the water containing high suspended sediment as well as high levels of total phosphorus is indicated by the presence of lower values of percent transmittance. The zone of turbid water apparently entered the reservoir as an interflow.

Relative good correlations were found between river discharge and several water quality parameters (see Table 1). Negative correlation coefficients reflect the more conservative parameters while positive values reflect constituents which generally increase during elevated discharge. Even though correlation coefficients for these two storm events are generally comparable, it should be pointed out that the relationships suggested by regression analysis are significantly different for each storm. It would appear that the specific relationship of a given parameter with river discharge may be highly variable, depending on the nature of the storm.

From these observations it is obvious that the chemical dynamics of a relatively undisturbed headwater stream are such that short term events may dominate the transport of certain substances, particularly phosphorus. Although the origin of the phosphorus cannot be tied to a particular section of the watershed, sequential relationships of other water quality parameters during the response of the river to a storm event suggest that its primary souce is not the grazing environment.

References

1. Johnson, Noye M., Gene E. Likens, F. H. Bormann, and Robert S. Pierce, Rate of chemical weathering of silicate minerals in New Hampshire, Geochimica et Cosmochimica Acta, 32, 531-545, 1968.

2. Cleaves, Emery T., Andrew E. Godfrey, and Owen P. Bricker, Geochemical balance of a small watershed and its geomorphic implications, Geological Society of America Bulletin, 81, 3015-3032, 1970.

3. Bilby, Robert E., and Gene E. Likens, Effect of hydrologic fluctuations on the transport of fine particulate organic carbon in a small stream, Limnology and Oceanography, 24, 69-75, 1979.

4. Standard Methods for the Examination of Water and Wastewater, APHA,AWWA, and WPCF, 14th Edition, Washington, D.C., 1975.

5. Methods for the Chemical Analysis of Water and Wastes, U. S. Environmental Protection Agency, Washington, D.C., 1974.

6. Paustian, Steven,J. and Robert L. Bechta, The suspended sediment regime of an Oregon coast range stream, Water Resources Bulletin, 15, 144-153, 1979.

7. Lee, G. Fred, Walter Rast, and R. Anne Jones, Eutrophication of water bodies: insights for an age-old problem, Environmental Science and Technology, 12, 900-907, 1978.

WATER USE FOR ELECTRIC POWER GENERATION

David M. Snyder, K.T. Mao, and Robert D. Glynn, Jr.*

A major demand for water in the future will be for cooling steam-electric generating facilities needed to meet the nation's increasing demands for electric power. Plans to provide water for power plant cooling must address economic, environmental, social, and legal considerations. Improved data from improved hydrologic forecasting will be vital in making difficult decisions regarding the allocation of water supplies for power plant cooling as well as the siting and design of such facilities.

This paper first describes power plant cooling water requirements and alternative cooling technologies. Next, significant environmental and social issues regarding the use of water for power plant cooling are discussed, followed by an assessment of projected national and regional requirements for cooling water. The paper concludes with a discussion of why hydrologic data are important in the planning of steam-electric generating facilities.

HOW WATER IS USED IN POWER PLANTS

Steam-electric power plants produce electricity using the Rankine Steam Cycle. In this cycle, water is fed to a boiler heated with coal, oil, gas, or nuclear energy. Steam from the boiler is used to rotate a turbine that is connected to an electric generator. Exhaust steam from the turbine is condensed and returned to the boiler to repeat the cycle (Figure 1). Cooling water is required to carry off the heat that must be removed in order to condense the exhaust steam.

In most systems, a separate water circuit is used to conduct cooling water between the condenser and the cooling system. In the case of "once-through" cooling systems, a continual supply of new, cool water is taken in from a large body of water. This water becomes heated as it passes through the condenser and is simply returned to the source. Other "closed-cycle" cooling systems recirculate water between the condenser and the cooling device. In these systems either the evaporation of a portion of the water or passage of the water through finned-tube heat exchangers (which operate on the same principle as an automobile radiator) are used to lower the temperature of the circulating water. Closed-cycle systems that use evaporative processes are called "wet cooling" systems. Other closed-cycle systems that use the flow of air

---

*Water Use for Electric Power Generation, by David M. Snyder, K.T. Mao, and Robert D. Glynn of Woodward-Clyde Consultants, San Francisco.

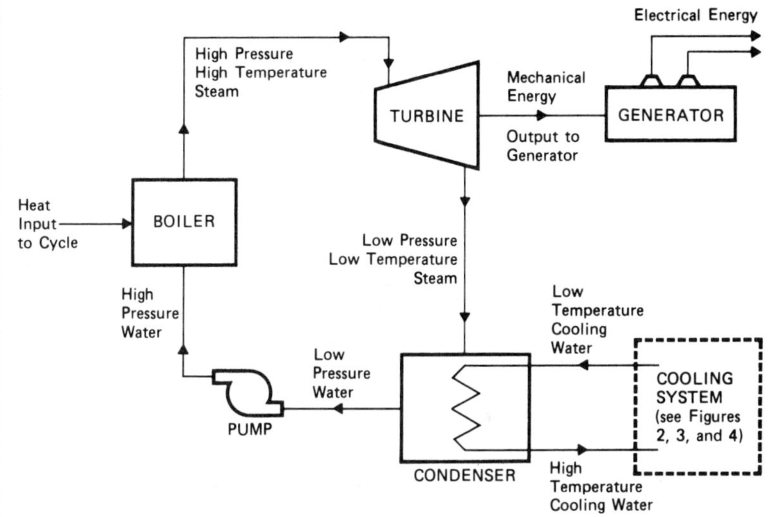

Figure 1. POWER PLANT WATER/STEAM CYCLE

over heat exchanger elements to cool the circulating water are called "dry cooling" systems.

In closed-cycle cooling systems that use evaporative processes it is necessary to provide "makeup water" to replace water that is evaporated. In addition, because evaporation of a portion of the circulating water results in the concentration of any dissolved or suspended solids that are present, a portion of the water must be periodically "blown down" and replaced with fresh water in order to maintain acceptable water quality in the system. Blowdown water can be returned to the water source; however, if the quality is unacceptable it might be disposed of in an evaporation pond. Alternatively, it is possible to treat blowdown water in reclamation facilities and return it to the cooling system, thereby reducing the amount of makeup water required.

The water flows in a closed-cycle cooling system are shown in Figure 2. The figure shows a storage reservoir that is included in some designs to supplement makeup water flows during periods when large flows are required or when the water source cannot support the necessary withdrawal rate.

Alternative cooling technologies will be discussed in detail later, but at this point it is important to note that, while all require withdrawals of water from some source, the amounts vary widely. Furthermore, "consumptive use" (water permanently removed from its source) also varies widely. In selection of a preferred type of cooling system, the choice frequently involves a tradeoff between one system that requires large water source withdrawals, but results in

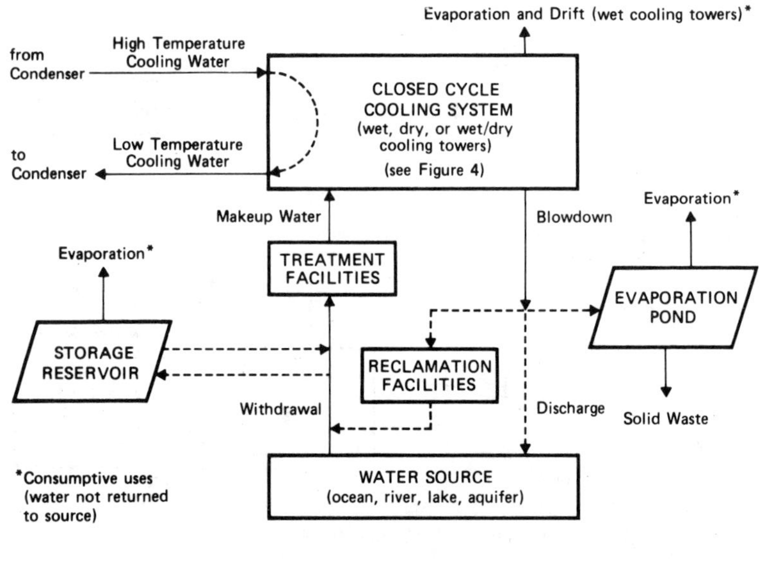

Figure 2. CONSUMPTIVE WATER USES IN POWER PLANT COOLING

very little consumptive use, or one which requires relatively small water withdrawals, but results in large amounts of water loss through evaporation.

WATER REQUIREMENTS FOR COAL AND NUCLEAR POWER PLANTS

This discussion will focus on coal and nuclear powered generating facilities for two reasons. First, these energy sources, which already produce the majority of electric power, are expected to comprise an even larger share of the nation's generating capacity in coming years (1). Second, the cooling water requirements for a single facility can be significant because generating plants of these types are typically quite large.

As mentioned above, water withdrawal and consumption rates for power plants can vary a great deal. This variability is a function of power plant thermal efficiency, type of cooling system utilized, environmental conditions at the plant site, and requirements for water for other than cooling purposes, such as pollution control at coal plants. Estimates of water requirements for 1000 MWe coal and nuclear power plants using various cooling system alternatives are given in Table 1.

As of 1975, approximately 73% of the generating capacity in the United States was cooled using withdrawals from navigable waters for once-through cooling (4). The current industry trend is moving away

TABLE 1. COOLING WATER REQUIREMENTS FOR 1000 MWe POWER PLANT

Cooling System	(a) Annual Withdrawals (acre-feet/year) Coal	Nuclear
1 Natural Lake or River (once-through)	377,000	772,000
2 2 Acre/MW Cooling Pond*	10,900 - 23,000	22,300 - 47,000
3 1 Acre/MW Cooling Pond*	9,000 - 18,500	18,500 - 37,900
4 Spray Pond*	7,200 - 13,900	14,800 - 28,500
5 Natural Draft Wet Towers*	6,700 - 10,500	13,700 - 21,500
6 Mechanical Draft Wet Towers*	4,600 - 13,200	9,500 - 27,000
7 Dry Towers	1,180	2,420

Cooling System	(b) Annual Consumption (acre-feet/year) Coal	Nuclear
1 2 Acre/MW Cooling Pond*	8,300 - 17,400	16,900 - 35,600
2 1 Acre/MW Cooling Pond*	6,800 - 14,000	14,000 - 28,600
3 Spray Pond*	5,400 - 10,500	11,100 - 21,500
4 Natural Draft Wet Towers*	4,700 - 8,000	9,600 - 16,300
5 Mechanical Draft Wet Towers	3,500 - 10,100	7,200 - 21,000
6 Natural Lake or River (once-through)	3,500 - 6,200	7,200 - 12,700
7 Dry Towers	1,180	2,420

* Evaporative cooling technologies.
Calculations based upon 75% capacity factors for both fuel types.
Thermal efficiencies: 38% coal, 33% nuclear.
In-plant heat losses assumed: 3% coal, 5% nuclear.
Cooling range assumed: 20°F.
For coal, 17% of waste heat dissipated directly to the atmosphere in the form of hot stack gases.
For coal, water requirements for sulfur and fly ash removal would add approximately 1,500 to 3,000 AFY.
Source: Adapted from (5).

from once-through cooling towards evaporative cooling technologies. This trend is a response to two major influences. First, Environmental Protection Agency standards require that all new power plants, and some existing units, install closed-cycle cooling systems. Second, the number of acceptable power plant sites possessing large available water supplies is diminishing. The important message contained in Table 1 is that, although the current trend will result in withdrawal and thermal discharge reductions, it will also increase the amount of water that is consumed for power plant cooling. The only apparent way to overcome this consequence would be widespread use of dry cooling technologies. Unfortunately, these technologies are not currently available for large-scale plants and are expected to remain too expensive to be used for the the time being, even for smaller facilities, except in a few cases where the additional capital and operating costs can be justified.

COOLING TECHNOLOGIES

Figures 3 and 4 show schematic representations of some of the alternative cooling systems listed in Table 1. Following are brief descriptions of how these systems operate.

Once-through (Figure 3a)

In once-through cooling systems water withdrawn from the source (any large body of water such as an ocean, natural lake, or river) is conducted through the main condenser. Water from the condenser is then returned to the source at a higher temperature, typically 20°F. warmer, than the intake temperature. A storage reservoir might be needed if the water source is a lake or river because water flow can be constrained by seasonal or drought conditions.

Lakes or ponds (Figure 3b)

Specially constructed lakes or ponds can supply water to the condenser in a manner similar to once-through cooling. The difference is that, in once-through systems, the major portion of the heat is dissipated through an increase in the temperature of the body of water receiving hot water discharged from the plant, while in cooling lakes or ponds the heat is dissipated to the atmosphere through evaporation. In cooling ponds, evaporative losses are large enough to require replenishment. The source of this makeup water might be an ocean, lake, river, groundwater aquifer, or, if available in sufficient quantities, treated sewage water. In practice, oceans and saline ground waters are usually considered undesirable because of salt concentration in the system and the adverse environmental effects from salts deposited downwind from the cooling pond.

Spray ponds are sometimes used if there is insufficient land area to construct a cooling lake of adequate size. In this system, heated water is sprayed into the atmosphere and is cooled by evaporation as it falls back into the pond.

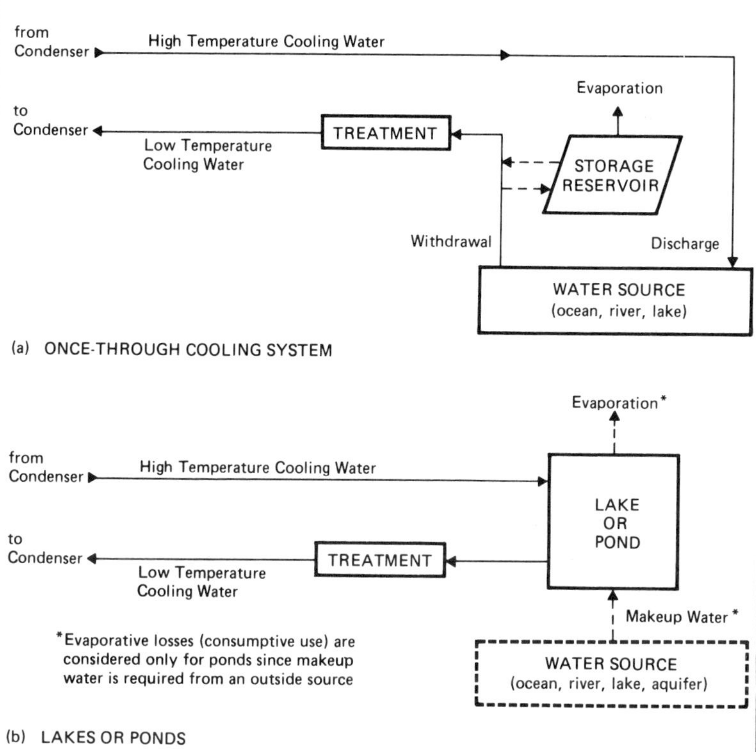

Figure 3. ONCE-THROUGH, LAKE AND POND COOLING SYSTEMS

Wet cooling towers (Figure 4a)

In these systems, heated water from the condenser cascades down inside the tower and comes into contact with unsaturated air flowing through the tower. Heat is removed primarily by the process of evaporation. Air is either conducted through the tower by natural convection, or is drawn through mechanically by large fans. In addition to evaporative and blowdown losses in wet cooling towers, there are drift losses due to small droplets of water carried away by the flow of air. Makeup water sources are the same as for cooling ponds.

Dry cooling towers (Figure 4b)

Dry cooling towers operate similarly to wet towers in that water can be circulated between the tower and the condenser, the only difference being that heat is dissipated into the atmosphere through the fins of tower cooling coils instead of by evaporation. Alternatively, exhaust steam can be piped directly to the dry cooling tower for condensation. Dry towers that circulate water to a main condenser are of the natural draft of mechanical draft types; mechanical draft designs are required if steam is piped directly to the tower. Small quantities of makeup water are still needed for blowdown replenishment in dry cooling systems.

Wet/dry cooling systems (Figure 4c)

Combinations of wet and dry towers can be described where water supplies are limited in order to avoid the higher cost of an all-dry system. In such a system the dry towers function throughout the year with the greatest operating efficiency occurring during the winter months. In warmer weather, during peak periods, the wet towers also function. Although Figure 4c shows a wet/dry system that has separate water circulation circuits, it also is possible to construct a wet/dry system in which both wet and dry towers are on the same circuit.

The ultimate decision regarding which cooling technology should be used at a specific plant depends on the interaction of a number of factors. These include cost, power plant characteristics, climatological conditions, how much water is available to be withdrawn for cooling purposes, and what level of consumptive use can be tolerated. The last two factors can be heavily influenced by environmental and social considerations.

ENVIRONMENTAL AND SOCIAL ISSUES

A number of adverse environmental impacts can result from the use of water for power plant cooling. Several of these impacts might be serious enough to warrant strict constraints on water use in particular situations. Furthermore, the use of water for power plant cooling might conflict with other competing uses.

Adverse environmental impacts from the use of surface water for cooling may result from either temporary or permanent removal of water from the source, reduction of water quality, heating of the water, or the ultimate disposition of water permanently removed from its source. As water is removed from a source, aquatic biota may be killed in the intake structure or during passage through the cooling system. The reduction in water level or flow that results from consumptive use can concentrate suspended or dissolved materials and other contaminants. Water level or flow may be reduced to such a degree that requirements for vegetation and wildlife cannot be met. Thermal discharges to a body of water might upset the existing ecological balance or exacerbate the effects of discharges from other sources. Finally, water from evaporative cooling systems evaporated or expelled as small droplets might affect local weather conditions by causing fog or icing to occur, or by salts deposited downwind from the power plant site.

Regarding ground water, a variety of problems might be encountered. When water is withdrawn from an aquifer where little or no recharge occurs, it is possible to deplete the stored supply or cause consoli-

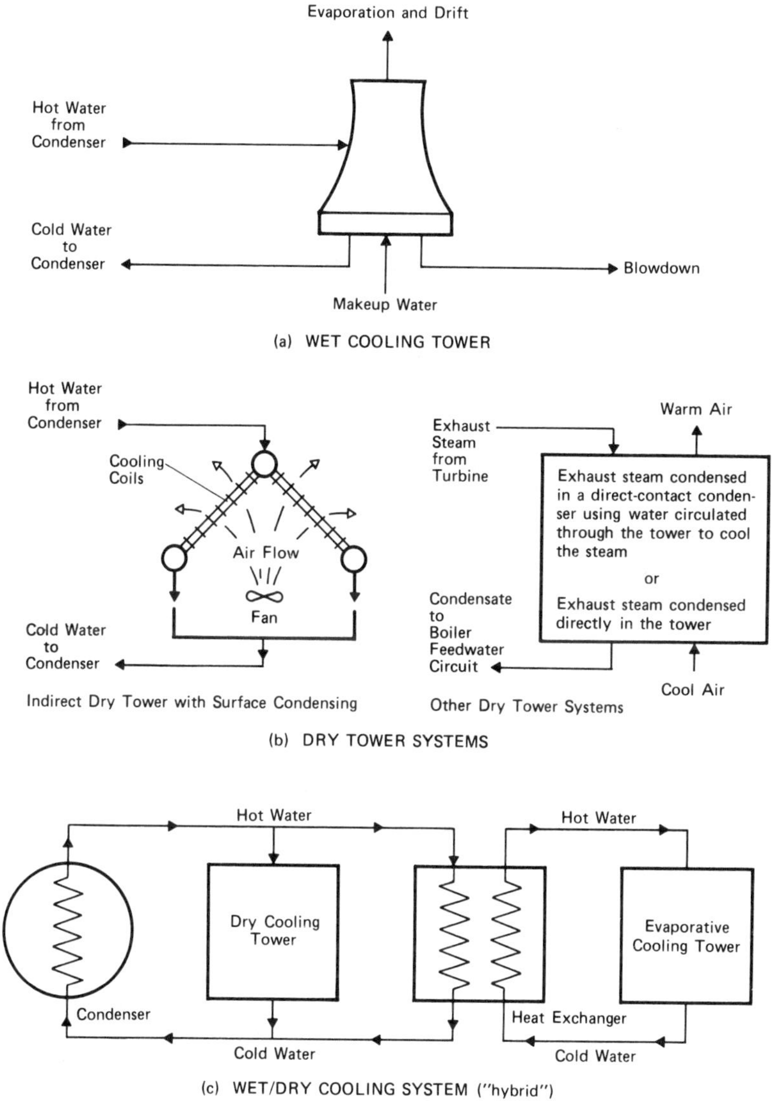

Figure 4. COOLING TOWERS

dation of the ground so that recharge is not possible. In loosely consolidated aquifers, subsidence is a potential problem when large amounts of water are withdrawn. In the case of stream-connected aquifers, it is possible to interfere with ground water flows which supply streams and springs. Fish and wildlife that depend on these streams and springs may be affected. Coastal ground water supplies may be affected by saline intrusion. No matter what source of water is used, if blowdown from a cooling system is disposed of in an evaporation pond, seepage from the pond can contaminate ground water supplies.

In 1975, the amount of water withdrawn for power plant cooling purposes was 22% of total withdrawals in the United States, second only to agricultural withdrawals which accounted for 40%. Other major withdrawals were for manufacturing (13%), and for municipal and domestic supplies (7%). In the same year, consumptive use for power plant cooling accounted for slightly more than 1% of the water consumed throughout the country. This was only the fifth largest consumptive use following agricultural (83%), municipal and domestic (7%), manufacturing (6%), and mining (2%) (6).

Strong competition for availabile water supplies will occur in many areas if the trend toward increased consumptive use in power plant cooling continues. Requirements for other energy development activities such as extraction, synthetic fuel production, and coal slurry pipelines will also contribute to the potential for conflicts over the use of available water supplies. One characteristic of this situation is that electric utilities will usually be able to easily outbid other competing users, primarily because it is frequently more cost-effective for a utility to pay high prices for water than it is to pay large capital and operating costs for water-conserving cooling technologies.

In some cases it might be possible to develop cooperative water use arrangements. In situations where evaporative cooling is required to avoid thermal discharges to a waterway, it might be feasible to divert heated water to aquaculture ponds or to greenhouses for crop cultivation. In other cases it might be possible to design cooling lakes or ponds so that they can also serve recreational purposes. In addition it is possible to use treated sewage water as the supply source (as planned for the Arizona Nuclear Unit near Phoenix).

Conflicts between electric utilities and other competing water users could become serious enough to require government intervention. This will be especially true where the diversion of water for power plant cooling purposes results in substantial losses of jobs or income. Additional legislation might be needed in some areas to supplement existing statutes and organizational structures, and specifically to refine the definition of water rights and the determination of allocation policy. Provisions for the conjunctive use and management of surface and ground water supplies may be particularly important as this could lead to more efficient water use and effectively increase available supplies (7).

PROJECTED ELECTRICAL DEMAND AND COOLING WATER REQUIREMENTS

The magnitude of environmental and social problems encountered in the provision of water for power plant cooling will depend on actual water requirements. This will be determined by factors such as electrical generating capacity requirements, in what areas the new capacity will be required, and what types of cooling technologies will be employed at these new facilities.

As natural gas and oil prices increase and supplies diminish, electrical energy will comprise an increasing share of United States energy consumption. Electrical energy made up 21% of the energy consumed in the United States in 1975 (6); depending on electrification of transportation this is projected to increase to somewhere between 43% and 63% by the year 2000 (9). Historically, electrical load growth for the nation has been 7% per year for the period 1955 to 1973. It is likely that for the next ten to twenty years the growth rate will be lower, perhaps 3 to 5% per year (8); some estimates, however, reach 7.7% per year (10).

Table 2 lists water requirement projections for steam-electric power generation as reported in The Second National Water Assessment by the Water Resources Council (WRC). These projections are based on growth rates for electric power production of 7.5% per year between 1975 and 1985, and 5.1% per year from 1985 to 2000. The estimates in Table 2 are likely to be somewhat high for two reasons. First, the Water Resources Council has assumed load growth rates near the top of the range of projections. Second, they have assumed that nuclear generation, which uses more water than does fossil generation, will account for a very high proportion of electric production (37% in 1985 and 65% in 2000).

It is useful to examine regional demands for the generation of electricity. In 1975, four of the twenty-one WRC regions generated 51% of the nation's electric power and are expected to continue to generate one half of the nation's power through the year 2000. These are the Ohio, South Atlantic-Gulf, Mid-Atlantic, and Great Lakes regions. Eight of the twenty-one regions are expected to increase their generation by 600% or more by the year 2000 (an average annual growth rate of 7.4% or more). These include the Upper and Lower Mississippi, Missouri, Arkansas-White Red, Texas Gulf, Great Basin, Alaska, and Carribean regions (Figure 5).

In a separate study by the Electric Power Research Institute, projections of water availability and demand were coupled with projections of requirements for new electric generating capacity on a regional basis to identify areas where potential water shortages might occur (11). The results of this study are shown in Figure 6.

THE ROLE OF IMPROVED HYDROLOGIC FORECASTING IN THE PLANNING AND DESIGN OF STEAM-ELECTRIC GENERATING FACILITIES

The preceding sections have described how water is used for power plant cooling and discussed some expected problems regarding provision

TABLE 2. WITHDRAWALS AND CONSUMPTION FOR STEAM-ELECTRIC GENERATION
Average Year (millions of gallons per day)

(a) Freshwater

Region	Withdrawals			Consumption		
	1975	1985	2000	1975	1985	2000
New England	1,263	1,069	375	21	18	167
Mid-Atlantic	7,463	7,130	4,657	103	224	644
South Atlantic-Gulf	12,768	12,912	13,952	153	722	1,857
Great Lakes	24,362	22,689	16,061	175	497	1,384
Ohio	21,022	21,008	10,574	324	656	1,692
Tennessee	4,799	5,738	4,581	42	231	417
Upper Mississippi	7,644	6,347	3,537	129	352	1,079
Lower Mississippi	4,175	9,313	16,687	54	118	291
Souris-Red-Rainy	82	23	0	1	0	0
Missouri	3,540	5,834	4,938	68	289	637
Arkansas-White-Red	498	1,026	1,012	89	237	457
Texas-Gulf	724	1,000	2,262	99	270	994
Rio Grande	34	16	10	18	9	5
Upper Colorado	101	157	201	38	106	151
Lower Colorado	68	150	154	63	134	126
Great Basins	33	65	82	3	42	52
Pacific Northwest	260	203	580	13	104	344
California	42	158	367	25	101	242
Alaska	26	17	22	0	2	5
Hawaii	0	0	0	0	0	0
Caribbean	0	0	0	0	0	0
Total: Regions 1-18	88,878	94,838	80,030	1,418	4,060	10,539
Total: Regions 1-21	89,904	94,855	80,052	1,418	4,062	10,544

(b) Saline Water

Region*	Withdrawals			Consumption		
	1975	1985	2000	1975	1985	2000
New England	4,831	14,604	10,161	47	108	293
Mid-Atlantic	16,997	28,634	26,208	96	305	813
South Atlantic-Gulf	7,009	14,976	20,239	62	233	904
Lower Mississippi	320	1,590	10,181	2	10	63
Texas-Gulf	4,139	4,950	10,371	37	45	199
Pacific Northwest	0	7	10,652	0	0	65
California	10,157	22,090	24,658	55	125	155
Alaska	67	44	420	0	0	2
Hawaii	1,085	2,005	3,637	3	8	20
Caribbean	1,124	2,337	2,288	9	35	149
Total: Regions 1-18	43,453	86,851	112,470	299	826	2,494
Total: Regions 1-21	45,729	91,237	118,815	311	869	2,663

* Eleven regions have no saline water withdrawals or consumption.
One million gallons per day = 1,120 acre-feet/year.
Source: Water Resources Council (6).

Figure 5. WATER RESOURCE COUNCIL REGIONS

of adequate water supplies to the electric power industry. The consequences of inadequate water supplies for power plant cooling systems include significantly increased costs for electricity and, if water shortages are severe enough to limit the growth of generating capacity, an increased brownout (voltage reduction) and blackout risk during peak demand periods.

Improved hydrologic data are needed in the planning and design of new steam-electric generating facilities for several reasons. Accurate regional and local water availability data are needed for use in the screening of geographic areas to identify candidate power plant sites. Characteristics of the available water supply strongly influence the choice and design of the preferred cooling technology. Uncertainties regarding water availability must be compensated for in the design of storage reservoirs and well fields.

In site selection it is necessary to choose a location where the available water supply is not only large enough to meet cooling system requirements, but also sufficiently reliable to minimize the risk of

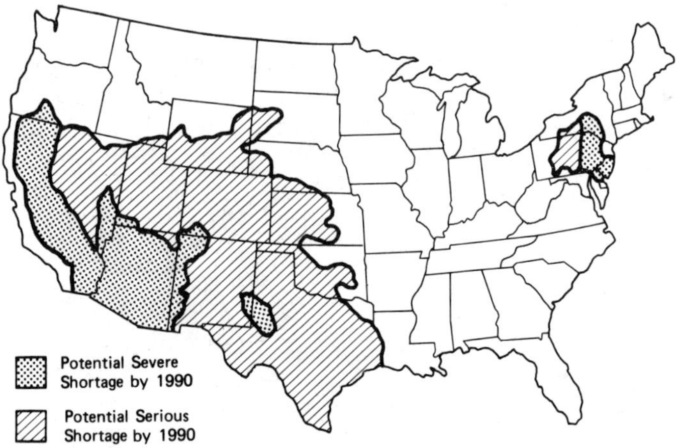

Figure 6. AREAS OF POTENTIAL WATER SHORTAGES

costly plant shutdowns due to lack of water. For nuclear power plants a reliable source of water is especially critical due to the need to maintain cooling of the reactor core in addition to steam condenser cooling. Improved prediction methods for low-flow characteristics of rivers, properties of ground water aquifers, and water quality conditions will be particularly useful in power plant siting studies.

Future assessments of the characteristics of water supplies for power plant cooling will require improved accuracy. As cooling systems with lower withdrawal rates are contemplated, smaller increments of water will become more important and the value of accurate information will increase. Better hydrologic information can result in savings through the reduction or elimination of the need for the very large reserve capacities often required in the design of storage reservoirs or well fields to compensate for uncertainties in the hydrologic data used in design.

As previously noted, the competition for available water supplies is likely to be intense in some areas. The availability of more accurate water supply characteristic assessments will also be helpful in the resolution of conflicts concerning allocation policies and water rights by making it easier to evaluate the adequacy of water supplies for competing uses.

The diligence of the hydrologic community in continuing to improve the state of its art will pay off as the need for electrical power in the United States continues to increase and the value of hydrologic data in the planning of generating facilities becomes more critical.

NOTES AND REFERENCES

1. In 1977, coal and nuclear facilities produced 46.4 and 11.8% of the electricity generated in the United States (2) By 1985 these percentages are projected to grow to 52 and 19% (3).

2. Energy Information Administration. 1978. _Quarterly Report_: Third Quarter 1978, _Energy Information Report to Congress_. Department of Energy, October.

3. Energy Information Administration. 1977. _Annual Report to_ Congress, Volume II 1977. Department of Energy.

4. Loftness, Robert L., D.Sc. 1978. _Energy Handbook_. Van Nostrand Reinhold Company.

5. Woodward-Clyde Consultants. 1976. _Summary of Predesignated Siting Study: Environmental Impact of the HTGR and LWR_, prepared for General Atomic Company. August.

6. Water Resources Council. 1978. _The Nation's Water Resources: The Second National Water Assessment_. U.S. Water Resources Council. April.

7. Comptroller General of the United States. 1977. _Groundwater: An Overview, Report to Congress_. Government Accounting Office, June.

8. Tucson Myers and Associates, Inc. 1977. _Water for Western Energy Development, 1977 Update_. Western States Water Council. September.

9. Electric Power Research Institute (EPRI) quoted in (8).

10. Federal Power Commission forecast cited in Winter, John V., and Connor, David A. 1978. _Power Plant Siting_. Van Nostrand Reinhold Company.

11. Water Quality Control and Heat Rejection Program. 1978. _Advanced Dry Cooling Tower Demonstration_. Electric Power Research Institute. August.

SESSION I - ANALYTIC TECHNIQUES IN HYDROLOGY - STATE OF THE ART

John J. Cassidy[1]

(Recap on Friday Morning). This session and the discussion following it raised the following general conclusions. No attempt has been made to summarize the individual papers since they are included herein. Instead particular emphasis has been placed on questions and discussion generated by the individual papers.

In flood hydrograph and peak-flow forecasting techniques, the most activity appears to be in the use of deterministic simulation programs such as HEC-1, TR-20, MITCAT, the USGS model, and SWMM for the simulation of single event floods. Efforts are being made by the U.S. Army Hydrologic Engineering Center (HEC), with NASA help, to develop techniques for obtaining land-use, soil-cover, field-moisture and soil-type information through remote sensing particularly through analysis of digital information obtained from NASA satellite tapes. This information is being used to start, or to attempt to start, the deterministic simulation models in an effort to simulate extreme events. According to Dr. Feldman, HEC would use the HEC-1 model for this purpose and are attempting to develop a feed-back process for use with HEC-1 by which corrections can be made to assumed soil parameters for improvement in calibration.

Little or no progress has apparently been made in relating rainfall recurrence-interval to flood recurrence-interval; nevertheless, the use of 100-year rainfall to generate a 100-year flood is common, particularly among agencies and firms facing the task of developing flood hydrographs where streamflow records are scarce or nonexistent. Simulation has been performed where an antecedent moisture condition is generated randomly giving rise to probabilistic, or pseudo-probabilistic, and not deterministic result.

Research contracts have been let by HEC to study the various analytical models for single-event generation and their relative value and effectiveness in flood simulation. One of the biggest needs in single-event analysis is a consistency of methodology for development of the PMF. As the state-of-the-art now stands great differences can arise in generated PMF hydrographs depending on the methodology used.

[1]Director, Washington Water Research Center, Washington State University, Pullman, Washington 99163.

Synthetic streamflow traces are generally generated primarily in two ways: Indirect from precipitation through use of a simulation model, and direct using a stochastic model to generate new streamflow values. There is some disagreement in methodology for generating stochastic traces but in general the Lag-1 Markov Model preserves "short-term" characteristics but substantial debate centers over just how "short" is "short-term." On the other hand the Fractional Gaussian Noise and Broken Line Models preserve long-term characteristics of variance and skew. The Fractional Noise model appears to be the most difficult to use from a computational standpoint but it does preserve both long- and short-term characteristics, apparently reasonably well.

The continuous simulation model (Hydrocomp, MITCAT, SACRAMENTO, etc.) seems to have developed the most usage in the generation of streamflow traces, probably because from a conceptual standpoint it is easiest to understand and has an apparent physical basis. Rainfall is sometimes generated, or at least filled in stochastically, and then the simulation model is applied to generate streamflow. There is general concern about disaggregating rainfall which has been generated stochastically, a process required in at least some simulation models, and it appears that streamflow could just as well be generated stochastically in a direct fashion.

Bayesion techniques have not been used in generating streamflow sequences, at least within the knowledge of the participants.

Records extended stochastically should not be used in a frequency analysis although this has apparently been done on occasion, particularly in design situations where data was seriously short.

The prediction of annual runoff in states such as California where a large percent is due to snowmelt is seriously limited by inaccuracies inherent in long-range weather forecasts. Snowmelt methodology in long-term forecasting is of much less importance early in the year than late in the year. As the water year progresses, the correctness of the weather forecast becomes relatively less important while accuracy in snowmelt methodology increases in importance. Long-range weather predictions remain inaccurate with inaccuracy increasing with range.

The science and art of simulating storm surges is well advanced. The NWS SPLASH and SLOSH Models give very good agreement with old storms and are widely used in predicting surges which may arise during actual storms. Data from old storms used for calibration is, however, very skimpy and may be unreliable. The National Weather Service Programs are easy to use and depend very strongly on the validity of storm input such as gradient of wind and wind speed. The programs are used readily during the hurricane season. However, the programs have not generally been applied to predict maximum surge heights arising from Probable Maximum Hurricanes, mainly because of lack of clear understanding and agreement relative to the independent parameters associated with such storms.

Considerable concern was voiced by several participants about the development of a probable maximum flood (PMF) and its general use as a design event. The National Dam Safety Act provides some guidance about the choice of a design event but the rationale is mostly subjective. Whether or not a project is designed for a PMF should be strongly governed by the individual project. In some cases the design can be accommodated readily without great expense. In others an adequate spillway may be very expensive or even impossible to develop. There is great need for consistency in PMF methodology and for a rational method of choosing a design flood.

SESSION II - ECONOMIC ASPECTS OF HYDROLOGIC FORECASTING

L. Douglas James*

In addressing the conference theme of "Improved Hydrologic Forecasting - Why and How," the second session pointed out how, in the language of the economists, the why of forecasting is the demand users have for forecast information and the how is the supply of forecasts being provided. In economic markets, a supplier whose product differs from the needs of users, loses sales and eventually goes out of business to be replaced by other suppliers who have a better product. In forecasting, the financial incentivies to do better are much weaker and are as likely as not ignored by forecasters who wonder why the public fails to heed their sage advice, just as the public wonders why government inundates them with technical gibberish.

The need is for the conscious effort to bring forecasting tools and their products into better agreement with user preceptions of ways to use better information for self improvement. This session looked at the problem from four viewpoints and came out of that look with what can be called the Benjamin-Martin-DeGraff-James laws governing forecasting effectiveness. Since the mnemonic BMDJ is much better as a tongue twister than as a memory aid, those of us who remain here will gladly blame it all on our chairman who organized and then left us. Our summary is contained in Knapp's four laws of forecasting:

1. The Benjamin Law: <u>Don't throw fancy statistical techniques at others before you understand them yourself because the bulky package is likely to land on your toe.</u> The world of statistics and probability contains many attractive toys to those who like to play numerical games, but few of us are impressed by someone else's toys. Statistics and probability theory provide forecasting practitioners some very powerful tools; but, when misapplied, these tools can produce some very misleading results. You may have difficulty following the mathematics of many of the techniques but should not worry about that because experts are around who can help you. The important point is to search out techniques that can provide more information from your data and distinguish them from techniques that can transform your data into nonsense.

2. The Martin Law: <u>Be brave enough to try new methods because it is easier to make a turn when you are walking the wrong path than when you are stuck in the mud.</u> Probably every one of us here has had the experience of seeing a new tool explained, becoming excited with its potential, going home and trying it, and getting exactly nowhere because

*L. Douglas James, Director, Utah Water Research Laboratory, Utah State University, Logan, Utah 84322

nothing worked out as it was described. We saw an application of operational hydrology to yield analyses from Berryessa and Shasta Lakes and may have seen aspects of the results that were unexpected theoretically. That, however, it not the point. What we need to remember is that all are better off as we practice what we learned and help others practice what we taught.

3. The DeGraff Law: Seek out and use unfamiliar inputs because telling others what you have done in ignorance makes them think you are ignorant. This conference spent a great deal of time on hydrologic probability but had only one paper on demographic and economic futures. These areas also have important stochastic elements, and considerable resources are available for forecasting use. Rules exist for combining joint probabilities and can be used to forecast probabilities for future flood peaks, water use, etc.

4. The James Law: Remember that others are not like you and very thankful that is so. Forecasts are too often presented to communicate with others who think much like to forecaster and delivered at the time of his convenience rather than at the time of user need. We need to look at user needs and user uses, particularly as we enter a period of emphasis on conservation and nonstructural measures. Who of us really knows how to forecast drought in the most effective way for getting our neighbors to conserve water?

Obviously, what we have in Knapp's four laws is an effort to help you remember what you heard this week. Many are going to go home, become involved in other activities, and forget. Don't forget too fast, however, because I don't ever want to see a group of this conference's participants looking ignorant with their sore toes hidden four feet deep in mud.

SESSION III - METEOROLOGY AND HYDROLOGY

Session Chairman: J. Paul Riley
Reporter: Roger A. Amisial*

Myths and Use of Hydrometeorology in Hydrologic Forecasting

by Gene Willeke

The author pointed out some issues associated with hydrologic forecasing which he identified as being myths, and underlined some potential uses of hydrometeorological methods in hydrologic forecasting. Most of the points brought out by the author involved the question of how to improve hydrologic forecasting.

Is there a physical maximum in nature? The question is related to the maximum probable precipitation and flood events and leads to the following question: Can a probability value be assigned to the physical maximum? A comparison between generalized PMP and Hersfield's frequency factor equation, indicates that the latter provides a lower estimate of PMP.

There seems to be no theoretical reason for considering hydrometeorological events as random accurrences, nor for classifying the hydrometeorological processes as stationary. Those issues are important when it is remembers that most frequency analyses used in forecasting are based on those assumptions. The author considers that the most reliable estimates of flood frequency can be obtained from precipitation events, if the analysis is based on areal precipitation computed from depth-area-duration curves.

Potential uses of hydrometeorological methods include the updating of generalized PMP maps, the estimation of maximum probable drought, the estimation of urban snow water content, and the short-term forecasting of flash flood parameters, such as direction, orientation, shape and speed of travel. The author also suggests that the use of PMP may be more justified for large basins than for small ones.

The Applicability of Present and Planned Spaceborne Observational Capabilities

by V. Ralph Algazi

The author addresses the question of whether remote sensing can be used to provide the data needed for hydrologic studies. He focused his presentation on current development in data acquisition by satellites and went on to describe future or programmed development. LANDSAT information depends on parameters such as resolution, frequency, and spectral bands. Current information being obtained with quite good accuracy from LANDSAT includes watershed physiography, drainage density,

*CDIAT-OAS, Merida, VENEZUELA

surveys of irrigated land, urban and suburban use, vegetation index, and turbidity of water.

Recently there has been much concern over the use of satellite information for specific applications such as calibration of hyrologic simulation models, and the estimation of soil moisture. Current efforts on improvements of satellite information include the development of micro-wave sensing to penetrate cloud cover.

The Benefits of Uncertainty: Probability in Short and Long Term Hydrologic Forecasting

by Rafael L. Bras

The author claims that the acceptance of uncertainty in hydrologic processes opens the door to the application of much probability theory which could not be used otherwise. Special emphasis is made on long-term forecasting, even in the case when data are not available, and on short-term forecasting when the prediction for the next period can be improved based on observation of the value of the antecedent period. This paper addresses the problem of how probability theory and optimal estimation theory can be used to improve long and short-term forecasting.

Long-term forecasting can be based on independent random variables such as rainfall intensity or effective rainfall, i_e, and rainfall duration, t_r, the joint distribution of which has been derived by Eaggleson. Dependent variables such as flow volume, or infiltration can be expressed as functions of i_e or t_r and their distributions derived. In particular, the kinematic wave equation and Phyllip's equation can be used to relate the independent variables with the flow and the infiltration volumes, respectively. Comparison of the results with studies realized at Colorado State University indicated that the method is reliable.

Short-term forecasting can be improved by coupling a conceptual model with the Kalman-Bucy filter. This enables an improved forecasting of the flow during the next period based on the flows predicted and observed during the immediate antecedent period.

Hydroclimatology - A Needed Cross-Discipline

by Ronald F. Kilmartin

The author proposed the study of the hydrologic processes on a global manner involving both the consideration of reconstructed data and the interpretation of the processes from a synoptic point of view. Long-term hydrometeorological data can be reconstructed based on long-term fluctuations in climate, and also based on synthetic data generation techniques which preserve various characteristics of the samples such as mean, variance, shewness, and Joseph effect.

On the other hand, the consideration of a synoptic climate may permit the establishment of a correlation between events occurring in different geographic areas. For example, it has been observed that a dry spell in a certain part of the earth has always been accompanied by wet spells in other geographic areas of the world.

Large and Small Scale Quantitative Precipitation Forecasting at the National Meteorological Center

by David Olson

The organization and activities of the National Meteorologic Center are briefly described. Basic forecast procedure include the reception of the data, their comparison with earlier forecasts, the use of forecast models, and an evaluation of the results. Emphasis was put on the scale of the storm which involves its areal extent and depth, moisture content, and velocity of upward motion. Those scale factors, together with meteorological feasutres, such as jet streams, convergence level, vorticity and center level, are very important when considering the accuracy of the forecast, and are generally responsible for low accuracies sometimes encountered in storm forecasting.

SESSION IV - NON-METEOROLOGIC DATA

W. R. Boggess*

The session was organized to present recent advancements in the development and use of paleoclimatic indicators for interpreting and extending hydrologic trends backward in time. Three of the four most commonly used methods were discussed. These included:

1. Pollen deposits in lakes, swamps and bogs; also algal spores and cysts;
2. Stratigraphic and geomorphic charactersitics of streams; and
3. Analysis of tree-ring series.

In the discussion of pollen analysis, the methods of collection, preparation, and pollen identification were emphasized. These points were illustrated by current work on sediment samples from a pond in northern California. Pollen profiles were developed for major plant species and/or species groups. These profiles, which indicate both the occurrence and relative abundance of plant species, can be used as indication of past environmental events and conditions. Limitations and disadvantages are:

1. Usually single samples because of amount of time and effort required for collecting, preparation and identification.
2. Lack of absolute dating control.
3. Essentially limited to low frequency events -- such as broad-scale climatic change.

Information that can be obtained by stratigraphic and geomorphic characteristics of stream channels and stream networks are largely related to the magnitude and frequency of hydrologic events -- e.g. flooding.

1. Except in arid regions the geometric properties of alluvial channels are highly correlated with floods that have a recurrence interval of 1-2 years. Channel bankfull capacity also is an indicator of high frequency floods.
2. Floodplain stratigraphy and sediments presents evidence of low frequency flood events. Such indicators are --
   a. Coarse gravel overlying fine sediments that represent a discontinuity in the normal "fining-upward" sedimentary sequence.
   b. Formation of point-bar sedimentation.
   c. These sediment characteristics can be combined with hydraulic "energy" equations to provide estimates of flood magnitude.

Dating requires the use of radiocarbon methods. Such dating of post-glacial alluvium in upper Mississippi River Valley indicate that flooding, erosion and sedimentation have occurred in an eposodic (non-random) fashion.

*W. R. Boggess, The University of Arizona, Laboratory of Tree-Ring Research, Tucson, Arizona 85721

Ring widths, especially from some of the long-lived tree species in the western USA provide a much more precise data base for reconstructing certain past hydrologic events than either of the other methods. However tree-ring chronologies generally do not extend as far back in the past. Except for the 7,000+ year Bristlecone Pine chronology, most tree-ring records are less than 400-500 years in length.

Examples discussed include:

1. Reconstruction of annual flows for the Upper Colorado River Basin -- especially at Lee Ferry, the point of the Colorado River Compact.
2. Reconstruction of large-scale drought occurrence (frequency, duration and severity) in the western USA.
3. Reconstruction of regional drought for Southern California.

Advantages of tree-ring chronologies include:

1. Precise dating.
2. Preservation of high frequency variations and low frequency variations -- a high frequency variation can be filtered out.
3. Adaptability of data to statistical analysis.

Reconstructions of Colorado River flows for the past 400 years aptly demonstrate the usefulness of such techniques for hydrologic forecasting. In this case the period of record upon which the Lee's Ferry Compact allocated water to the Upper and Lower Basins was shown to be the wettest period during the 400 years covered by the tree-ring reconstructions. These projections have been confirmed by flow records since the time the Compact was fromulated.

SESSION V - POWER FORECASTING

By C. B. Cecilio*

A total of five papers were presented. Three papers addressed the problem of power forecasting during the drought; one addressed forecasting for water supply and one paper addressed power forecasting utilizing a conceptual model.

The papers discussed a variety of problems encountered in preparing forecasts. They ranged from the inability of older regression equations to provide positive answers to computer models unable to work under the extreme event of the drought. One major complaint was that there was no other available data which could be used as a basis to forecast a drought. The 1976-1977 drought years were the worst in history and therefore left several forecasters with no historical basis of assumptions to compare future forecasts.

There was a general consensus among several people in the audience that forecast procedures could actually have been improved if a simulation model was used. The old-fashioned regression equations could still be used but the present state-of-the-art in system modeling could produce a much faster and much more accurate prediction of long and short term forecasts. A model minimize gross assumptions such as a median rainfall. All readily available information can easily be accessed and manipulated. Better rule curves of operation to conserve water can be developed and perhaps take advantage of the available storage in the reservoirs. In general better utilization of water available can be obtained.

Of the four papers on flow forecasting only one marketing agency took advantage of a simulation model. The rest used multiple regression equations. Although the remaining three California agencies admitted having a simulation model, none were utilized during the drought as a management tool. It was pointed out that one large utility company could have optimized their complex watershed system with a watershed simulation model as a management decision tool.

Decision making manager's tools of watershed simulation model are already available in varying complexity. There are the stochastic models for planning and the deterministic models for operation. It is just a matter of convincing managers to adopt these modern tools in their decision making. The oldest model has been in existence for 18 years.

*C. B. Cecilio, Pacific Gas and Electric Company, San Francisco, CA

The power forecasting session produced an "ad hoc" session in the afternoon. The session described the potential benefits that could be derived in watershed simulation models.

SESSION VI - CHANGING NEEDS IN HYDROLOGIC FORECASTING

W. P. Henry*

The session on changing needs in hydrologic forecasting consisted of three papers:

- Changing Needs in Urban Flood Forecasting, by L. Scott Tucker,
- Contamination in The Hydrologic Data Base, by Larry Ferral, and
- Let's Look at The Legal Implications of Stream Flow Forecasting, by Thomas Stetson.

All three papers commented on WHY we need improved forecasting and offered suggestions on HOW to improve hydrologic forecasting. Let's look first at why we need improved hydrologic forecasting.

The speakers gave three solid reasons why improved forecasting techniques are needed. In urban areas subject to flooding, the time for either evacuation of residents or placement of flood mitigation measures is short. In many cases, there is less than an hour to react effectively to a forecast of imminent flood danger. With such a short time available for action, increased accuracy of forecasts using existing techniques or better forecasting techniques to identify problems earlier are needed.

Hand-in-hand with the short time available in urban flood forecasting is the requirement for accuracy in flood forecasts.

Dealing with the public, a flood forecasting agency must preserve its credibility. Too many flood warnings issued will cause the public to ignore future warnings. If too few flood warnings are issued, the flood forecasting system has failed in its mission. Thus, in this particular role, the need for accuracy in hydrologic forecasting is high.

In addition to the shorter time and higher accuracy requirements for hydrologic forecasts, the legal implications of inaccurate forecasts were discussed. These legal implications can be time consuming to resolve, expensive, and downright unpleasant. Scott Tucker discussed some of the legal liabilities his District would have in failure to warn

*W.P. Henry, Manager of Industrial Projects, CH2M HILL, San Francisco, CA.

residents of impending floods. Tom Stetson described 22 years spent on one legal problem, the majority of which is attributable to an initial poor hydrologic forecast.

This session marked the first time the word "lawyer" appeared in the conference. It came up again during the sessions on flood plain management and environmental concerns. The legal arena is one in which we hydrologists, like most other professionals including geotechnical engineers, will find ourselves more and more. For that reason, our hydrologic forecasts must be improved to make sure that our work is systematic, thorough and accurate.

Let's move now to the question of how to improve hydrologic forecasting and break that subject into two parts:

- What technique can we use to improve our current forecasting efforts?
- What types of forecasts do we need over and above those currently produced?

On the first point, Larry Ferral described problems with using the published data base. He showed in one area that a 5-percent variation in precipitation gave a 15 to 25 percent variation in flows computed. During the discussion period Dolores Taylor described a rain gage with a slow clock which has been thoroughly investigated because the time of stoppage of rainfall is a key point in a murder trial. She commented on the tremendous amounts of work that was done to take the data and come up with as accurate a record as possible. We wondered whether that data would have been as thoroughly evaluated if it were to be used only in an hydrologic analysis, and not in a murder investigation.

Larry Ferral suggested that our data can be evaluated through the use of simulation models that have been verified for an area. All agreed that the data required for use in verifying models was in short supply.

Two types of forecasts were suggested as being required in the near future. Scott Tucker suggested that real-time forecasting techniques are needed in the urban flooding problem, and that these techniques need considerable improvement before they can be reliably applied. Tom Stetson felt that complex operational forecasting techniques would be applicable in the Colorado River System and in other complex water resource systems.

SESSION VII - FLOODING AND FLOOD PLAIN MANAGEMENT

David Dawdy *

How would improved hydrologic forecasting improve flood plain management? I don't think that it necessarily will. Certainly none of the panelists gave improved forecasting a high priority as a need.

George Phippen started out by describing the dynamic nature of flood plain management. Changing policies represent changing government attitudes. Public attitudes and expectations are changing also. This dynamic nature implies that the goals and, thus, the methodology to achieve those goals are changing--not rapidly, but constantly. That fact alone means that improved hydrologic forecasting probably never will hold the attention of the program managers.

Scott Tucker struck a common note by stressing that FIA's net impact is positive, even though its policies create problems. Those troublesome policies were touched on by others also: prediction of flooding based on present conditions only, flood mapping based on political boundaries, inadequate mapping detail for local needs, etc. His most telling point, however, was the overall lack of a Federal commitment to urban areas. In particular, Federal commitment to hydrologic data collection and analysis in urban areas is almost totally lacking. The feds create a demand for information but do not supply it to meet that demand.

Bill Johnson presented several character sketches of those involved in hydrologic forecasting and flood plain management. An important message was that the only happy character was "ignorance." If you want to be happy, don't know any more than necessary. The character who gets the job done carries a sword--to cut red tape or in case someone talked back?

Earl Jones made a case for flexibility for site specific characteristics in order to achieve uniformity of treatment in flood plain management. Uniformity of regulations may not result in uniformity of treatment or impact. Further, the appeals process places a burden on the community or the individual. If there are valid technical areas for challenging FIA findings, the community or individual must identify them and bear any cost involved.

Brian Mrazik closed by discussing where FIA is today and where he sees it going. There still are outstanding problem areas for determination of flooding potential. These include coastal area flooding, alluvial fan flooding, mudflows, influence of structures (levees in particular), storage and velocity aspects of floodways, headwater

*Dames & Moore, Washington, D.C.

flooding and drainage (the federal-local cost allocation included), the effects of predicted future urbanization (insurance needs may differ from regulation needs), and the overall social, environmental, and economic impact of the National Flood Insurance Program. The Program is supposedly a response to users' needs. Who are the users, what do they want, and how can FIA give it to them? Thus, FIA is still playing a game of catch-up. This partly is a measure of the size of the problem, and partly a result of the dynamic nature of the problem, as mentioned by George Phippen.

Brian divided FIA standards into two categories--primary and secondary. Examples of primary standards are the choice of the 0.01 flood and the permissible 1 foot rise in flood elevation as a result of encroachment on the flood plain. These probably will not change. They are technically defensible and are almost impossible to change now. Exceptions are permissible, but the onus of proof of need is on the community. An example of a secondary standard which could change is dry floodproofing (traditional FIA) against wet floodproofing (a new alternative for FIA).

In summary, where is Flood Plain Management today, as seen by the panel? Any program which impacts on a public reacts to the attitudes of that public. That creates a dynamic, ever-changing program. Scott Tucker expressed a user attitude, and outlined several points he felt needed changing. What mechanism is there for FIA and other involved Federal agencies to consider and react to such suggestions? In particular, should a Federal agency create such an intense demand for data, as has FIA, and yet have no policy for Federal collection of that data? What are the trade-offs between uniformity of regulations (easy to manage) and uniformity of treatment (the ultimate objective)? Is improved hydrologic forecasting or social forecasting the answer to an advance of the present state-of-the-art? Questions without answers--or at least easy answers? Perhaps that is, as usual, the state-of-the-art.

SESSION VIII - ENVIRONMENTAL APPLICATIONS

Session Chairman:
Arliss D. Ray
Environmental Consultant
EMANCO Inc.
Houston, Texas

Recorder:
Michael B. Sonnen
Consulting Engineer
Walnut Creek,
California

Introduction and Summary

The Thursday evening session was an appropriately culminating experience for the conference on several grounds:

1) It was characterized by adaptability to circumstances, occasioned by the misfortune of two of the three scheduled speakers not being able to attend for last-minute reasons. And adaptability to circumstances was one of the important lessons learned about forecasting during the rest of the conference.

2) Nevertheless it was focused by Arliss Ray on environmental assessment, another important application of forecasting science, (flood plain management being the one emphasized in the morning).

3) It was specifically narrowed to the very particular topic of electric power plant water needs (and hence water forecast needs) by David Snyder.

4) Each of these two presentations was heightened all the more by the discusisons that followed, and through them the purpose of the conference was made all the more clear, and Bill Johnson's admonition in the morning session was made the more poignant, which was that forecasting is only a part of the whole process of addressing human needs. In the discussions the questions were largely and significantly removed from forecasting algebra and details and were directed more fundamentally to "what is the problem?" It became clear at this point that the problem encompasses forecasting processes as part of a whole. Indeed it became clear that forecasting may be an important part, but perhaps even a minor part, of a process for defining and solving human problems.

Hydrology and Environmental Assessments -- by Arliss Ray

The modern study of water in the environment -- Hydrology -- is not an old science. It is less than three professional generations old. The study of the environment itself as it may be impacted by man's social and public or private works programs is even younger. Consider that the National Environmental Policy Act of 1969 (NEPA) is actually not yet one decade old. It follows from those short histories that concepts, techniques, procedures, and perceived requirements in these areas are not yet well established or even sensible, in the best sense of the word. (Consider Johnson's earlier image of the four elephants supporting the earth.)

Nonetheless, NEPA and the guideline for implementing it, even if they change and are changed yet again, are serious social statements by people and their representatives, attempting to assert that science will be mustered to serve the needs of human kind, which is a departure from and a significant step away from the past in which science was selectively and occasionally invoked by policy (makers) to serve policy makers' (politicians') porkbarrel needs, which may by happenstance have filtered down to serve a human need.

NEPA means people got into the act. And they are going to stay in it. And if they need a hydrologic forecast, they are going to want it to be more accurate, more defensible, more understandable, more believable than ever before. Therefore, the challenges to hydrologic forecasters are clear: they must derive better methods, better data bases, and better means of communicating results. People and private industries, and opponents and proponents are demanding that now. The solution of human problems makes those things important; NEPA demands them; and scientific research by this group and others will make them possible.

Water Use Implications of Electric Power Generation -- by David Snyder

Electric power generation is a significant water withdrawer from the environment. It ranks second behind agriculture in this regard. Electric power generation is not, however, a significant consumptive user of water on a percentage basis. This must mean that large quantities of cooling water are returned to the environment undiminshed in quantity, if altered in physical properties. Indeed that is exactly the case.

The production of electric power through steam-driven turbine generators is inextricably (apparently) connected with enormous cooling water requirements. The need for such plants is continuing to grow. The areas of greatest growth in demand are in the American southwest, precisely the areas of least available supply of water. This, then, is a large, significant, socially and economically compelling human problem. "Improved data from improved hydrologic forecasting will be essential." In several southwestern power plant siting project examples, "hydrology has run the project."

Among the factors influencing or controlling cooling technology selection from plant to plant are: 1) cost, 2) power plant characteristics, 3) climate, 4) water available for withdrawal, 5) a tolerable level of consumptive use, and 6) environmental and social considerations.

(One harkens back to earlier reports this week that quantitative precipitation forecasts beyond one day are iffy, although the National Weather Service experts are considering expanding capability to perform a 3-day forecast.)

Additional forecasts needed in the siting and cooling technology decision areas are: flow requirements to be met, water quality availability and constancy, ecologic forces and populations, thermal discharge behavior, fog or icing conditions, flooding potential, ground-water depletion, land subsidence, phreatophyte usage of available supplies, saline water intrusion, contamination of aquifers by seepage, and others. These are problems of human beings, needing solution in part by forecasters. When two or more human beings meet to argue these questions, scientific forecasts and data bases will be the weapons of resolution. Hydrologic and water quality forecasters will have to develop those still needed.

Discussions

While the two presentations were about the needs for hydrologic forecasting in the environmental area, the discussions interestingly enough were all about the environment and its problems, rather than hydrologic methods.

Following Ray's talk there were questions and colloquies about the National Pollutant Discharge Elimination System permit process; industrial plants, oil and gas works, and nuclear power plants and a possible backlash against environmental controls of these; endangered species such as the snail darter; and procedureal regulatory and legal matters by which environmental problems, though sticky, can and will be solved.

In the discussion following Snyder's presentation, much of the interest was not on how hydrology could help to solve the problems posed, but dwelt on the (not very realistic) possibilities for retrofitting or redesigning steam power plants. That is, the sense of the meeting was decidely one of trying to solve that human problem, even if hydrologic forecasting did not appear to be the most direct and fruitful way. The group, having decided it was jointly and severally incompetent to deal with the thermodynamic redesign of the power plant machinery, pressed on to suggest other avenues of solution. These included alternate uses of the heated water or the heat in it; alternate sources of available supply, such as agricultural return water and treated sewage; and finally, but seemingly only as a grudgingly-admitted last resort, which was curious for this group, the potential of mathematical models for water quantity and quality that might perhaps be useful in the solution of these problems was raised.

And so the meeting had come full circle, but the exercise had given the topic its proper perspective.

CONFERENCE REGISTRANTS

David P. Adam
U. S. Geological Survey

V. Ralph Algazi
Univ. of California-Davis

Roger A. Amisial
CIDIAT-OAS

Jack R. Benjamin
Engr. Dec. Anal. Co. (EDAC)

William R. Boggess
University of Arizona

Rafael L. Bras
Mass. Inst. of Technology

John J. Cassidy
Washington State University

Catalino B. Cecilio
Pacific Gas & Electric Co.

Donald Charlton
Tippetts-Abbett-McCarthy-Stratton

Sandford S. Cole
Engineering Foundation

Robert F. Collins
U. S. Army, Corps of Engineers

Kenneth D. Corbin
Corps of Engineers

Norman Crawford
Hydrocomp

David R. Dawdy
Dames & Moore

Thomas N. Debo
Georgia Tech.

Henry Degraff
U. S. Dept. of Commerce
Bureau of Economic Analysis

R. Larry Ferral
NWS River Forecast Center

Quentin L. Florey
Off. Water Res. & Tech.

Delbert D. Franz
Hydrocomp

Gary J. Freeman
Pacific Gas & Electric Co.

Robert L. Goodell
Delaware River Basin Comm.

William R. Gordon
Bonneville Power Adm.

Roderick F. Hall
Sierra Hydrotech

William P. Henry
CH2M Hill

L. Douglas James
Utah State University

Chester P. Jelesnianski
NOAA-NWS

Donald T. Jensen
U. S. Bureau of Reclamation
E&R Center

William K. Johnson
Hydrologic Engrg. Center

D. Earl Jones, Jr.
Dept. Housing & Urban Dev.

R. W. Julian
U. S. Bureau of Reclamation

Ronald F. Kilmartin
International Engrg. Co.

Jerry W. Knapp
CH2M Hill

James C. Knox
University of Wisconsin

Orley H. Lindgren
Entropy Limited

Billy E. Martin
Bureau of Reclamation

Harold W. Meyer
U. S. Bureau of Reclamation

John Mickelson, Jr.
Espey, Houston & Assoc. Inc.

Brian R. Mrazik
Federal Insurance Adm.

David Olsen
National Meteorological Center

George R. Phippen
U. S. Army Corps of Engineers

John M. Power
Environment Canada

Jack F. Rasmussen
St. Louis District Corps of Engrs.

Arliss D. Ray
EMANCO Inc.

J. Paul Riley
Utah State University

Maurice Roos
California Dept. of Water Resources

David Snyder
Woodward-Clyde Consultants

Michael B. Sonnen
2227 Banbury Place

Richard E. Stein
East Bay Muncipal Util. District

Thomas M. Stetson
Stetson Engineers, Inc.

Charles W. Stockton
University of Arizona

Dolores B. Taylor
Ventura County Flood Control District

John M. Tettmer
Los Angeles County Flood Control District

L. Scott Tucker
Urban Drainage & Flood Control District

Brian Van Weele
P.B.Q. & D. Inc.

Bi-Huei Wang
Harza Engineering Company

Ronald B. Wigle
James F. Mallaren, Ltd.

Gene E. Willeke
Miami University

Subject Index

Author Index